Automation, Collaboration, & E-Services

Volume 2

Series editor

Shimon Y. Nof, Purdue University, West Lafayette, USA
e-mail: nof@purdue.edu

About this Series

The Automation, Collaboration, & E-Services series (ACES) publishes new developments and advances in the fields of Automation, collaboration and e-services; rapidly and informally but with a high quality. It captures the scientific and engineering theories and techniques addressing challenges of the megatrends of automation, and collaboration. These trends, defining the scope of the ACES Series, are evident with wireless communication, Internetworking, multi-agent systems, sensor networks, and social robotics – all enabled by collaborative e-Services. Within the scope of the series are monographs, lecture notes, selected contributions from specialized conferences and workshops.

More information about this series at http://www.springer.com/series/8393

Shimon Y. Nof · Jose Ceroni
Wootae Jeong · Mohsen Moghaddam

Revolutionizing Collaboration through e-Work, e-Business, and e-Service

Springer

Shimon Y. Nof
PRISM Center & School of IE
Purdue University
West Lafayette Indiana
USA

Jose Ceroni
School of Industrial Engineering
Catholic University of Valparaíso
Valparaíso
Chile

Wootae Jeong
Korea Railroad Research Institute
Uiwang 437-757
Republic of South Korea

Mohsen Moghaddam
PRISM Center & School of IE
Purdue University
West Lafayette Indiana
USA

ISSN 2193-472X ISSN 2193-4738 (electronic)
Automation, Collaboration, & E-Services
ISBN 978-3-662-45776-4 ISBN 978-3-662-45777-1 (eBook)
DOI 10.1007/978-3-662-45777-1

Library of Congress Control Number: 2014958088

Springer Heidelberg New York Dordrecht London

Printed on acid-free paper

Springer International Publishing AG Switzerland is part of Springer Science+Business Media (www.springer.com)

This book is dedicated to our colleagues of the PRISM Center and PRISM Global Research Network for their invaluable contributions in revolutionizing collaboration through engineering augmentation.

Foreword: Augmenting Us to Collaborate Better for Further Progress

We know the difference between communication and collaboration. When we communicate --- we hope somebody is listening to us and paying attention: Our family, friends, students, co-workers, our boss, the BIG boss...and they probably expect the same from us on the other end of the (wireless) connection.

When we collaborate --- well, now, there is some work to be done, some efforts resulting in expected outcomes. And with farther reach, scaled-up networks, more parties and team-workers and customers and partners, there are more information, bigger data, more work. Deluge of expectations. Where are the limits? We have forever been trying to collaborate better. We learned to collaborate with machines, as drivers of vehicles; as pilots of planes; as operators of equipment. Trusting them and their built-in automation, and at the same time (even unknowingly) also collaborating with the many scientists, engineers, technicians and programmers that teamed up to create and maintain them.

But collaborating with other people is still a challenge. Through the ever larger communication pipelines, technology enables greater communication flows, deeper inter-networking, closer than ever and faster interactions, all promising so much, yet we are swamped and overloaded. This calls for engineering solutions to also automate and improve the collaboration support systems, significantly by cyber with intelligent control algorithms.

So there come the engineered augmentation of collaboration, and this book. Many useful contributions, conferences and workshops have added up to enable this movement. Among them, and from our perspective, we remember the NATO Advanced Research Workshop on Integration: Information and Collaboration Models (Il Ciocco, Italy, 1993), where we evaluated emerging theories of collaboration from earlier years; the EURO XIV meeting in Jerusalem, Israel (1995), the first time we presented recent developments of collaborative e-Work; and the first time we presented CCT, the Collaborative Control Theory, at the 2002 IFAC World Congress: Challenges of a New Millennium, in Rotterdam, The Netherlands. And there are many other inventive contributors and innovations, many of them included in this book.

The evolving tools and techniques of augmenting collaboration, which are the subject of this book, have been developed and applied in multiple industries and applications, with proven success. Generations of research students and scholars in our PRISM Center at Purdue University, and many colleagues over the PRISM

Global Research Network, are partners in pioneering, developing and implementing this exciting field of knowledge. We thank all of them. In particular, we would like to thank our PRISM Center former and current research students, (in addition to co-authors Jose Ceroni, Wootae Jeong, and Mohsen Moghaddam): Chwee Beng Ang, Pornthep Anussornnitisarn, Jorge Avila Soria, Pat Banerjee, Thomas Bellocci, Radhika Bhargava, Glenn Candranegara, Daniel Castro Lacouture, Abhijit Chaudhury, Jianhao Chen, Xin Chen, Tze Chao Chiam, Caludia Chituc, Surenjan De, Prabhu Devadasan, Yael Edan, Keyvan Esfarjani, Ed Fisher, Yashuharu Fujii, Gavin Furtado, Richard Gurecki, Daud Hanna, Chin Yin Huang, Sunjay Jayaram, Ezhil Kanagaraju, Howard Kang, Nitin Khanna, Chang Ouk Kim, Hoo Sang Ko, Hong Li, Marco Lara, Hanan Lechtman, Doo Young Lee, Geoffrey Lenart, Jiaxi Li, Hee Jong Lim, Yan Liu, Oded Maimon, Diana Milner, Amit Mukherjee, Richard Nash, Yukiko Nishibori, Engin Ozsoy, Karianne Prytz, Anurag Puranik, Jose Peralta, Colleen Phillips, Ignacio Puig, Venkat Rajan, Parbati Ray, Rodrigo Reyes Lavalle, Benhard Rembold, Robert Remski, Andrew Robinson, Pierre Sauvaire, Manuel Scavarda, Cristy Sellers, Hyesung Seok, Michael Shaw, Karna Singh, Itshak Tkach, Lina Uribe, Nitin Vallapuneni, Juan Velasquez, Graeme Warren, Neal Widmer, Robert Wiegner, Robert Wilhelm, NaRaye Williams, James Witzerman, Tetsuo Yamada, Chao Lung Yang, Sang Won Yoon, Lu Zhang, Hao Zhong, and the many post docs and visiting scholars. We are also grateful to Dr. Tom Ditzinger, who has patiently guided us throughout the development and completion of this book.

In the future, requirements will continue to intensify for more effective, timely, and trustworthy collaboration; it is essential for addressing scientific, technological, and social challenges, advance and sustain our life quality. We need to continue collaborating well for further progress. The collaboration revolution is a vital capacity of civilization's vision with the cyber and knowledge revolution. May this book and the ACES book series serve as a contribution in this exciting journey.

Autumn 2014 (S. Korea and USA); Spring 2014 (Chile)

Shimon Y. Nof
W. Lafayette, IN USA

Jose A. Ceroni
Valparaiso, Chile

Wootae Jeong
Seoul, S. Korea

Mohsen Moghaddam
W. Lafayette, IN USA

Contents

Chapter 1
Definitions, Scope, and Significance

Abstract. Engineered, augmented collaboration goes beyond the availability of cellphones, teleconferences, and other communication devices. When those enable human to human communication, it increases work overload and engages the human parties, usually preventing them from full participation in other assignments or meetings. But when cyber support, operating at a parallel cyber space, provides streamlined and timely information analyses and optimized workflows, collaboration activities can be fulfilled more efficiently and effectively, and with better quality of outcomes. This chapter discusses why we collaborate, and introduces definitions, scope, and significance. The objective here and in the rest of the book is to explain how collaborative e-Work, "electronic work", the foundation of e-Business and e-Service ("electronic business" and "electronic service", respectively), can be engineered, streamlined, and optimized; how effective, efficient and sustainable collaboration can be designed and guided successfully.

1.1 Why Do We Collaborate?

We all learn to collaborate very early in life: With our mother, then with family and friends. Then we collaborate in school projects, games and sports; with team workers and managers. Our common goals guide us to support and complement each other, share, coordinate, and work together to achieve those goals. We even learn to collaborate with competitors (when it makes sense). While for most of us collaboration is a necessity, for many it is also a source of joy.

Why? Because we know that we cannot always do it on our own.

But can collaboration processes among people, machines, and software agents be engineered, improved, systematized, and learned? Is it necessary to achieve a better collaboration process in order to produce better outcomes, and how can that advantage be evaluated? In other words, is there collaboration intelligence?

We believe that the answer to the above questions is yes. The goal of this book is to describe and explain how collaboration is being revolutionized through better engineered collaboration processes. Collaboration has always been among humans and for humans, but over time it has evolved. It has also encompassed collaboration among:

S.Y. Nof et al., *Revolutionizing Collaboration through e-Work, e-Business, and e-Service*,
Automation, Collaboration, & E-Services 2, DOI: 10.1007/978-3-662-45777-1_1

- Humans and machines – drivers and their cars; pilots and their aircraft.
- Humans and computers – designers and their computer aided design; children and computer games.
- Internet parties – sellers and buyers; servers and clients; social networks.
- Cell phones – to communicate, coordinate, share and exchange with all of the above, with other humans, machines, computers, and the Internet.

The evolution is clear: We can collaborate with more. More people can and want to interact and collaborate with us. We can reach farther and faster…and more. More contacts, more information and knowledge. It may mean better quality of life, but it may also mean greater overloads and stress, and less joy. So what can be done?

Figure (1.1-a) shows the collaboration among humans and with other entities. Figure (1.1-b) shows the same collaboration, but with some of the collaboration process functions being engineered and augmented, to be handled in a parallel cyber space by cyber support, through algorithms, protocols, and collaboration software agents. Thus, collaboration can be designed and managed, alleviating humans from being overloaded and enabling us to focus on the purpose of collaboration, not on its details and overhead.

Engineered, augmented collaboration goes beyond the availability of cellphones, teleconferences, and other communication devices. When those enable human to human communication, it increases work overload and engages the human parties, usually preventing them from full participation in other

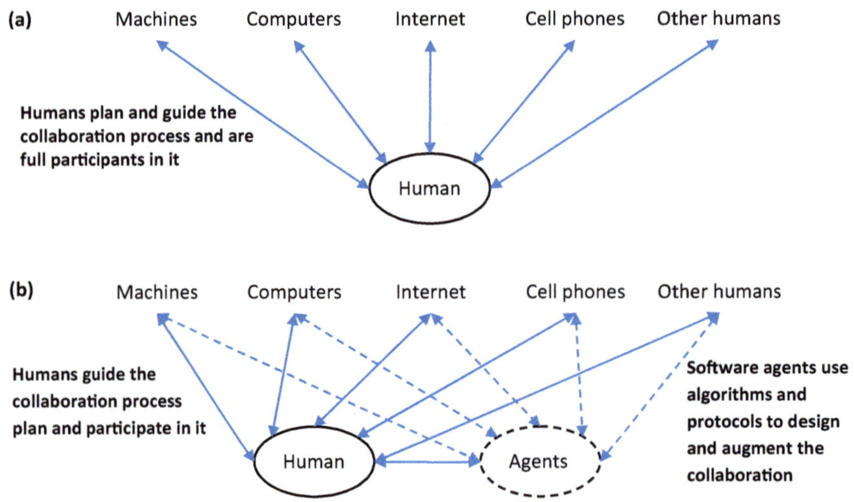

Fig. 1.1 Collaboration: (a) Humans and other entities use cyber technologies to collaborate; (b) Revolutionized Collaboration – Humans and other entities use cyber technologies to collaborate, while the collaboration process itself is augmented by cyber functions.

assignments or meetings. But when cyber support, operating at a parallel cyber space, provides streamlined and timely information analyses and optimized workflows, collaboration activities can be fulfilled more efficiently and effectively, and with better quality of outcomes.

This book explains how collaborative e-Work, "electronic work", the foundation of e-Business and e-Service ("electronic business" and "electronic service", respectively), can be engineered, streamlined, and optimized; how effective, efficient and sustainable collaboration can be designed and guided successfully.

1.2 Definitions and Scope

All researchers and practitioners involved in the design and operation of production, manufacturing, and service systems have realized a substantial evolution over the years; a transformation from monolithic self-reliant systems to highly distributed *systems of systems*. The transformation is driven by the emerging needs for distribution of intelligence and autonomy, from the levels of nano and micro systems all the way to global enterprise networks, with the purpose of achieving more sustainable and dependable systems. Such emerging systems, however, entail higher levels of complexity due to the increasing frequency and impact of *interaction* and *collaboration* among distributed entities in pooling their information, resources, and responsibilities to create more added values. That is, those collaboration processes need to be engineered and optimized. *Collaboration engineering*, as the focal theme of this book, refers to the development of tools and techniques to go beyond better communication technologies (e.g., teleconference, remote file sharing) and provide significantly improved collaboration support by cyber through algorithms, protocols, and collaboration support software agents.

This chapter reviews the fundamentals of collaborative e-Work (henceforth, "e" stands for electronic), as the main enabler of this systematic revolution. We will show throughout this book that the potential benefits, opportunities, and sustainability of emerging e-Systems such as virtual manufacturing, e-Healthcare, automated inspection, e-Supply, e-Production, e-Collaboration, e-Logistics, and other e-Activities will not materialize without the design of effective e-Work. The "four-wheels" of collaborative e-Work, their respective 15 e-Dimensions, and their role in e-Business and e-Service are explained and illustrated. Examples of e-Work, e-Manufacturing, e-Logistics, e-Business, and e-Service are also provided to help readers comprehend the depth and breadth of on-going efforts, potential inhibitors to revolutionizing such e-Systems, challenges and emerging solutions.

1.3 e-Work, e-Systems, and e-Activities

The landscape in which everyday work and personal activities are accomplished by humans, computers, machines, and organizations has evolved significantly in recent years. Advances in information and communication technologies (ICT) over the years have revolutionized traditional work disciplines to telecommuting and remote work, and now e-Work. In this context, e-X is no longer the same as X. From a production point of view, specifically, work has been reshaped by e-Work. Just as work, business, commerce and service are different from each other so are e-Work, e-Business, e-Commerce and e-Service. So what are the differences? To understand the differences, let us first formally define e-Business, e-Commerce and e-Service; they are the most substantial instances of "e-Systems" today. Also, let us understand what e-Work is, and how they relate to it:

- *e-Business*: is the integration of a company's business, including products, procedures, and services, over the Internet (Rosen, 2002). A more comprehensive definition provided by Browne et al. (2003) states that e-Business is about conducting business both internally and/or externally by electronic means, e.g., over the Internet, intranet or extranet and that it involves not only buying and selling but also operating automated, efficient internal business processes servicing customers and collaborating with suppliers and business partners (Figure 1.2).

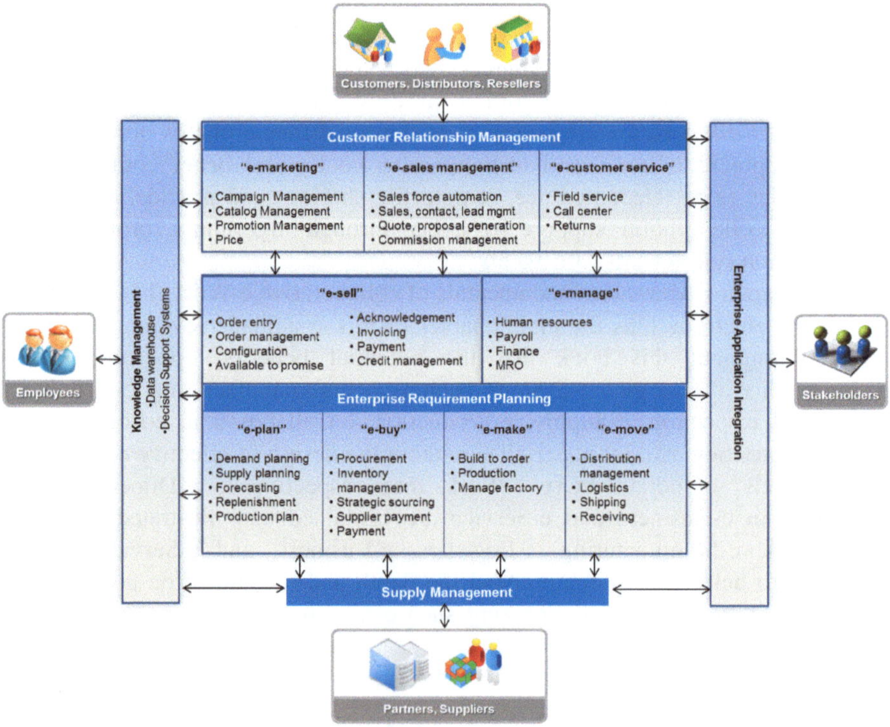

Fig. 1.2 e-Business application framework.

- *e-Commerce*: covers the range of online business activities for products and services, both business-to-business and business-to-consumer, over the Internet (Rosen, 2002). Besides businesses, individual consumers are increasingly taking advantage of e-Commerce. For instance, more than 50% of all travel reservations made in the USA are now made via the Internet rather than travel agencies. Just in the USA, e-Commerce has grown from US$ 57.4 billion in 2000 to US$ 658.4 billion by 2012, i.e., around 100% increase per year (see Figure 1.3).

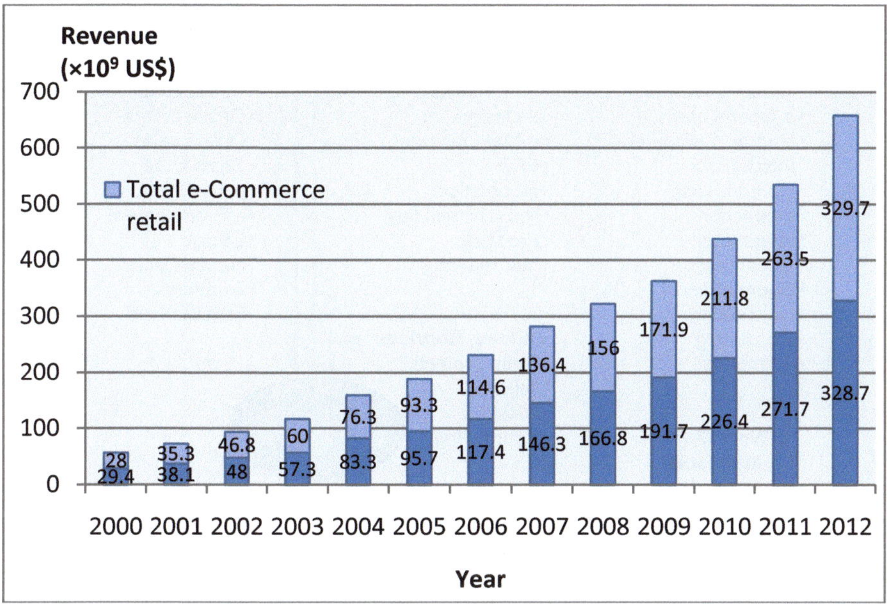

Fig. 1.3 Business-to-consumer e-Commerce revenue by service and retail sectors for 2000–2012 in US$ (Courtesy of the US Department of Commerce, US Census Bureau).

- *e-Service*: is the provision of services over electronic networks such as Internet, intranets or extranets without its scope being limited to service organizations but rather encompassing all enterprises, even those that manufacture goods and which require the development and implementation of sound service practices over electronic networks.

The key point in the definition of all e-Systems is the "e" aspect, the reliance on the Internet and ICT. A classic activity cannot be copied *as-is* to an e-Activity. Substantial transformations are indeed required. "e-Activities", in essence, refer to activities based on and executed through electronic information and communication systems, supported by the Internet and ICT. As a simple example, compare the search activity performed by three different generations of workers: A) search through manuals, usually papers, B) search through a database by queries, and C) search through the Web by a browser. The evolution of traditional search activity to an e-Activity is obvious. But clearly there is a significant logical, organizational, and technological augmentation, and perhaps there are fewer similarities than differences between traditional search and e-Search. In this context, e-Work provides a foundation to support the e-Activities in any e-System, in general (see Figure 1.4).

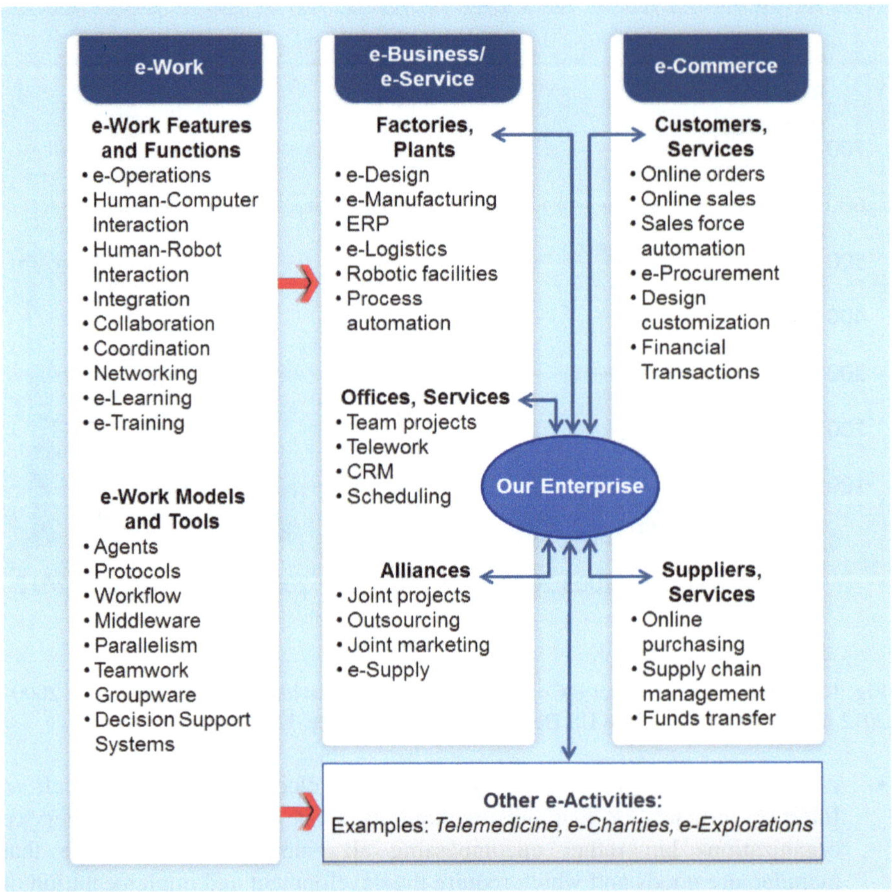

Fig. 1.4 e-Work as the foundation for e-Business, e-Service, e-Commerce and other e-Activities.

e-Work is a collection of collaborative, computer-supported, and communication-enabled e-Activities, e-Operations, e-Functions, and e-Support systems that enables other e-Systems and e-Activities (Nof, 2003); e.g., e-Business, e-Commerce, e-Manufacturing, e-Healthcare, e-Training, e-Logistics, v-Factories, v-Enterprises and many more (Figure 1.5). e-Work relies on the use of computer support and ICT and, at the same time, necessitates *collaboration engineering* to effectively design and support the interactions among humans, computers, and machines. The revolutionary impact of e-Work is legitimately characterized by the following quote (Nof, 2003):

> "*As power fields, such as magnetic fields and gravitation, influence bodies to organize and stabilize, so does the sphere of computing and information technologies. It envelops us and influences us to organize our work systems in a different way, and purposefully, to stabilize work while effectively producing the desired outcomes*".

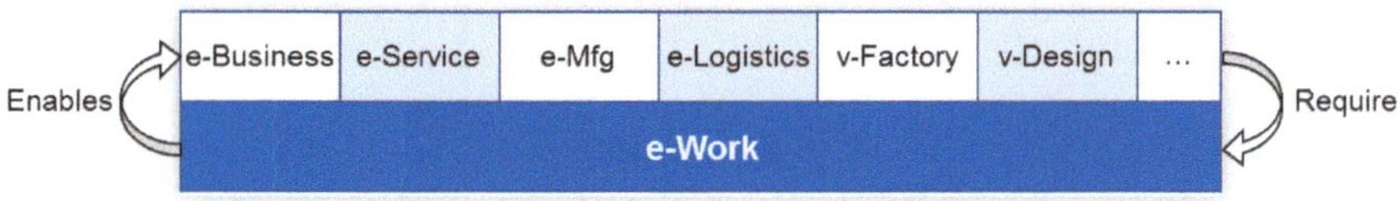

Fig. 1.5 Scope of collaborative e-Work as the common foundation of e-Activities.

1.4 Collaborative e-Work *vs*. Automation, Tele-Work, and Robotics

Before discussing the fundamentals of collaborative e-Work, it is necessary to highlight how it relates to automation, tele-work, and robotics. We begin our comparison by formally defining each concept:

- *Automation*: implies "*operating or acting, or self-regulating, independently, without human intervention*" (Nof, 2009). The root of the term "automation" is a Greek work "automatos", which mean acting by itself, or by its own will, or spontaneously. Automation refers to purposeful combinations of machines, tools, devices, installations, and systems, as platforms performing a certain set of activities without human involvement.
- *Tele-work*: or "tele-commuting", refers to any work arrangement in which workers, in general, do not physically commute in a central place and instead, perform a given set of work or action from some distance away. The term "tele" is also a Greek word (τηλε), which means offsite or remote (Basañez and Suárez, 2009).
- *Robotics*: refers to "*the science and technology of designing, building, and applying robots, computer-controlled mechanical devices, such as automated tools and machines*" (Nof, 2009). The term robotics was first coined by Isaac Asimov, science-fiction author and scientist, in 1941: "*Since physics and most of its subdivisions routinely have the "-ics" suffix, I assumed that robotics was the proper scientific term for the systematic study of robots, of their construction, maintenance, and behavior, and that it was used as such*" (Asimov, 1985).

There is an overlap between the definitions of automation and robotics. They both rely on the use of automatic control, based on four major components: platform, autonomy, process, and power source. In robotics, however, a robot is often considered a machine, thus the platform is mostly a machine, a tool or device, or a system of tools and devices. In contrast, automation, by definition, is not restricted to motion and mobility, and encompasses other areas beyond robotics such as software, decision-making, planning and optimization, process automation, office automation, enterprise resource planning automation, and e-Services. Thus, robotics is a subset of automation (see Figure 1.6).

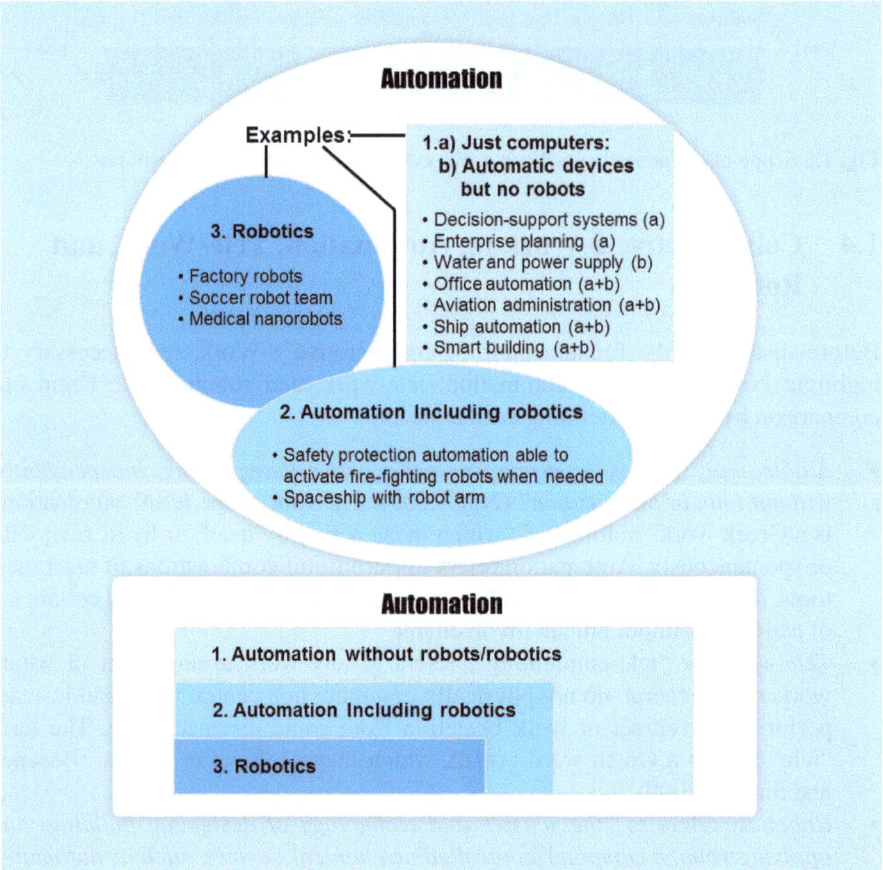

Fig. 1.6 The relation between robotics and automation: The scope of automation includes applications: (1a) with just computers, (1b) with various automation platforms and applications, but without robots; (2) automation including also some robotics; (3) automation with robotics.

With the evolution of computers and ICT, new challenges also emerged for better "interaction" and "integration of islands of automation" to systems of systems. Nowadays, all e-Activities rely on computer support and ICT, and require collaboration in their inherent interactions between machines, humans, and computers; e.g., tele-robotics, remote services, and telework. Advances in ICT have enabled new levels of work augmentation, e.g., tele-communication and computer internetworking, as important enablers and supporting technologies for e-Work and collaborative control. This unique functional property, i.e., "collaboration", is the main building block differentiating collaborative e-Work from the concepts of automation, tele-work, and robotics.

1.5 Collaboration *vs*. Coordination and Cooperation

Often the terms "collaboration" and "coordination" are used interchangeably; but even though they are closely related, they do exhibit significant differences. Formal definitions (Nof, 2003; 2006; 2007; Camarinha-Matos and Afsarmanesh 2006; 2008) help clarify the scope and breadth of each term by placing them in a broader context, including closely related terms such as networking, mobility, and cooperation. It is important to highlight the differences between all these terms, as their implications to the design of e-Work are essential (Figure 1.7):

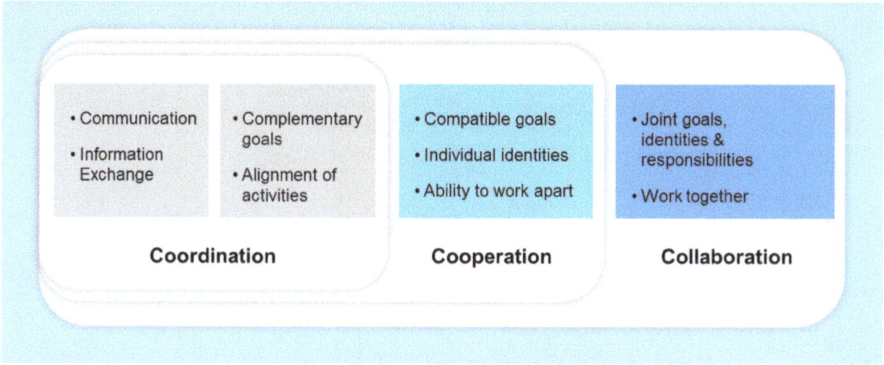

Fig. 1.7 Coordination, cooperation and collaboration level of interaction.

- *Coordination*: It means the use of communication and information exchange to reach mutual benefits among parties by working harmoniously.
- *Cooperation*: Involves all the aspects of coordination but also incorporates a resource-sharing dimension to support goal achievement. Commonly, cooperation will exhibit a division of labor component among all participants and therefore the aggregated value is the result of adding the "individual" parts.
- *Collaboration*: It is the most demanding of the three processes, where all involved parties share information, resources and responsibilities to jointly plan, implement, and assess the set of activities required to achieve a common goal, thus jointly creating value added.

1.6 Fundamentals of Collaborative e-Work

Over the years, computers and advances in communication technologies and related areas have enabled us to significantly improve the quality of service and products, as well as the level of customization. Specifically, *collaborative e-Work* is revolutionizing the capabilities of e-Systems. By leveraging existing and well-established work, business and service models, theories, and solutions with the new electronic world, it is possible to augment capabilities of distributed systems to meet their emerging needs. The four knowledge clusters or "wheels" of collaborative e-Work, i.e., 1) e-Work, 2) Integration, Coordination and Collaboration, 3) Distributed Decision Support Systems, and 4) Active Middleware are shown in Figure 1.8. In this section, the 15 e-Dimensions (*e-D*) corresponding to the four knowledge clusters are briefly discussed along with some practical applications. We will elaborate on different dimensions of collaborative e-Work in the next chapters.

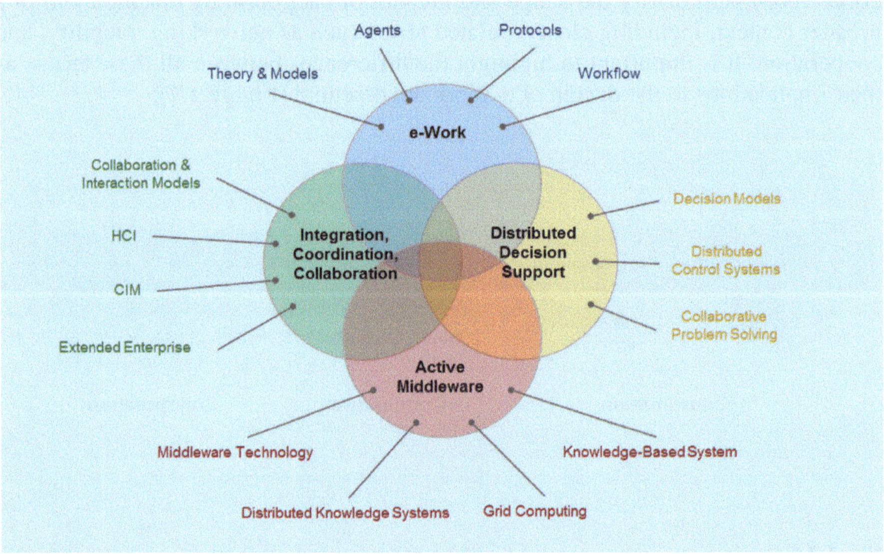

Fig. 1.8 The four wheels of collaborative e-Work and the 15 e-Dimensions.

1.6.1 Wheel 1: e-Work

e-D 1: *e-Work theory and models*: provide the foundation for augmenting the abilities of (1) e-Workers to accomplish their e-Activities, (2) organizations to accomplish their goals, and (3) agents (e.g., human, computers, robots) to collaborate.

e-D 2: *Agents*: are independent and autonomous programs, which operate and execute specific tasks under certain protocols, and are responsible for handling expected and unexpected events. Agents enable the automation and integration of flexible, adaptive, and heterogeneous activities, components, and tasks in a network. There has been a drastically increasing trend in design and development of agent-based systems over the past three decades (Figure 1.9). We will comprehensively discuss agents, agent-based systems, and basic and emerging agent-based optimization and control architectures in Chapter 4. Note that agents usually also include the programmable and cyber-controlled machines included in e-Work.

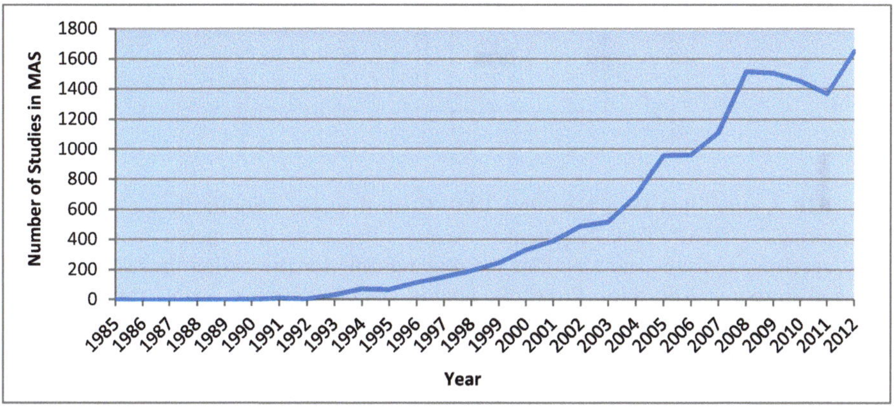

Fig. 1.9 Number of studies in design and development of multi-agent systems (MAS) in different years (Source: Engineering Village).

e-D 3: *Protocols*: provide efficient and effective distributed platforms for autonomous agents to interact and accomplish their tasks to achieve mutual benefits and common goals. Various types of coordination protocols exist, depending on the application. Examples include (1) dependence-based coordination between tasks, (2) decision- and time-based protocols to adapt requirement changes and complex scenarios that require decisions beyond coordination. A main requirement of such protocols is to allocate and administrate certain activities or tasks among distributed autonomous entities. Simple protocols, however, may encounter difficulties in decision-making in complex and dynamic collaborative systems. Task administration protocols (TAP) provide a foundation for protocols to cope with such difficulties and conflicts (see Figure 1.10), through task requirement analysis, shared resource allocation, and synchronization.

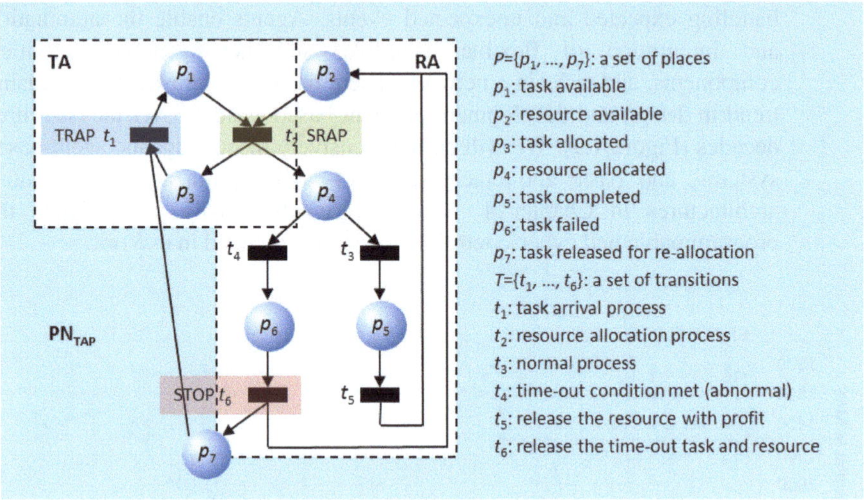

Fig. 1.10 A simplified Petri-Net (PN) TAP; Control scope: prioritization, allocation, and monitoring (1) TRAP (Task Requirement Analysis Protocol): is triggered upon arrival of a task; receives, analyzes and identifies the requirements for processing the task, (2) SRAP (Shared Resource Allocation Protocol): reports tasks to resource agents through certain bidding processes, (3) STOP (Synchronization and Time-Out Protocol): monitors the processes and checks for time-out conditions.

e-D 4: *Workflow*: enables scalability, availability, and reliability of processes through systematic assignment and control of activities in distributed and dynamic environments. Workflow technology, as an emerging area in software development, (1) provides user-friendly interfaces for editing, improving, and executing process models, especially in e-Business, (2) enables information exchange between homogenous and heterogeneous information systems through service standardization, and (3) enhances circulation of information, and service integration with standard web-based protocols. The approach depends on systematic modeling of the required work, interactions, input-output relations, required data and resources, conditional functions, etc. Depending on the application, workflow management systems are generally classified into three categories of production, administrative, and ad hoc workflows. An example of an inter-enterprise workflow management system is depicted in Figure 1.11.

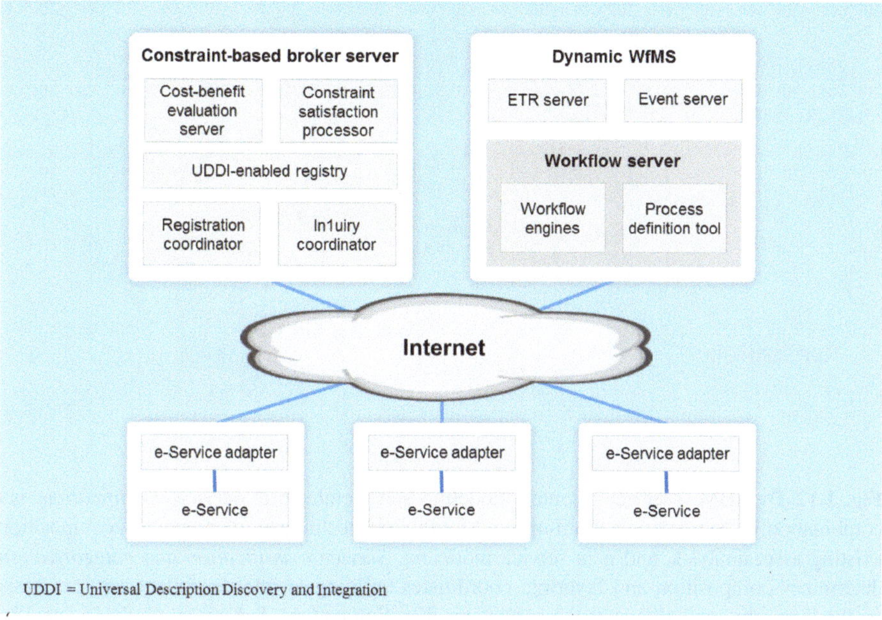

Fig. 1.11 Dynamic workflow management system for a collaborative network of enterprises; standardized integration and interfacing the data resources, processes, and services regarding different scenarios.

1.6.2 Wheel 2: Integration, Coordination, and Collaboration

e-D 5: *Collaboration and interaction models*: identify and model relationships between information resources and processing agents, and leverage this knowledge to improve performance of collaborative service-oriented business and software systems. Service-oriented modeling applies certain modeling and interaction languages understandable for both the business and the IT domains. Service-oriented architectures (SOA) improve interaction and collaboration and enhance agile re-configurability, modularity, collaboration, and open communication among distributed business and service entities (see Figure 1.12). Among the foremost areas being influenced by higher level of integration, coordination and collaboration are e-Business (Park, 2003; Li et al., 2005), e-Healthcare (Malamateniou and Vassilacopopulos, 2003; Weerakkody and Ray, 2003), transportation (Krajewska and Kopfer, 2006), and construction (Han et al., 2007).

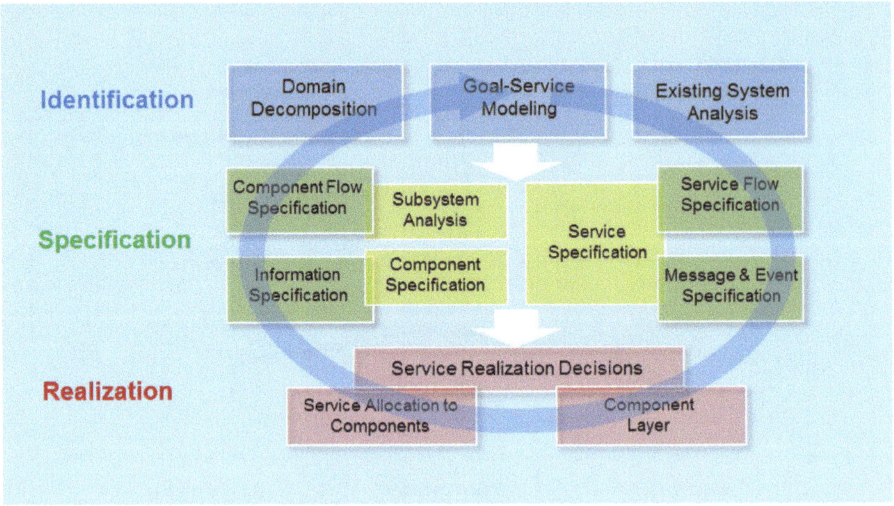

Fig. 1.12 The IBM® service-oriented modeling and architecture; *Service identification*: is a combination of top-down, bottom-up, and middle-out techniques of domain decomposition, existing asset analysis, and goal-service modeling. *Service classification and categorization*: determines composition and layering, coordinates building of interdependent services based on the hierarchy, and alleviates the service proliferation pattern. *Subsystem analysis*: specifies the interdependencies and flow between the subsystems. *Component specification*: specifies the details of the components implementing the service, e.g., data, rules, services. *Service realization*: decides whether the software that realizes a given service must be selected or custom built (adapted from Arsanjani, 2004).

e-D 6: *Computer-integrated manufacturing (CIM)*: enables harmonized integration and high level of collaboration in the entire product lifecycle from design to customer service. Product customization, agility and responsiveness, product and process innovation, sustainability, and global distribution of competition are among the major challenges of modern businesses. Evolution of CIM has been in line with those challenges (see Figure 1.13), e.g., emergence of integrated-enterprise systems (Figure 1.14), RFID (Radio-Frequency Identification), intelligent products, and more.

e-D 7: *Human-computer interaction (HCI)*: refers to systems, platforms, and interfaces that enable interactions between people and computers and support their roles as workers, learners, and researchers. In line with emergence of new technologies, the roles of humans in today's workplace and the impact of HCI continue to evolve. HCI applications are as wide-ranging as the applications of computers, in general, in different areas. For instance, virtual manufacturing is an HCI technology, which helps managers and decision-makers to better interact with complex work environment, distributed agents, models/protocols, parameters, and decision alternatives (Figure 1.15-a). Human-robot interaction is another aspect of HCI, as a multidisciplinary field between artificial intelligence, robotics, behavioral sciences, and HCI, which enables tele-operation and remote control of intelligent machines and robots via visual and haptic devices (Figure 1.15-b). HCI applications in e-Service are also plentiful. Microsoft recently released Surface, an interface that enables users instant access to information and entertainment via their hands and gestures, while improving collaboration and empowering consumers in the decision-making process (Figure 1.15-c).

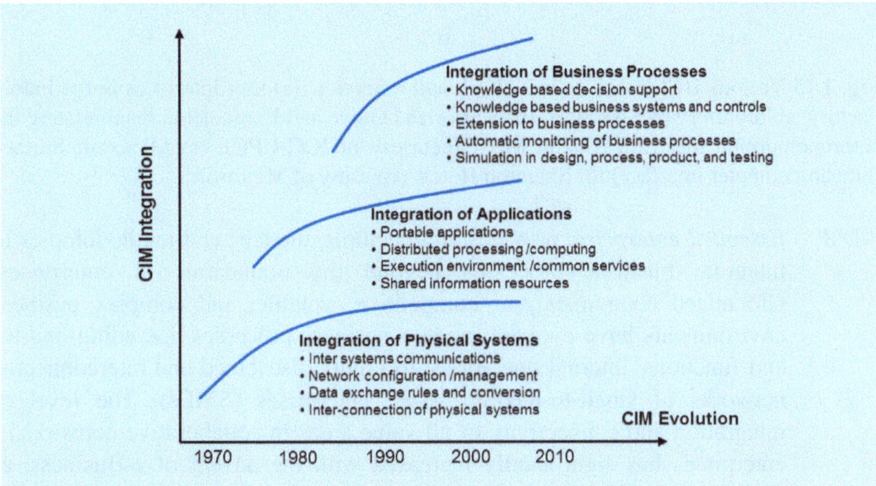

Fig. 1.13 Evolution of CIM (adapted from Nagalingam and Lin, 2008).

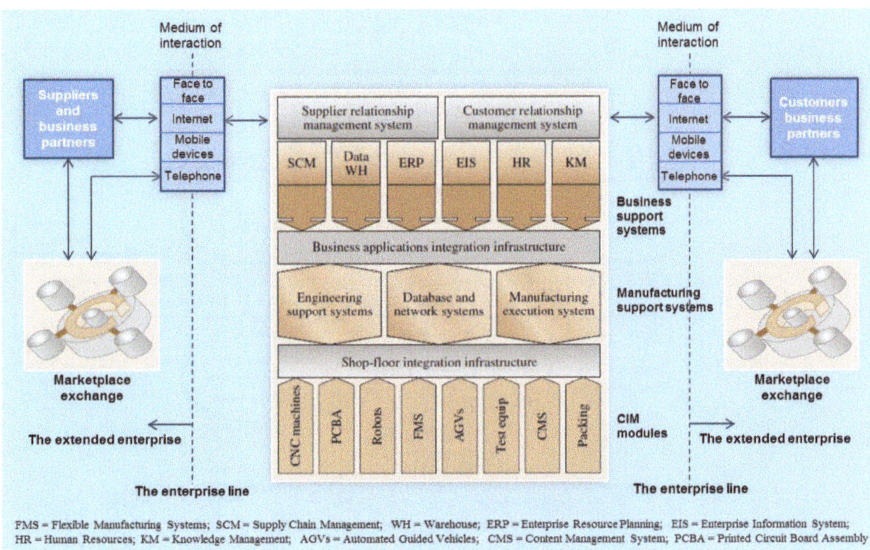

Fig. 1.14 The integrated enterprise system architecture; integration of critical success factors for development of different stages/components of an integrated-enterprise system (adapted from Ho and Lin, 2004).

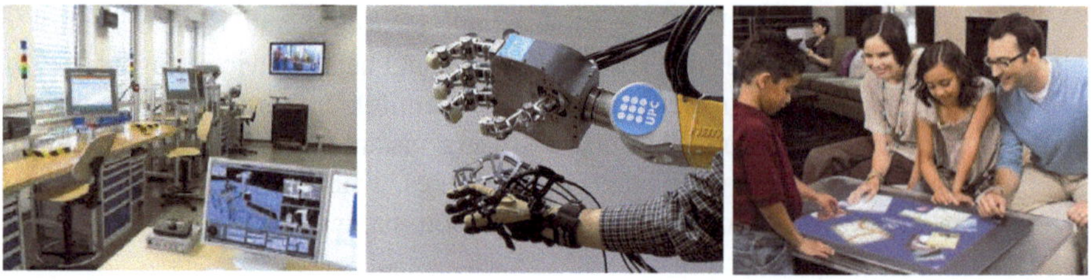

(a) (b) (c)

Fig. 1.15 Various HCI instances in e-Work and e-Service: (a) user interfaces in the Future Factory lab (courtesy of SAP AG), (b) Sensorized Glove and Exoskeleton manipulating the anthropomorphic mechanical hand MA-I (courtesy of IOC-UPC), (c) Microsoft Surface (human-computer interface) for Sheraton Hotels (courtesy of Microsoft).

e-D 8: *Extended enterprise*: provides architectures, models, and methodologies to integrate business processes beyond the boundaries of enterprises. Globalized economies and competitive, volatile, and complex business environments have evolved modern business processes, including end-to-end functions, internal operations, to highly distributed and interconnected networks of small-to-medium sized enterprises (SMEs). The level of integration and connectivity of all value areas in collaborative networks of enterprises has significantly increased with the advent of e-Business, e-Commerce, e-Services, and globalization. The evolution of traditional enterprises into extended enterprises, as defined by a project cluster due to the European Information Society Technologies, is shown in Figure 1.16.

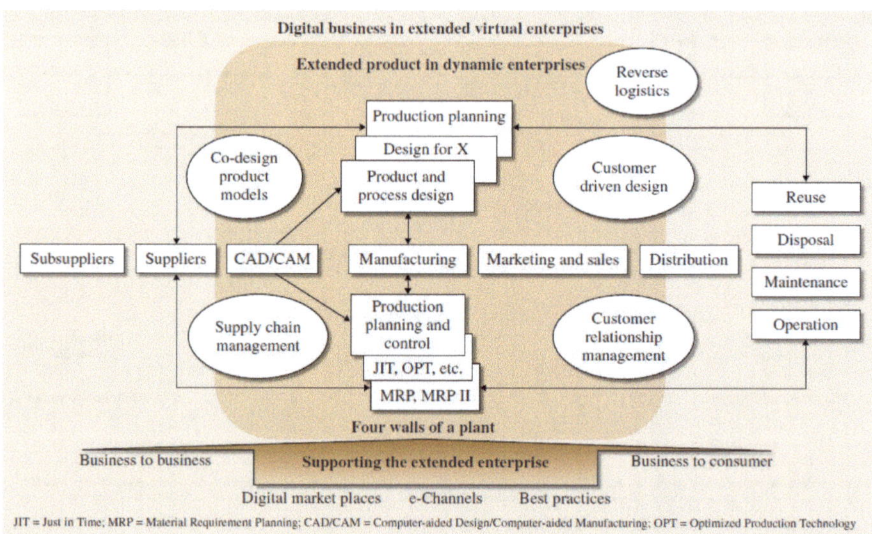

Fig. 1.16 Extended products in dynamic enterprises architecture (after Browne et al., 2003).

1.6.3 Wheel 3: Distributed Decision Support

e-D 9: *Decision models*: refer to general-purpose knowledge representation languages with normative status and well-understood computational properties of decision-modeling formalisms and algorithms (Nof, 2006). Decision models enhance quality and effectiveness of distributed decision support systems through integrating computer tools and methodologies in implementation of decision theories. With the increasing complexity of enterprise management information systems, advent of *big data*, and demand for real-time responses, human intuition and expert judgments are being replaced with collaborative intelligence relying on powerful predictive models, e.g., data mining and pattern recognition. For instance, the online quality information system, shown in Figure 1.17, enables access to real-time data and information regarding distributed locations, enhanced with active agents, to make complex quality and deployment costs decisions.

e-D 10: *Distributed control systems*: enable higher degree of autonomy of distributed entities (e.g., agents, supply network parties) to negotiate for the assignment/allocation of tasks and their respective individual and group incentives. Distributed control systems include dynamic processes where autonomous controllers are distributed throughout the entire system, and each component, or sub-system, is controlled by one or more controllers. In manufacturing, for example, intelligence, autonomy, proactivity, and fault-tolerance in the distributed shop floor devices (e.g., intelligent sensors) is improved by the use of cooperative objects (COBS). COBS are intelligent physical objects capable of sharing information, cooperating, and acting as elements of a community, with strong roots in ubiquitous computing, embedded systems, wireless sensor networks, and strong internet and communication infrastructures (Figure 1.18). Besides manufacturing, distributed control systems have variety of applications in geographically distributed systems, e.g., electrical power grids and power plants, water management systems, oil refining systems, sensor networks, pharmaceutical manufacturing, and cement process.

e-D 11: *Collaborative problem solving*: enables information exchange between loosely coupled entities (where less mutual benefits are acceptable or expected by collaborators) and interactions to improve harmony and consistency in the decision-making process and the overall results. Distributed entities, in general, can interact, learn from the others' actions, create their own actions and decisions, share knowledge, request resources, and access to the required information to fulfill their ongoing tasks.

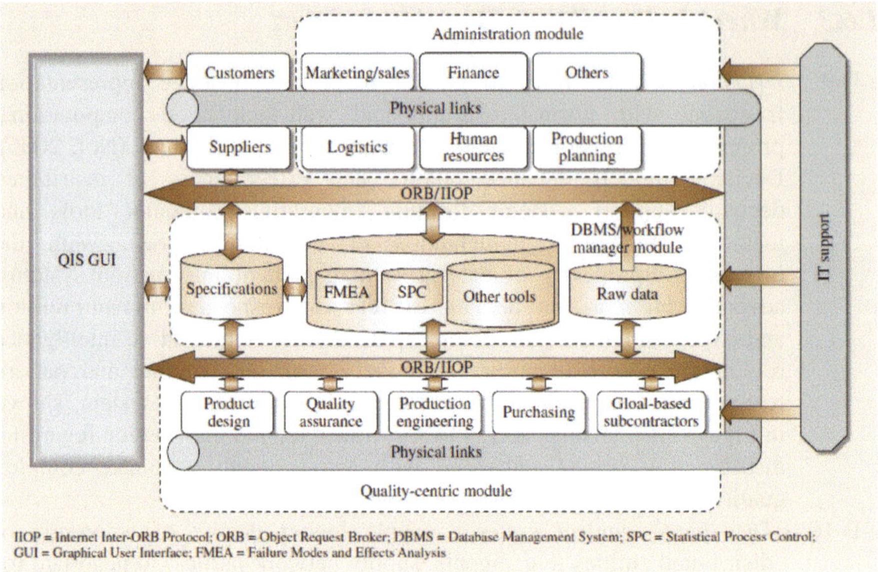

Fig. 1.17 The online quality information system reference architecture (after Jiao et al., 2007).

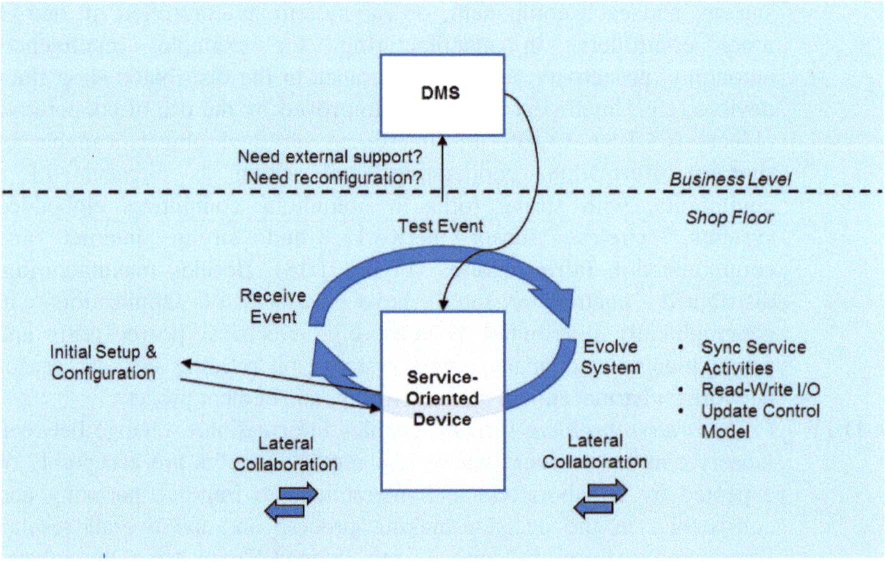

Fig. 1.18 Cross-layer (vertical) collaboration between the COBS (shop floor) and the business level via web service enabled middleware, and cross-domain (lateral) collaboration between homogenous/heterogeneous COBS, requesting/providing decision information from MES/ERP/DMS (adapted from Colombo et al., 2010).

1.6.4 *Wheel 4: Active Middleware*

e-D 12: Middleware technology: is a layer of services that enables interoperability among distributed and homogenous/heterogeneous operating systems, and network and application layers. Active middleware consists of six major components classified into two categories (see Figure 1.19):

- Active – (1) multi-agent systems, (2) workflow management systems, and (3) protocols.
- Supportive – (4) decision support systems, (5) modeling tools, and (6) databases.

The roles of active components are to enhance proactivity and adaptability (multi-agent systems), integrate and automate execution of complex processes (workflow management systems), and provide platforms for the agents to interact and automate the workflow (coordination protocols). The supportive components, on the other hand, support the active components through providing information, rules, and specifications.

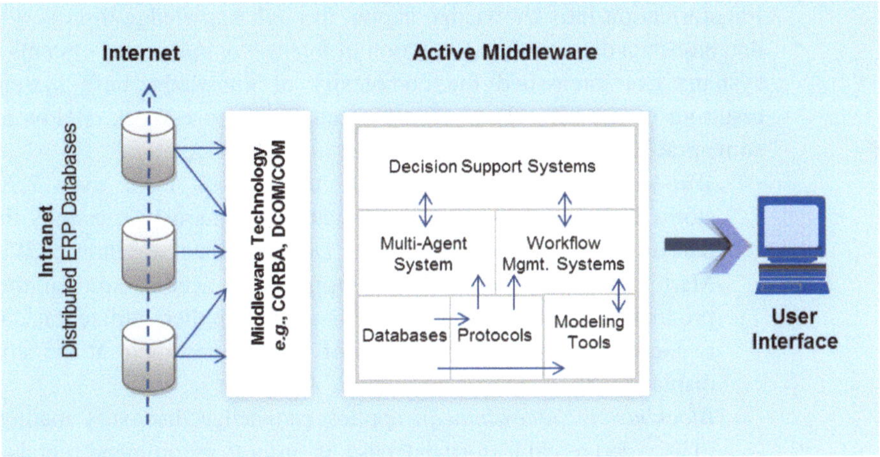

Fig. 1.19 Active middleware architecture; formulation, description, decomposition, and allocation of problems; synthesis of the results among distributed/autonomous entities; enabling applications to interact; conflict resolution; group decision-making.

e-D 13: *Distributed knowledge systems*: are techniques and standards to design distributed information and knowledge systems in order to sustain high reliability, integrity, and quality of service over large-scale and highly distributed networks of entities, e.g., search robots, semantic webs. Distributed knowledge systems are collections of autonomous knowledge-based objects, known as *knowledge agents*, which interact and work with each other to evolve the knowledge required for improving decisions or solving queries, based on their respective abilities.

e-D 14: *Grid computing*: refers to *super virtual computers* that enable rapid sharing and aggregation of loosely coupled, heterogeneous, and geographically dispersed computing resources in a virtual (v-) environment, for solving large-scale and data intensive computing applications. A computational grid provides dependable, pervasive, consistent, and inexpensive access to high-end computational resources and enables its users to perceive heterogeneous resources as homogeneous. In e-Work and e-Business, grid technology is a primary prerequisite for creation of v-Factories and v-Organizations for supporting distributed agents and integration of technological, methodological, and organizational aspects.

e-D 15: *Knowledge-based systems*: provide the capabilities to mine, refine, construct, and extract information from a variety of sources, enabling its transformation into knowledge capital through knowledge discovery in databases via data mining. Evolution of *internet of things* and systems of systems has intensified the complexity of knowledge-base systems resulting in big data, as an emerging challenge to e-Work. Below are some practical challenges in e-Business and e-Service:

- *Wal-Mart*: captures point-of-sale transactions from over 2,900 stores distributed in six countries and continuously transmits this data to its massive 7.5 Terabyte *Teradata* data warehouse. Wal-Mart allows more than 3,500 suppliers to access data on their products and to perform data analysis. The suppliers utilize this data to identify the purchasing behavior of the consumers at the store display level.
- *Blockbuster Entertainment*: applies knowledge discovery methods to its video rental history database, in order to recommend rentals to individual customers.
- *American Express*: suggests products to its customers through analyzing their monthly expenditures using knowledge discovery methods.
- *National Basketball Association*: applies knowledge discovery methods along with image processing of basketball games, in order to help coaches orchestrate plays and their tactics/strategies.

1.7 Examples

Examples of collaborative e-Work systems are plentiful. They began through computing and communications, for instance for cooperative design (e.g., Klein, 1993) and models of material flow in manufacturing (e.g., Huang and Nof, 1999). Throughout this book, we will discuss various examples and several emerging applications. As an introduction and to provide an introductory overview, four real-life examples of e-Systems are selected and described in this section.

1.7.1 Example 1: Holonic Manufacturing Systems (HMS)

HMS is an emerging approach for resolving emerging issues in modern manufacturing systems concerning semi-customized products and mass customization, networked manufacturing to increase efficiency and competitiveness, and reduce risk, efficient coordination and collaboration, high flexibility and agility, and collaborative control of distributed agents (see Chapter 4 for more details). Figure 1.20 shows an example of HMS, where different intelligent agents (holons) for resource, products, and orders are deployed and collaborate with each other for process planning. This HMS is composed of the following components (Hsieh and Chiang, 2011):

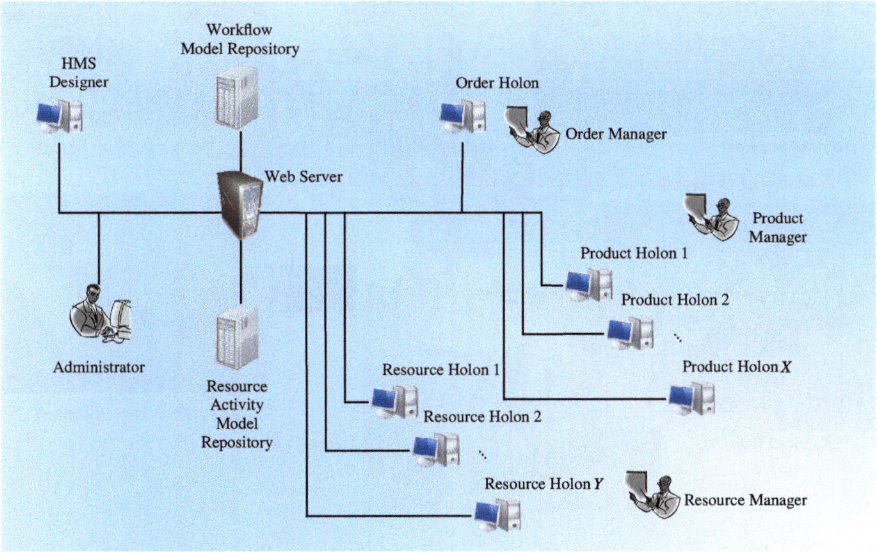

Fig. 1.20 Problem solving architecture using holons (after Hsieh and Chiang, 2011; with permission from Elsevier B.V.).

- *Workflow model repository*: captures the workflows of product holons.
- *Resource activity model repository*: captures the activities of resource holons.
- *HMS Designer*: constructs the holons in HMS, activates negotiation of holons, and monitors the system operation and established contracts.
- *Administrator*: manages the account of users.
- *Order manager*: creates and updates order holons.
- *Product manager*: defines and creates product holons.
- *Resource manager*: defines and updates resource holons.

1.7.2 Example 2: e-Production & e-Logistics

Integration of manufacturing and logistics with services is among the most prominent objectives of emerging e-Work systems. It relies on the maturity and experience gained in integration and in collaborative control, as well as better understanding of fundamental processes, including bioprocesses. An example of such integration is the modern mining technology deployed by the CODELCO (Corporación Nacional del Cobre de Chile or, the National Copper Corporation of

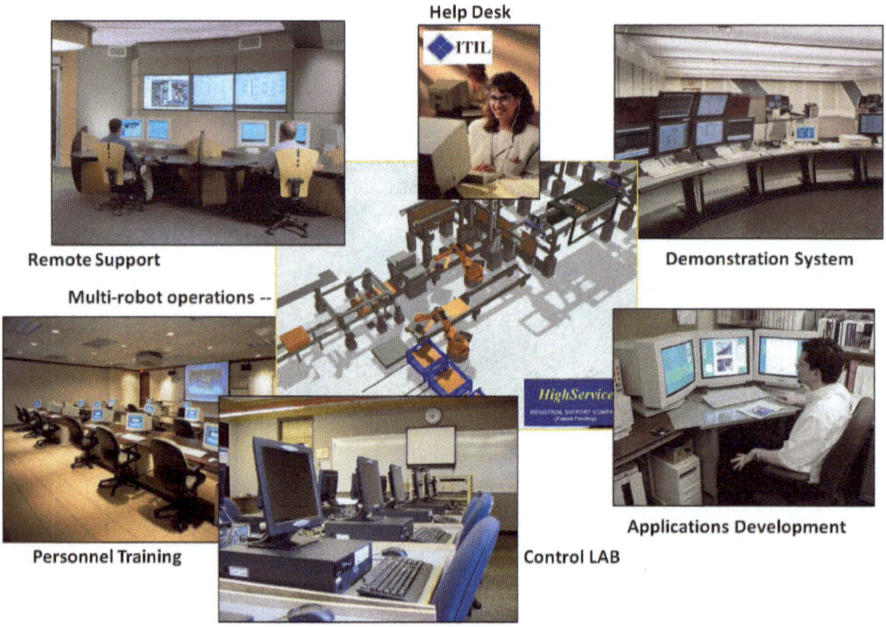

Fig. 1.21 e-Production; automation and control for pre-processing, mining, manufacturing, logistics and services (courtesy of CODELCO, Chile).

Chile). The bioprocess enriches copper ores in-site prior to mining, packaging and delivering, through integration of e-Production and e-Logistics (see Figure 1.21). The process is performed by intelligent and autonomous mining trucks, which collaborate through a GPS network (see Figure 1.22).

Fig. 1.22 (a) Wireless links autonomous mining, (b) High speed, large bandwidth wireless communication, high precision GPS for equipment position, and navigation control, (c) Autonomous driver-less mining trucks.

1.7.3 Example 3: Virtual Factory

Virtual manufacturing integrates and synthesize manufacturing environments to facilitate decision-making and control at different levels. Virtual manufacturing environments provide more powerful interface for managers and decision-makers to better interact with the shop floor, and the decision models/algorithms. Virtual Factory Framework, initiated by the European Commission 7th Framework Program (FP7), aims at providing a conceptual framework to simulate complex and dynamic behavior of collaborative manufacturing agents, their design, control, evaluation, and reconfiguration (see Figure 1.23). Digital factory is an alternative terminology, defined by VDI (association of German engineers), which refers to "*a comprehensive network of digital models, methods and tools, including simulation and 3D/VR visualization, which are integrated through continuous data management*" (Zulch and Stowasser, 2005).

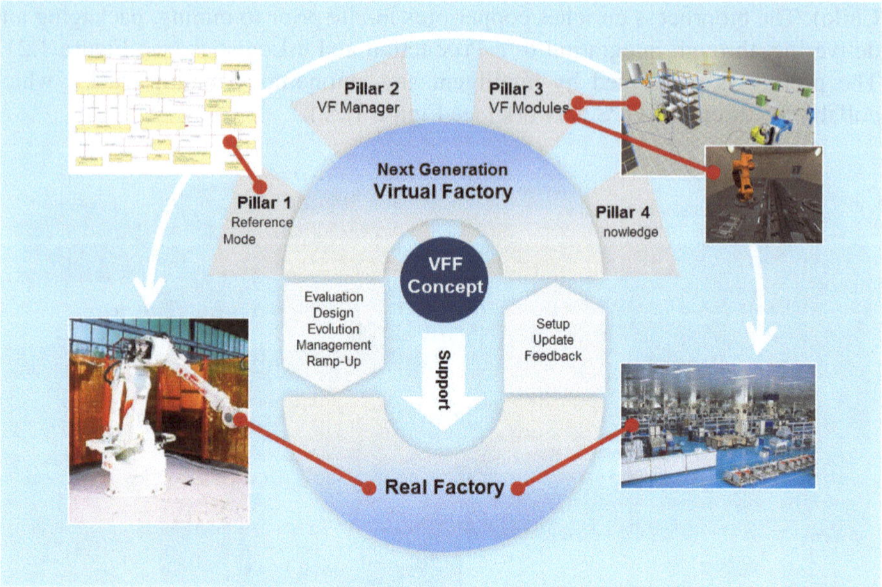

Fig. 1.23 The Virtual Factory Framework; (1) *Reference model*: factory data models for common representation of factory objects, (2) *Virtual factory manager*: acts as a server supporting I/O communications among the decoupled modules that require accessing the factory data model, (3) *Virtual factory Modules*: decoupled functional software tools responsible for services such as factory design, performance evaluation, and management, (4) *Knowledge*: Supports modeling and provides information and comprehension on the modeling phenomena (adapted from Pedrazzoli et al., 2007).

1.7.4 Example 4: e-Models in e-Commerce and e-Service

e-Models refer to the work-methods by which an enterprise provides its products and services to create value and generate revenues to sustain its profit and reputation. e-Models enhance global connectivity, reduce costs (e.g., communication, transaction, agency), improve interactivity, flexibility, and customization, and enable dynamic pricing and inter-enterprise interaction and collaboration. Examples of e-Models include (but are not limited to):

- *Virtual storefront*: sells products and services online (e.g., online marketing).
- *Marketplace concentrator*: combines the information from several providers and shares (or sells to) with the clients.
- *Online exchange*: bid-ask systems, multiple buyers and sellers.
- *Information broker*: provides information (e.g., prices, models) of products and services.
- *Transaction broker*: buyers view rates, terms from various sources.

Table 1.1 Examples of e-Models.

e-Model	Storefront	Market concentrator	Auction	Name-your-price	B2B exchange	Information broker
Example	Target	PriceGrabber.com	eBay	Priceline	Dell	fatwallet.com
Model mix	Information broker	Information broker	1. Market concentrator 2. Information broker	1. Market concentrator 2. Information broker	Storefront	None
Channel	Dual	Internet only	Internet only	Internet only	Dual	Internet only
Revenues	Product sales	Commission Advertisement	Commission Advertisement	Commission Advertisement	Product sales Providing service	Advertisement
Web Sevice	1. Product info 2. Store info 3. Search engine 4. Account management	1. Product info 2. Compare prices 3. earch engine 4. Compare products	1. Search engine 2. Seller rating 3. Bidding info	1. Search engine 2. Booking tour info	1. Product info 2. Search engine 3. Online technical support 4. Access customized DB	1. Product list 2. Link to source 3. Search engine 4. User forums
e-Work	HCI	Integration	Coordination	Coordination	Collaboration	Networking
e-Services	Online orders	e-Marketing	e-CRM	e-Supply	e-Logistics	e-Marketing
Advantages	1. Product tangibility 2. Visibility of the Store 3. Reduce customer inputs 4. Customer satisfaction	1. Product tangibility 2. Lower price 3. Easy to find a product 4. Help customers find favorite products	1. Easy to find a product 2. Easy to avoid fraud 3. Help the buyer bidding price 4. Guide the novice to use the site	1. Easy to find a deal 2. Help traveler find, reserve and plan hotel and car 3. Guide the novice to use the site	1. Product tangibility 2. Easy to find a product 3. Faster responses and prompt support	1. The latest info 2. Alternative products/sellers 3. Detailed info on products and services, views and feedback available

- Auction: electronic clearinghouse of products; shares prices and updates in response to the demand (Reverse auction: buyer sets the price and submits to multiple sellers).
- *Digital product delivery*: sells software, movies, music, photos, other digital products online.
- *Content provider*: creates revenue through providing clients content for a fee and advertising.
- *Online service provider*: provides service, support for hardware, and software products.
- *Business-to-business exchange*: automation of purchase and sale transactions from business to business, where many buyers and sellers are present.

Table 1.1 summarizes some best practices from different e-Models and compares their unique characteristics, e.g., channels, revenues, web services required, and the corresponding e-Systems.

1.8 Significance and Summary

As emerging e-Work systems are not the same as traditional work systems, their corresponding design considerations and criteria are also different. That is, the transformation from centralized systems to networks of intelligent and autonomous agents has brought up new design criteria and measures beyond classic measures such as productivity, efficiency, and effectiveness. The emerging criteria, henceforth called "e-Criteria", are basically anchored in the emerging characteristics of e-Work concepts; systems of systems, internet-of-things, as well as ever-increasing innovations in computers and ICT. In this context, although traditional measures continue to be valid, they need to be supported by emerging e-Criteria to enhance effective design of collaborative e-Work systems. Common e-Criteria, which reflect achievable, tangible and measurable benefits of e-Work, e-Business, and e-Service systems are:

1) *Integrability*: Ability to integrate data from distributed agents, thus increasing its usefulness. The two dimensions of information integrability are
   - *Connectivity*: Type, level, and quality of ICT-supported connections between distributed operating systems, application layer, and network layer, and
   - *Behavioral patterns*: Willingness to share necessary information.
2) *Agility*: Ability of a system, from individual agents to the whole manufacturing network, to respond and adapt to changes in real-time.
3) *Scalability*: Capability of a process, system, network, or economy to handle increasing amount of works/tasks and adapt to that growth.
4) *Reachability*: Ability and efficiency of interconnections and interactions between individual agents in distributed networks.

5) *Viability*: The ratio between costs of operating and sustaining distributed agents and the rewards gained from their service.
6) *Autonomy*: Level of delegation of authority, task assignments, and decentralization in distributed networks of entities, i.e., how active are agents, protocols, and other e-Work dimensions.
7) *Dependability*: Probability of a task to be successfully executed, requiring
   - *Availability*: Readiness of the system for service
   - *Reliability*: Continuity of service
   - *Sustainability and safety*: Avoiding catastrophic consequences, and
   - *Integrity and maintainability*: Eliminating improper alterations of information.
8) *Resilience*: a.k.a. *transformability* or *adjustability*, the ability to survive the unforeseen circumstances, risks, disruptions, and high impact events.

In addition, there are other tangible and intangible benefits. Design to achieve the benefits enabled by e-Work and optimized e-Systems (e.g., e-Business, e-Service) will require satisfying respective emerging e-Criteria. The challenges to designers of e-Systems are how to design e-Work models to represent the four wheels and the 15 e-Dimensions, and how to implement and benefit from the e-Work models to address the aforementioned (or other) e-Criteria. Collaborative control theory (CCT) has been developed over the last decade to support effective design of collaborative e-Work, e-Business, and e-Service systems through several design principles (Velasquez and Nof, 2009).

The theoretical foundation and design principles of CCT, emerging principles, and collaboration support systems to enable successful implementation are discussed in Chapter 2. Then, Chapter 3 reviews strategies and techniques for economic justification and rationalization of such systems with respect to various e-Criteria. After introduction of the fundamentals and rationalization, Chapter 4 discusses some basic and emerging models and architectures for optimization and control in collaborative e-Work systems, followed by a comprehensive discussion on collaborative e-Work tools in Chapter 5; decision support systems, coordination protocols, and software agents. Thereafter, we show how collaborative e-Work models and concepts can be applied to the design and development of products and services (Chapter 6), and production/manufacturing systems and supply networks (Chapter 7). Chapter 8 is specifically devoted to two emerging CIM technologies, sensors and RFIDs, and their applications in e-Production, e-Manufacturing, and e-Service. In Chapter 9, we review applications of collaborative e-Work in various e-Service industries, performance analysis, monitoring, and optimization, and basic design principles for e-Service. Applications of collaborative e-Work in the automation and augmentation of education through e-Learning and e-Training are reviewed in Chapter 10, along with basic design principles and performance monitoring/evaluation techniques. Emerging trends and challenges, and roadmaps to the design of next-generation collaborative e-Work systems are presented in Chapter 11.

References

Arsanjani, A.: Service-oriented modeling and architecture. IBM Technical Library (2004), http://www.ibm.com/developerworks/library/ws-soa-design1/ (November 09, 2004)

Anussornnitisarn, P., Nof, S.Y.: e-Work: The Challenge of the Next Generation ERP Systems. Production Planning and Control 14(8), 753–765 (2003)

Asimov, I., Looking Ahead, Foreword to Handbook of Industrial Robotics (S.Y. Nof, Ed.), John Wiley & Sons, New York (1985)

Basañez, L., Suárez, R.: Teleoperation. In: Nof (ed.) Springer Handbook of Automation, ch. 27, pp. 449–468 (2009)

Browne, J., Higgins, P., Hunt, I.: e-Business principles, trends and visions. In: Gasós, J., Thoben, K.-D. (eds.) e-Business Applications Technologies for Tomorrow's Solutions, pp. 3–16. Springer, New York (2003)

Camarinha-Matos, L.M., Afsarmanesh, H.: Collaborative networks: Value creation in a knowledge society. In: Wang, K., Kovacs, G., Wozny, M., Fang, M. (eds.) Proc. PROLAMAT 2006, IFIP Int. Conf. Knowl. Enterp. – New Challenges. IFIP, vol. 207, pp. 26–40. Springer, Boston (2006)

Camarinha-Matos, L.M., Afsarmanesh, H.: Concept of collaboration. In: Putnik, G., Cunha, M. (eds.) Encyclopedia of Networked and Virtual Organizations, Hershey, Pennsylvania (2008)

Cassidy, A.: A Practical Guide to Planning for e-Business Success: How to e-Enable Your Enterprise. CRC, New York (2002)

Colombo, A.W., Karnouskos, S., Mendes, J.M.: Factory of the Future: A Service-oriented System of Modular, Dynamic Reconfigurable and Collaborative Systems. In: Artificial Intelligence Techniques for Networked Manufacturing Enterprises Management. Springer Series in Advanced Manufacturing, pp. 459–481 (2010)

Han, S.H., Chin, K.H., Chae, M.J.: Evaluation of CITIS as a collaborative virtual organization for construction project management. Autom. Constr. 16(2), 2–199 (2007)

Ho, L.T., Lin, G.C.I.: Critical success factor framework for the implementation of integrated-enterprise systems in the manufacturing environment. Int. J. Prod. Res. 42(17), 3731–3742 (2004)

Hsieh, F.S., Chiang, C.Y.: Collaborative composition of processes in holonic manufacturing systems. Computers in Industry 62, 51–64 (2011)

Huang, C.Y., Nof, S.Y.: Models of Material Handling and Robotics. In: Brandimarte, P., Villa, A. (eds.) Modeling Manufacturing Systems: From Aggregate Planning to Real Time Control. Springer (1999)

Jiao, R.J., Pokharel, S., Kumar, A., Zhang, L.: Development of an online quality information system for e-Manufacturing, J. Manuf. Technol. Manag. 18(1), 36–53 (2007)

Klein, M.: Integrated support for cooperative design coordination: Managing processes, conflicts, and memories. In: Nof, S.Y. (ed.) Information and Collaboration Models of Integration, pp. 435–459. Kluwer Academic Publishers (1993)

Ko, H.S., Nof, S.Y.: Design and application of task administration protocols for collaborative production and service systems. International Journal of Production Economics 135, 177–189 (2012)

Köksal, G., Batmaz, I., Testik, M.C.: A review of data mining applications for quality improvement in manufacturing industry. Expert Systems with Applications 38, 13448–13467 (2011)

Krajewska, M.A., Kopfer, H.: Collaborating freight forwarding enterprises. OR Spectrum 28(3), 3–301 (2006)
Li, H., Fan, Y., Dunne, C., Pedrazzoli, P.: Integration of business processes in Web-based collaborative product development. Int. J. Comput. Integr. Manuf. 18(6), 452–462 (2005)
Malamateniou, F., Vassilacopopulos, G.: Developing a virtual patient record using XML and web-based workflow technologies. Int. J. Med. Inform. 70(2/3), 591–598 (2003)
Nagalingam, S.V., Lin, G.C.I.: CIM-still the solution for manufacturing industry. Robotics and Computer-Integrated Manufacturing 24, 332–344 (2008)
Nof, S.Y.: Modeling of e-Work with TIE, a teamwork integrator evaluator. Research Memo 99-20 (School of Industrial Engineering, Purdue University) (1999)
Nof, S.Y.: Design of effective e-Work: Review of models, tools and emerging challenges. Production Planning and Control, Special Issue on e-Work: Models and Cases 15, 8–681 (2003)
Nof, S.Y.: Collaborative e-work and e-manufacturing: challenges for production and logistics managers. Journal of Intelligent Manufacturing 17, 689–701 (2006)
Nof, S.Y.: Collaborative control theory for e-Work, e-Production, and e-Service. Annual Reviews in Control 31, 281–292 (2007)
Nof, S.Y.: Automation: What It Means to Us Around the World. In: Nof (ed.) Springer Handbook of Automation, ch. 3, pp. 13–52 (2009)
Park, N.: Collaboration and integration of the shared process system with workflow control. Prod. Plan. Control 14(8), 743–752 (2003)
Pedrazzoli, P., Sacco, M., Jönsson, A., Boer, C.R.: Virtual Factory Framework, Key enabler for future manufacturing. In: Cunha, P.F., Maropoulos, P.G. (eds.) Digital Enterprise Technology, pp. 83–90. Springer, US (2007)
Rosen, A.: The e-Commerce Question and Answer Book. American Management Association, New York (2002)
Shen, W., Hao, Q., Yoon, H.J., Norrie, D.H.: Applications of agent-based systems in intelligent manufacturing: An updated review. Advanced Engineering Informatics 20, 415–431 (2006)
Su, S.Y.W., Meng, J., Krithivasan, R., Degwekar, S., Helal, S.: Dynamic inter-enterprise workflow management in a constraint-based e-Service infrastructure. Electron. Commer. Res. 3(1/2), 9–24 (2003)
Velásquez, J.D., Nof, S.Y.: Collaborative e-Work, e-Production, and e-Service. In: Nof, S.Y. (ed.) Springer Handbook of Automation, ch. 88, pp. 1549–1576 (2009)
Weerakkody, G., Ray, P.: CSCW-based system development methodology for health-care information system. Telemed. J. e-Health 9(3), 273–282 (2003)
Zulch, G., Stowasser, S.: Editorial: The Digital Factory: An instrument of the present and the future. Computers in Industry 56, 323–324 (2005)

Review and Challenge Questions

1. Is collaboration a means, or a goal, or both? Explain and illustrate the participants and their collaboration roles (a) relative to your own work experience. (b) for the case of (i) a teacher; (ii) an artist; (iii) a farmer; (iv) an assembly robot; (v) a truck driver.

2. Consider the work of a design engineer working with a computer system. Identify in this case: (i) the e-Work; (ii) the e-System; (iii) the e-Activities. For each, explain your answer.
3. Repeat Question 2 for (a) a dentist; (b) a dancer; (c) a restaurant server; (d) a bank worker; (e) an airplane pilot; (f) a store manager.
4. For the cases in Questions 2 and 3, (a) explain the role of collaboration; (b) Are there any circumstances under which they can work and function effectively without collaboration?
5. Think: Can there be any work that can be accomplished well without e-Work? Explain and illustrate.
6. It is said that Work $\neq$ e-Work. Using the four wheels of collaborative e-Work, explain and illustrate this inequality.
7. Prepare a table containing a column for each of the following business e-models: (1) storefront model, (2) market concentrator model, (3) auction model, (4) name-your-price-model, (5) B2B exchange model, and (6) another model of your choice. For each model (column), find and list one or two examples of e-Businesses (e-organizations) that operate under the corresponding e-model. Visit the Web site of each of the organizations that you have selected.
8. Answer briefly the following questions (in the table rows and below the table of Question 7):

   (a) Model Mix: Do the companies operate following one or more of the defined e=business models (e.g., storefront and auction)? If so, which models do they implement?

   (b) Channel: Are the companies Internet-only companies, or brick-and-mortar businesses?

   (c) Revenues: How do the companies generate, or plan to generate revenue?

   (d) Web services: In each column, list 4-6 e-Activities that provide unique value to the client/visitor/reader. Below the table, explain briefly what these e-Activities are, and what the advantage of providing them on the Internet is.

   (e) e-Work functions: For the organization(s) you listed in each column, illustrate one function of e-Work that can specifically improve this particular e-Activity. Below the table, explain why it can improve. (Select a different e-Work function for each column.)

   (f) e-Service: For each column, illustrate one key function of e-Service unique to this model.

   (g) Access: Explain the difference between the Internet, extranet, and intranet, and illustrate if any are used in each of the six models.
9. (a) What are the four main domains (wheels) of knowledge needed for the design of effective collaborative e-Work? Can e-Work be effective without them? With just some of them?

   (b) Explain briefly what the 15 e-Dimensions are, and their purpose. Illustrate them with the four examples at the end of the chapter.

(c) Describe three to five functional challenges in e-Production, e-Business, and e-Service, and illustrate how each of the four domains (wheels) would be applied to overcome such challenges.

10. Consider library search for books in three eras: (i) manual only, before computers; (ii) supported by computer database catalogs; (iii) Internet era. For each era, identify and explain:
    a) Who collaborates with whom, and the purpose of their collaboration(s);
    b) What are the advantages of e-Work, if any, in each era? Disadvantages?
    c) Rank the library search for a particular book during each era for (i) Speed of search completion; (ii) quality of the search outcomes (from 1 being slowest or poorest, to 3 being fastest or best).
    d) Repeat Question (c) for a library search, this time for a group of several books written on the same topic.
    e) Suggest other metrics to compare the library search in the three eras, and rank them too.
11. Repeat Question (10) for driving navigation in three eras: (i) paper printed maps; (ii) GPS based navigation; (iii) GPS based navigation supported by crowd sourcing of the current roads' conditions.
12. Review the chapter and Questions (1) through (11) and explain several examples of (i) improved performance and outcome quality by e-Work just by including better available communication; (ii) improved performance and outcome quality by collaborative e-Work including engineered collaboration. (An example for (i) – a designer applying computer file and communicating with customers to consider design changes. An example for (ii) – same example, with software agents reviewing the proposed changes and alerting for design conflicts or cost and time conflicts.

Lab Questions

13. Obtain access to an e-Work system. Examples: A search engine; a Computer Aided Design system; MERP (Mobile ERP, by MBE Simulations), or similar. Read the tutorial and explain briefly: (i) What kind of e-Business or e-Service does it represent? (ii) What collaboration functions does it support? (iii) What e-model does it represent? (iv) How does it compare with the four case examples at the end of the chapter?
14. Comparison between e-Work systems: Surf and select your "e-Work Pair" -- two similar web based systems or companies that are in the same type of business or service; write their URLs. (For example, two different search engines; two different ERP/CRM e-Learning systems; or two organizations selling real-estate, travel, cars, cameras, etc., or two universities, state governments, etc.) After spending the same time (no more than 15 minutes) in each site, compare the two Web sites along five criteria (hint: a comparison table is preferred):

(i) Usefulness of information you can obtain; (ii) variety of web services they offer; (iii) ease of use; (iv) availability of search tool(s); (v) "cluttered" or "attractive" design. Be specific and illustrate each comparison with specific and relevant examples. (Note: We will cover in later chapters detailed methods, models and criteria for e-Work design comparisons; simple yet thoughtful answers are expected here.)

Chapter 2
Design with Collaborative Control Theory

Abstract. Recent developments in the theoretical foundations of e-Work, the four wheels and the corresponding 15 e-Dimensions, are the major enablers of emerging e- Business and e-Service. In line with the need for decentralization of work, emerging e-Systems are increasingly influenced by e-Work and applications of agent-based systems. The opportunities provided by e-Work, however, are always associated with emerging challenges and complexities, e.g., the scalability of workflow, information, and task overload. The intrinsic advantages, opportunities, and sustainability of the emerging e-Systems and e-Activities will not materialize without effective design for e-Work through collaboration engineering. The objective of this chapter is to introduce and discuss several design principles of collaborative control theory (CCT), and their application for better understanding and addressing these emerging requirements and challenges. Implementation of the CCT design principles, in turn, requires efficient design of effective software agents and programming frameworks, enabling the automation and integration of flexible, adaptive, responsive and heterogeneous e-Systems. These design issues will be addressed further in Chapter 4, and software engineering design and programming-oriented implementation of collaborative e-Work will be discussed in Chapter 5 in detail.

2.1 CCT Design Principles for Collaborative e-Work

In this section, fundamental principles for the design of collaborative e-Work, which have emerged over the years as useful results of CCT, are described. Their common objective is to augment the e-Work, and the achievements of e-Systems, and enable individual agents, from nano and micro objects all the way to global enterprises to accomplish their individual and organizational common goals.

Collaboration is known to be an essential means for effective design and control of e-Work. Collaboration enables all involved entities of decentralized e-Systems to share their resources, information, and responsibilities, such that mutual benefits are obtained. The amount and level of collaboration, however, may vary from minimal information sharing, to fully collaborative networks. Consider, for simplicity, a "bag handling system" as an example of a collaborative

S.Y. Nof et al., *Revolutionizing Collaboration through e-Work, e-Business, and e-Service*,
Automation, Collaboration, & E-Services 2, DOI: 10.1007/978-3-662-45777-1_2

system, where three robots are assigned to unload bags from a conveyor and load them on a truck. (We will discuss various basic and advanced material handling systems in Chapter 7). Suppose the conveyor delivers the bags to the robots at variable times, sometimes several bags appear together. Which of the three robots should handle the next arriving bag? If they all attempt to pick it up at the same time, collisions will occur, perhaps even causing some damage. If none of them picks it up, bags will fall and pile up. For heavy bags, two robots, or all three need to coordinate the handling by sharing the load. For three trained human workers assigned to the same job, we could assume that they will collaborate well together. It may not always be true, especially if the tasks are relatively more complex. We use this simple but basic example throughout our discussions on different CCT principles (more complex assumptions will be made later on).

In collaborative e-Work terminology, the bag handling system described above is a network of three intelligent and autonomous agents; their *task* is to pick up bags from the conveyor and place them on the truck, and their *mutual goal* is to improve the efficiency of the system in terms of cycle time; the time consumed to unload a certain number of bags from the conveyor and load them on the truck. The type of collaboration in this system can be one or a combination of the following (Nof, 1999):

- *Mandatory collaboration*: Two or more agents (robots, in this case) must process a task together, simultaneously. The collaboration between robots is mandatory in handling the *heavy* bags.
- *Optional collaboration*: One agent may or may not be sufficient for processing all tasks. In our example, instead of deploying three small robots (that are not capable of handling heavy bags on their own), a single bigger robot can be used, which is capable of handling all types of bags without requiring any help from any other robot. (We will see, however, that this scenario is not preferred, especially when several bags arrive simultaneously).
- *Concurrent collaboration*: Two (or more) agents process portions of the same task (i.e., handle the same bag), such that the overall task completion time is reduced. This type of collaboration does not have any contribution to the bag handling system. That is, if the collaboration is not mandatory for processing a given task (i.e., one robot is sufficient for handling a bag), it is not optimal to assign multiple robots to it. Nevertheless, this type of collaboration makes sense if the handling time is reduced by adding another robot (in case there is a free robot available in the system at that time).

Based on the above definitions, the type of collaboration that we prefer for the robotic bag handling system is mandatory collaboration, i.e., collaborate if required.

2.1.1 *Collaboration Requirement Planning (CRP)*

The first step necessary is to rely on planned collaboration, as opposed to ad hoc, unplanned collaboration. Effective collaboration requires advanced planning and feedback loops, as targeted by the CRP principle. CRP is composed of two phases:

CRP-I: Plan Generation

In CRP-I, a detailed requirement plan is generated, based on the work objectives and available resources, to answer the following questions: *With whom*, *how*, and *when* to collaborate. CRP-I generates a requirement plan based on a Collaboration Requirement Matrix (CRM), representing the mutual relationships between the *tasks' requirements* and the *available resources*. The entries of CRM represent 1) performance capabilities of resources in processing different tasks, 2) resource constraints, and 3) resource-task compatibility. Let T and R denote the set of tasks and the set of resources, respectively. The CRM can then be expressed through the following mapping in $\mathbb{R}$:

$$CRM : A(R)\times T \rightarrow \mathbb{R},$$

where $A(R)$ denotes the set of resources allowed for processing each task, and $CRM(r,t)$ represents the normalized score of resource r for processing task t, taking into account performance capabilities, resource constraints, and compatibilities. Accordingly, if a resource r is unable to process a task t, then $CRM(r,t)=0$. Similarly, if a resource r is the best fit for processing a task t, then $CRM(r,t)=1$. Let N denote the number of elements in R. Then, $A(R)$ is a subset of the power set (excluding the null set) of R, i.e., the maximum number of elements in $A(R)$ is $2^N - 1$. In practice, due to external collaboration limitations and/or physical limitations between collaborating resources, set $A(R)$ is much smaller than the power set.

In the bag handling example, each bag, which may have a certain *type* (type may be defined in terms of size, weight, dimension, etc. of bags) represents a specific task (t), and each particular robot or *team* of robots represents a specific resource. In our example, the teams may be composed of one, two, or three robots. That is, having robots ROB_1, ROB_2, ROB_3, there are $N = 7$ different teams (i.e., resource groups) as follows: $r_1 = \{ROB_1\}$, $r_2 = \{ROB_2\}$, $r_3 = \{ROB_3\}$, $r_4 = \{ROB_1, ROB_2\}$, $r_5 = \{ROB_1, ROB_3\}$, $r_6 = \{ROB_2, ROB_3\}$, $r_7 = \{ROB_1, ROB_2, ROB_3\}$. To illustrate the CRM, consider three tasks (i.e., bags) t_1, t_2, and t_3, which respectively require one, two, and three robots to be handled. In this case, CRM is a 7×3 matrix as follows. (Note that for simplicity, we have assumed that all robots are similar):

$$CRM = \begin{bmatrix} 1 & 0 & 0 \\ 1 & 0 & 0 \\ 1 & 0 & 0 \\ 1 & 1 & 0 \\ 1 & 1 & 0 \\ 1 & 1 & 0 \\ 1 & 1 & 1 \end{bmatrix}.$$

The above CRM implies that all resources (i.e., teams of robots; $R = \{r_1, r_2, \ldots, r_7\}$) can process the first task t_1, task t_2 can only be processed by the last four resources (i.e., teams of two or three robots), and task t_3 can only be processed by the last resource (i.e., team of three robots). Note that since all robots are similar, the entries of CRM happened to be binary (Entries of CRM may not be all binary if the robots are different). The plans for processing tasks t_1, t_2, and t_3 by resources R, e.g., resource allocation, sequencing, etc., can then be generated based on the CRM. CRP-I is a simplified version of Task Requirement Analysis Protocol (TRAP) in TAP, the task administration protocol. (We will discuss TAP in Chapter 9 in detail).

CRP-II: Plan Execution & Revision

CRP-II executes the *plan* generated by CRP-I in real-time and revises the plan following spatial and temporal challenges, changes, and constraints. The purpose of CRP-II is to assign or reassign tasks to resources based on their mutual scores in the CRM. CRP-II identifies all existing plans for each task, resolves the conflicts between the plans, and supports collaboration within a plan through a best-fit search protocol. The search is performed over specific state spaces, representing the partial plans (i.e., for a specific set of tasks). The state level represents the number of tasks included in the plan, where $\varnothing$ level denotes an empty state. The states in the open list are ordered following the *best-first* discipline (which may be performed based on different criteria). The first (i.e., the best) state in the list is selected for expansion (if required) at the beginning of the search iterations. If it contains all tasks, then no expansion is required; the optimal global plan is obtained and thus the search stops. Otherwise, the state is expanded through merging it with the plans associated with the additional tasks. Let *OPEN* represent the open list of the states. The *best-fit search protocol* is then executed as follows:

Step 1: $OPEN = \varnothing$.

Step 2: Find all tasks with no precedence constraints. Update *OPEN* to the best-first order of the set of all plans for such tasks.

Step 3: Select the first element of *OPEN*. Call this element *FIRST*.

Step 4: If *FIRST* includes all tasks, report *FIRST* as the global optimal plan, and stop. Otherwise, continue.

Step 5: Find the set of feasible tasks whose predecessors are already included in *FIRST* (and also *OPEN*) inclusion in the current state will not violate the system's constraints. Call this set *TASK*.

Step 6: Find all plans *P* associated with each task *t* in *TASK*, merge *P* into *FIRST*, and generate the successor states *SUCCESSOR*.

Step 7: Compute the cost for each element in *SUCCESSOR*.

Step 8: Merge *SUCCESSOR* into *OPEN* according to the best-first order.

Step 9: Return to Step 3.

In the bag handling example, the potential plans generated for pickup and loading need to be updated continuously following the best-fit search protocol. Each plan may define which bag is picked by which team of robots, when, and in what order. (Note that each individual bag represents a specific task). That is, each plan defines the sequence of bags on the conveyor and their arrival time and the formation of the teams of robots, accordingly, and assigns each bag to a specific team for pickup and loading on the truck. Naturally, the plans must be updated over time as new bags arrive in the system. In case the sequence of loading is important (e.g., big/small bags first), the generation and updating of the plans through CRP become even more complicated.

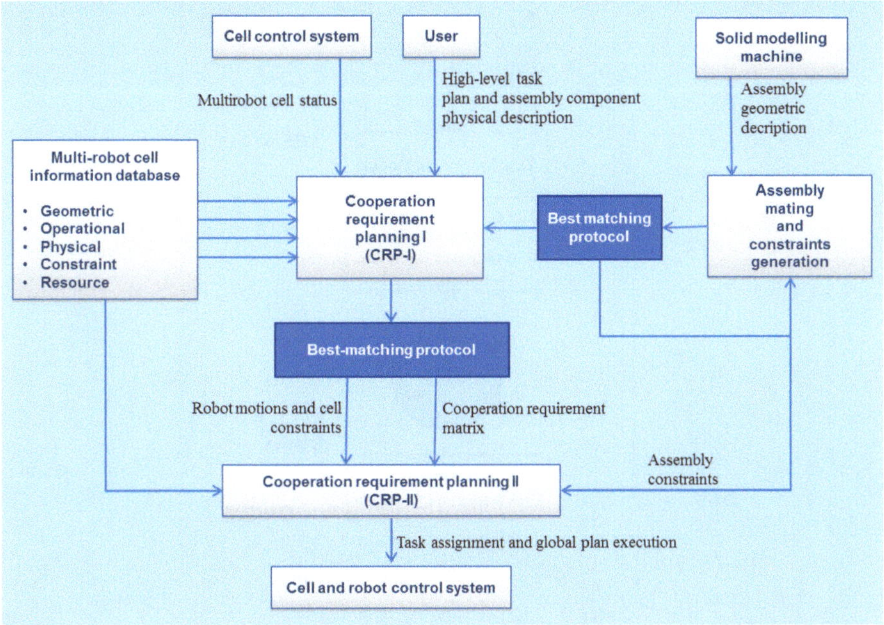

Fig. 2.1 The CRP architecture (best-matching protocol-enabled) for multi-robot assembly; *Geometric information*: specifies the work-cell layout and robot kinematic capabilities; *Operational information*: specifies the non-geometric processing capabilities; *Physical information*: specifies physical attributes.

CRP is the foundation for Shared Resource Allocation Protocol (SRAP) and Synchronization and Time-Out Protocol (STOP) in TAP. Fig. 2.1 shows an application of CRP principle in a multi-robot assembly system, coupled with best matching protocol (described later on in this chapter), which aims to enhance the flexibility and reliability of multi-robot cells. The framework shown in Fig. 2.1 can be used for cell design and optimization, multiprocessor collaboration, and integrated process planning and scheduling. Real-time applications of CRP requires deployment of powerful learning mechanisms (e.g., neural networks, machine learning) and adaptive planning to further enable timely adaptive modifications based on the CRP-I--CRP-II feedback loops. We will discuss it further in the context of intelligent collaboration in Chapter 4.

2.1.2 e-Work Parallelism (EWP)

The EWP principle exploits the fact that in any e-System, as a distributed network of agents (from nano-sensors to business and service enterprises), some activities

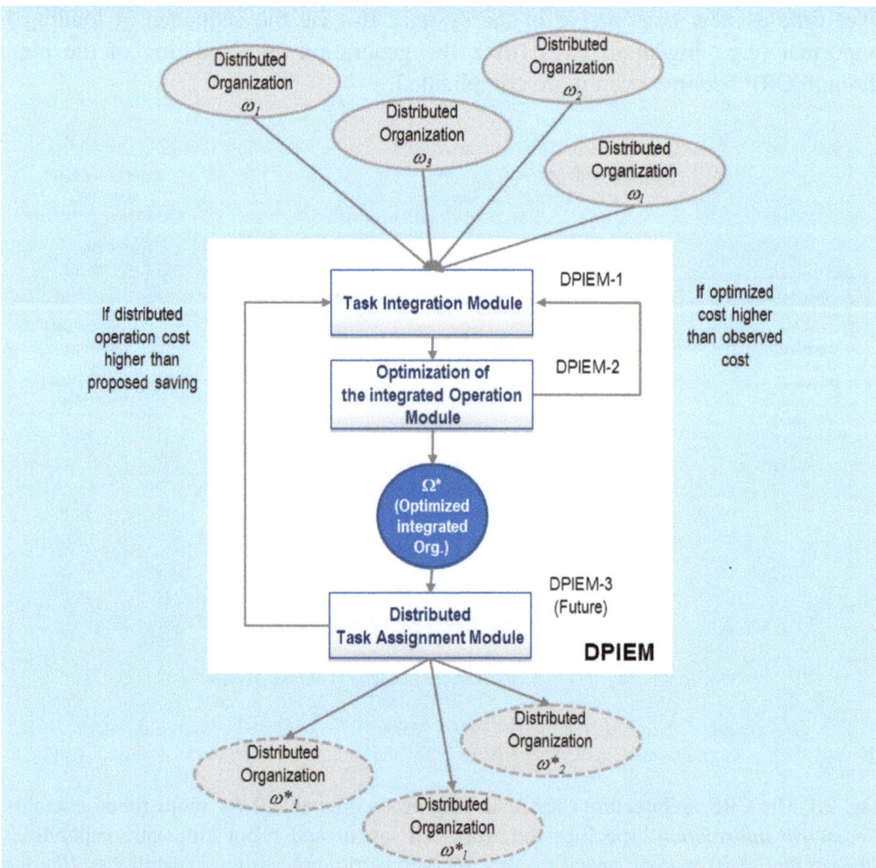

Fig. 2.2 The DPIEM architecture.

can/should be performed in parallel. In e-Work, the principle of EWP has deeper implications due to the distributed nature of the e-Systems and the fact that interactions take place through human and software workspaces, including 1) human-human interactions, 2) human-machine and human-computer interactions, and/or 3) machine-machine and computer-computer interactions. EWP is planned and performed through the Distributed Parallel Integration Evaluation Model (DPIEM), which optimizes task-based integration of workflow considering information availability, task dependence, and operational constraints. The DPIEM is composed of three main modules as described below (Fig. 2.2).

Task Integration Module (DPIEM-1)

In this module, the tasks corresponding to different distributed agents ω_a are identified and analyzed in order to generate an integrated agent Ω and the respective task graphs G_Ω (see Fig. 2.2). Task integration is performed through the following steps:

Step 1: *Task dependency analysis* – identifies the precedence relations between the tasks, at both distributed (within an agent ω_a) and integrated (Ω) levels. According to the tasks' information, the analysis of tasks' dependencies classifies the task data elements into four main categories (see Fig. 2.3):

X: Data elements required during the execution of task
Y: Data elements required and changed during the execution of task
Z: Data elements generated and further required during the execution of task
W: Data elements generated during the execution of task *t*

Accordingly, two tasks t_1 and t_2 can be executed in parallel if they do not have flow dependencies and output dependencies (Fig. 2.3). In other words,

$(X_{t_1} \cup Y_{t_1} \cup Z_{t_1}) \cap (W_{t_2} \cup Y_{t_2} \cup Z_{t_2}) = \varnothing$, (No flow dependency from t_1 to t_2)

$(W_{t_1} \cup Y_{t_1} \cup Z_{t_1}) \cap (X_{t_2} \cup Y_{t_2} \cup Z_{t_2}) = \varnothing$, (No flow dependency from t_2 to t_1)

$(X_{t_1} \cup Y_{t_1} \cup Z_{t_1}) \cap (X_{t_2} \cup Y_{t_2} \cup Z_{t_2}) = \varnothing$, (No concurrent outputs)

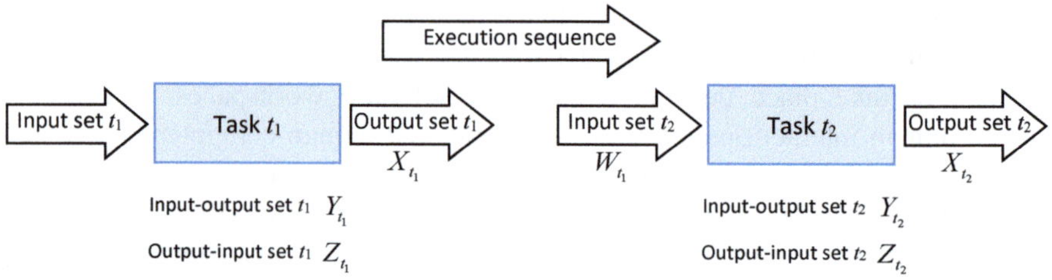

Fig. 2.3 Data element set for the dependency analysis.

Step 2: *Task integration* – integrates and assigns tasks to distributed agents ω_1 and ω_2 under three different situations regarding the products/services produced or delivered:

(a) *Identical products/services*: In this situation, two cases are considered: (i) the distributed agents perform the same process (the integrated set (union) of tasks is the same as the set of tasks taking place by either of the distributed agents), (ii) the distributed agents perform at least one task that is not the same for both agents. In case (ii), two alternatives are considered; to either modify or not modify the uncommon tasks, taking into account the costs and rewards associated with the task modification. Fig. 2.4-(a) shows the task graph G_Ω corresponding to agents ω_1 and ω_2 with dissimilar processes and non-modifiable tasks.

(b) *Complementary products/services*: This situation requires certain collaboration mechanisms among agents. The primary information required for the integration of the tasks between agents ω_1 and ω_2 is the sequence of the processes corresponding to each agent (Fig. 2.4-(b)).

(c) *Unrelated products/services*: In this situation, the execution sequence of the tasks between agents ω_1 and ω_2 is unrelated. The integration process is then performed according to the dependencies of the tasks' inputs and outputs in each individual agent (Fig. 2.4-(c)).

To better understand the mechanics of DPIEM-1, let's take a look at the example of robotic bag handling system again. DPIEM-1 analyzes the dependencies between the tasks (i.e., bags arriving on the conveyor), then integrates and allocates them to the distributed agents (i.e., teams of robots) that are dynamically created. Recall that in this example, there can be seven teams composed of one, two or three robots. In this case, the dependencies must be defined with respect to two factors and their relationship: (1) Arrival times of the bags; (2) Possibility of forming the qualified teams. Then, integration and sequencing can be performed based on the defined dependencies, in a way that the generated plan is feasible. For instance, two *heavy* bags (that require teams of two or three robots) cannot arrive at the same time, since the available resources are not enough to process them simultaneously, and as a consequence, one of them will fall and pile up (This issue requires effective communication between the *task agents*, as described later on). Hence, the sequence (and the arrival times) of the tasks in the integrated task graph, which is type (a) (see Fig. 2.4), must be consistent with their types along with the availability of resources in the integrated agent $\Omega = \{ROB_1, ROB_2, ROB_3\}$. Depending on the type of the task, specific robots/teams of robots may be called from Ω, where teams of the same size can interchangeably process each other's allocated tasks, e.g., $r_4 = \{ROB_1, ROB_2\}$ may process a task associated with $r_6 = \{ROB_2, ROB_3\}$.

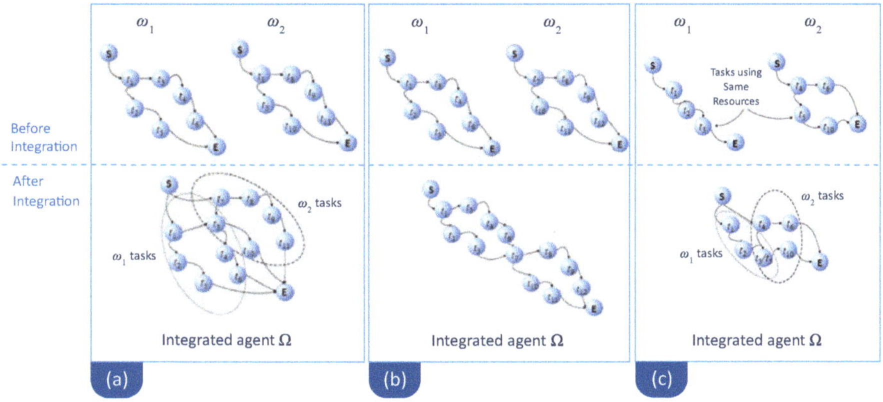

Fig. 2.4 Integration of task graphs for producing (a) identical (different processes; non-modifiable tasks), (b) complementary, and (c) unrelated products/services.

Task Parallelism Optimization Module (DPIEM-2)

This module considers the economic aspects of EWP through optimizing Φ; the total cost of the integrated agent Ω. The total cost Φ is composed of two main components (see Fig. 2.5):

(1) *Communication cost* (Γ): refers to the non-operational costs of executing tasks in each node of the integrated graph G_Ω, considering the information, parts, materials, tools, etc. being shared among the task agents. The communication cost is composed of three elements:

(a) *Sending the information by the source task*:

$$\gamma^1_{t_i,t_j} = \max\left\{1, \frac{p_j}{p_i}\right\} c^i_{st} + L\left(\frac{1}{p_i}\right) c^i_{tr}, \tag{2.1}$$

(b) *Network delay*:

$$\gamma_{t_i,t_j}^2 = \frac{L}{p_i \max\{1, p_j/p_i\}} c_n, \tag{2.2}$$

(c) Receiving the information by the target task

$$\gamma_{t_i,t_j}^3 = \max\left\{1, \frac{p_i}{p_j}\right\} c_{st}^j + L\left(\frac{1}{p_j}\right) c_{re}^j, \tag{2.3}$$

where

p_x	# of parallel agents available to task t_x	c_{st}^x	Startup cost at t_x
c_{tr}^x	Transmission cost at t_x (per byte)	c_{re}^x	Receiving cost at t_x (per byte)
L	Message length (in bytes)	c_n	Network cost (per byte)

The communication cost between tasks t_i and t_j is then formulated as follows:

$$\Gamma = \gamma_{t_i,t_j}^1 + \gamma_{t_i,t_j}^2 + \gamma_{t_i,t_j}^3. \tag{2.4}$$

(2) *Operational cost* (Π): refers to the costs associated with the execution of tasks in each node of the integrated graph G_Ω, considering the "*critical path*". According to the *Amdahl's law*, the execution cost of a task has two major components: serial and parallel. The serial component must be processed by one agent, while the parallel component can be executed by two or more agents. Then, the completely serial task execution cost is formulated as follows:

$$\pi_{t_i} = \left(\alpha_{t_i} + \frac{1-\alpha_{t_i}}{p_{t_i}}\right)\rho_{t_i}, \tag{2.5}$$

where

α_{t_i}	The serial portion of t_i ($0 \le \alpha_{t_i} \le 1$)	ρ_{t_i}	Reliability of the agents executing t_i

The operational cost for the task graph is then computed by adding the execution cost of the tasks in the *critical path* of the graph:

$$\Pi = \sum_{\substack{\text{Critical}\\ \text{path}}} \left(\pi_{t_i} + \max_i \left\{ \gamma^2_{t_{pr_i}, t_i} \right\} \right), \tag{2.6}$$

where t_{pr_i} denotes a predecessor of task t_i.

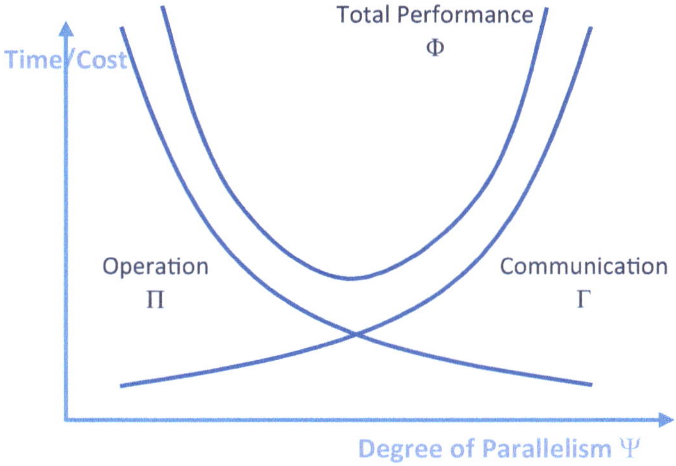

Fig. 2.5 Trade-off between the communication (Γ) and operational (Π) costs and the behavior of the total cost (Φ) of the integrated agent Ω.

A function over a task graph that yields the maximum number of its concurrent (parallel) agents needs to be defined to obtain the optimal *degree of parallelism* (DOP). DOP provides more information than the number of agents, considering alternative parallel implementations (see Ceroni and Nof, 2002). DOP, denoted by Ψ, can be calculated for distributed and integrated task graphs, where Ψ^* denotes the optimal DOP for an integrated task graph G_Ω. Interestingly, DOP is sensitive to the *gap* between the numbers of agents available to each task. As shown in Fig. 2.5, the communication cost is higher for higher DOP. However, according to Eqns. (2.1) to (2.3), if the number of agents available to each task is the same (regardless of its value), the respective communication cost (Γ) is the minimum, and even decreases for higher numbers of agents available to each task given that the number is equal for all tasks. Indeed, finding the optimal DOP Ψ^* is equivalent to finding the optimal number of agents available to each task, considering both the communication and operational costs, along with the cost of agents that has been discussed here.

Consider the task graph corresponding to tasks t_1 to t_6 (before integration) shown in Fig. 2.4-(a), and suppose each task represents a bag arriving on the conveyor in the robotic bag handling system described earlier. Table 2.1 presents the required information for each task. The number of available (parallel) agents in the system for the tasks that require one or two robots is three (i.e., r_1 = {ROB_1}, r_2 = {ROB_2}, r_3 = {ROB_3} for the bags that require one robot; r_4 = {ROB_1, ROB_2}, r_5 = {ROB_1, ROB_3}, r_6 = {ROB_2, ROB_3} for the bags that require two robots), while there is only one team available to the tasks that require three robots (i.e., r_7 = {ROB_1, ROB_2, ROB_3}). The communication cost between tasks t_3 (as transmitter) and t_4 (as receiver), for instance, can then be calculated as follows (see Eqns. (2.1)-(2.4)):

$$\gamma^1_{t_3,t_4} = c^i_{st} + \frac{L}{3} c^i_{tr}, \text{ (Cost of sending information by } t_3\text{)}$$

$$\gamma^2_{t_3,t_4} = \frac{L}{3} c_n, \text{ (Network delay cost)}$$

$$\gamma^3_{t_3,t_4} = 3c^j_{st} + Lc^j_{re}, \text{ (Cost of receiving information by } t_4\text{)}$$

$$\Gamma = c^i_{st} + 3c^j_{st} + L\left(\frac{c^i_{tr} + c_n}{3} + c^j_{re}\right). \text{ (Communication cost)}$$

The overall communication cost can be obtained by calculating the communication costs between each pair of tasks in a similar manner as shown above, indeed by defining proper values for the cost and message length parameters.

Table 2.1 Tasks information; Robotic bag handling system

Task	t_1	t_2	t_3	t_4	t_5	t_6
# of robots required	1	2	2	3	1	2

Calculation of the operational cost for the robotic bag handling example requires some additional assumptions. The processing time of each task is equal to the time consumed to carry a bag from the conveyor to the truck. It is natural that a bag that requires only one robot is carried faster that the heavier ones that require two or three robots. In addition, heavier bags consume more resources (two or three robots) than the lighter bags. Therefore, the critical path, in our example, is the path composed of the tasks that require more resources (i.e., t_1-t_3-t_4-t_6). The other issue is that although some tasks have several alternative agents to process these tasks, they can only be processed by one of them. That is, after a team of (one or two) robots picks up a bag, no other team interferes with the process and the bag has to be carried and placed on the truck by the same team. Hence, the tasks have no parallel portion (i.e., $\alpha_{t_i} = 1,\ \forall i = 1,\ldots,6$). Accordingly, the operational cost can be calculated as follows (see Eqns. (2.5) and (2.6)):

$$\Pi = \rho_{t_1} + \rho_{t_3} + \rho_{t_4} + \rho_{t_6} + \frac{L}{3} c_n.$$

The numerical value of the operational cost can then be defined by assigning proper values to the reliability, cost, and message length parameters.

Distributed Task Assignment Module (DPIEM-3)

This module provides an upper bound for the allocation of tasks to distributed agents. DPIEM-3 deals with the practical-operational aspects of EWP, taking into account real-time costs that cannot be included in the optimization module DPIEM-2. The additional real-time costs are mainly concerned with 1) reallocation, and/or 2) modification of tasks. Another source of variability in the realized total costs is the complexity of interactions among the distributed agents, e.g., dynamically changing economic conditions, barriers, and conflicts that may diminish the integration efficiency. These drivers of change are complex, by nature, which makes them difficult to quantify. In general, the EWP principle comprises five important features, as the enablers of collaborative e-Work support systems:

(1) Formulation, decomposition, and allocation of e-Work parallelism design and integration problems.
(2) Enabling the applications to communicate and interact under TAP.
(3) Re-synchronization of independent agents to act coherently in problem solving and decision-making.
(4) Enabling agents to reason, interact, and collaborate with other agents if required.
(5) Development of conflict and error resolution, diagnostics, and recovery strategies.

2.1.3 Keep It Simple, System (KISS)

Any e-System can be as complicated as need be, as long as it is as simple as possible for human participants' interactions. This principle builds on and embraces traditional human-computer and human-automation usability design principles (Albers and Still, 2012; Guy, 2013), and goes beyond them. Two examples are described to illustrate the KISS principle.

(1) *The continuously improved e-System*: Suppose a complex computer operating system governs an e-System. It is well designed for usability purposes, yet it keeps being updated, repaired and improved, sometimes even fixed daily. Such improvements and updates can be complicated, but as much as possible should be designed to refrain from forcing their users to keep learning new ways of working with the revised version.

(2) *The evolutionary learning e-System*: This example is similar to the previous one, with one key difference: It is designed with its own internal machine learning mechanisms to improve itself as it evolves with its environment.

In addition to the need for human interaction with e-Systems that are typically complex and not well understood (in terms of their internal mechanisms) by their human users, imagine the additional workload, confusion, and cost when those e-Systems keep changing. KISS is, therefore, considered especially a central design principle for effective collaborative e-Work. This principle states that e-Systems must be as complex as needed in their cyber sides, as long as the e-Work support system is enabled to work autonomously, in parallel to and supportive of humans, and subject to the humans' inputs and instructions. At the same time, for human e-Workers, the e-System must be the simplest to comprehend and operate, in order to minimize the need for repeated and costly human retraining.

Followed by the EWP principle, KISS incorporates internal parallel e-Functions in the design upon releasing new versions and advanced functions into the e-System. In other words, regardless of the complexity of the e-System and its functions, it should be simple for the human users, the e-Workers, to work with the e-Work support systems.

2.1.4 Error Prevention and Conflict Resolution (EPCR)

This principle deals with the detection of errors and conflicts among collaborating agents, and the costs associated with resolving the detected errors and conflicts. Naturally, any system that cannot overcome effectively its errors and conflicts will get out of control and eventually collapse. In general, the rates of errors and conflicts among agents are proportional to the rate of interactions and the number of active collaborating agents. Hence, effective collaboration requires timely detection and resolution of errors and conflicts as economically as possible. Errors and conflicts are formally defined as follows:

- *Error*: is any input, output or intermediate result in an e-System, which does not meet the predefined specifications, expectations, or comparison objectives. An error is defined as follows:

$$\exists E\left[\omega_a(t)\right], \quad \text{if } s_{\omega_a} \xrightarrow{\text{Dissatisfy}} \kappa_r(t), \tag{2.7}$$

  where

E	Error	$\omega_a(t)$	Agent a at time t
s_{ω_a}	State of agent a at time t	$\kappa_r(t)$	System constraint r at time t

- *Conflict*: refers to the inconsistencies between the goals, plans, tasks, and e-Activities of collaborating agents. A conflict is defined as follows:

$$\exists C\left[\Omega(t)\right], \quad \text{if } s_{\Omega} \xrightarrow{\text{Dissatisfy}} \kappa_r(t), \tag{2.8}$$

where

C	Conflict	$\Omega(t)$	Integrated (network) agent at time t
s_{Ω}	State of integrated agent Ω at time t	$\kappa_r(t)$	System constraint r at time t

Errors and conflicts are different but related. The key difference is that an error may involve only one agent, while a conflict involves at least two agents. An error may cause a conflict (or a series of conflicts) and vice versa. For instance, a machine breakdown (error) could shut down a workstation feeding an assembly line (conflict), or: a collision between AGVs, Automated Guided Vehicles, carrying raw materials (conflict) could lead to unexpected delays in a set of jobs processed by a machine (error). Three basic methods for detection of errors and conflicts, and the respective performance evaluation metrics are presented in Table 2.2.

Table 2.2 Classification of error and conflict detection methods [Chen and Nof., 2007].

Method	Description	Metric example
Violation of specifications	Specifications of the system; defined by rigid ranges that need to be satisfied	*Detection latency*: time between occurrence and detection
Violation of expectations	Characteristics of the system; more flexible than the specifications	*Coverage*: the percentage of errors and conflicts detected
Violation in comparisons	External standards and benchmarks used for comparison with the current system	*Performance cost*: overhead caused by detection

In the example of bag handling system, an error could be the failure of a robot in reaching to the conveyor at the right time (which will cause the congestion of the bags on the conveyor, and thus, a conflict), or dropping the bag while carrying it towards the truck. Examples of potential conflicts in this case, on the other hand, could be the collision of two (or more) robots (e.g., AGVs) during their movements between the conveyor and the truck, improper communication between the task agents (to synchronize their respective parameters such as arrival time), or failure of one or more robots in interacting and collaborating with the other peers while picking up/carrying heavy bags (that require two or more robots).

In general, the EPCR principle is composed of eight consecutive functions, or modules, starting from detection (Table 2.2) all the way to resolution and exception handling, as described below (see also Fig. 2.6).

(1) *Detection*: Searching for existing errors and/or conflict.
(2) *Identification*: Classifying the observation as an error or a conflict.
(3) *Isolation*: Determining the exact point of an error or a conflict in the system.
(4) *Diagnostics*: Determining the type, magnitude, time, and cause of the out-of-control status.
(5) *Prognostics*: Analyzing, predicting, and preventing propagation (cascading) of an error or a conflict.
(6) *Error recovery*: Removing or mitigating the effects of an error.
(7) *Conflict resolution*: Resolving a conflict.
(8) *Exception handling*: Managing exceptions, i.e., constructive deviations from the process.

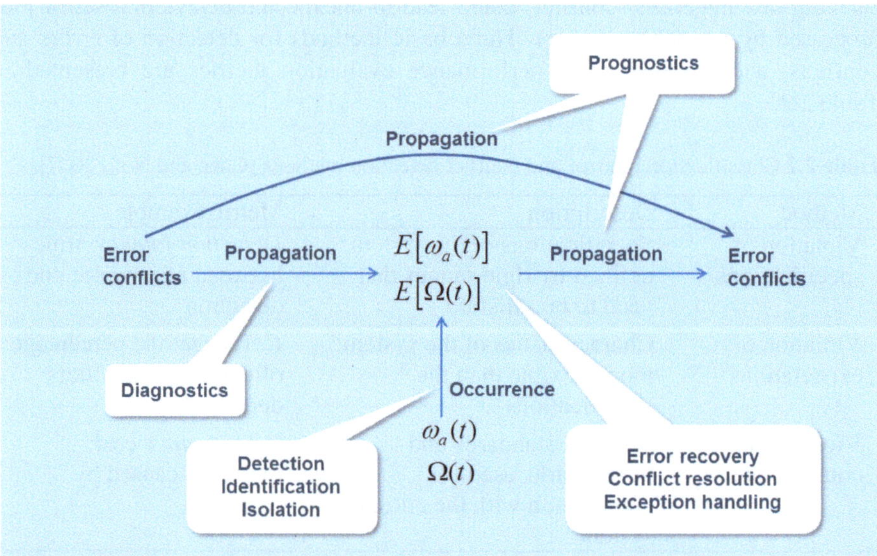

Fig. 2.6 Error and conflict propagation and the eight consecutive functions, from detection to recovery, resolution, and handling.

The total cost of EPCR including costs of each module is proportional to the rate of *human intervention* in the e-Work environment. That is, in a manual system (i.e., human operators), the frequency of errors and conflicts and thus the cost of EPCR lead to unbounded exponential growth, which in turn causes the e-System collapse. In contrast, with negligible human intervention, and with automated handling of errors and conflicts, the total cost of EPCR can be guaranteed to reach an upper bound, resulting in higher sustainability of the e-System.

2.1.5 Collaborative Fault Tolerance (CFT)

The CFT principle (aka, Fault Tolerance by Teaming) is based on fault-tolerant and feed-forward collaborative control protocols. The purpose is to achieve higher efficiency and reliability from a *network of weak* agents (e.g., micro- and nano-sensors) rather than a single stronger agent. The idea of the CFT principle lies in the fact that a collaborative team of agents generally yields better results compared to more powerful but individual agents. The latter may fail, and without backup, the system risks failure. In this context, collaboration engineering is a must. The reliability of an e-System is improved through application of multiple redundant agents, in case agent (ω_a) errors and/or network (Ω) failures are likely. For instance, networked micro-/nano-sensors improve the fault tolerance of a wireless sensor network, relying on the redundant sensory information that compensates faulty or erroneous readings by individual sensors (Fig. 2.7).

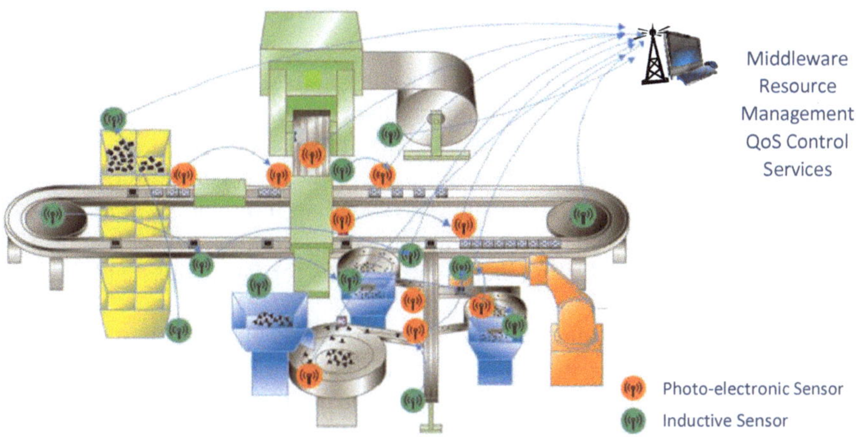

Fig. 2.7 Industrial application of a wireless micro-sensor network with fault-tolerance capabilities (i.e., protocols) to aggregate weak sensor signals. (FSN – Facility Sensor Network; MidFSN – Middleware for it; QoS – Quality of Service).

In the robotic bag handling system, for example, another alternative besides using three "small" robots (that are not able to handle heavy bags on their own) is to deploy one "big" robot capable of handling all types of bags. Even if the big robot is also more reliable than each individual small robot, the reliability of the bag handling system with one big robot is practically lower than the case of having thee collaborative robots. The reason is that, in practice, robots' failure (e.g., breakdowns) is a Poisson process (i.e., *birth-death process*), and therefore, the failure of multiple robots "at the same time" is much less probable than the failure of one single robot (this gap increases drastically as the number of collaborative robots increases). Therefore, the system with one big robot is relatively more prone to failure than the system with multiple collaborative small robots. Additionally, the collaborative robotic system increases the efficiency and throughput of the system through parallelizing the process (handling multiple bags simultaneously). Furthermore, the collaborative robotic system outperforms the alternative with one big robot in terms of *resource utilization* and *flexibility*. When the big robot is handling small bags, it utilizes a small portion of its capacity while consuming more energy to move its bigger structure. Also, in periods where only small bags arrive in the system with a low arrival rate, the collaborative system can adapt itself to this change by turning one (or more) of the robots into the *standby mode* to save energy, while this option is not available to the system with only one robot (issues of flexibility and scalability; see Chapter 4).

Agents interact with each other through *communication protocols* (see Chapters 4 and 5 for details). For example, in a wireless sensor network, micro-/nano-sensors interact with each other through *point-to-point*, *flooding*, *gossiping*, and *minimum transmission energy* protocols. The reliability and fault tolerance of the network of agents, however, are not guaranteed by these communication protocols. They are commonly designed to handle certain specific deficiencies in terms of speed, efficiency, and utilization of the agents. The design and control considerations concerning collaborative fault tolerance are addressed by the fault-tolerance timeout protocol (FTTP) which comprises the following seven steps (see Chapter 8 for more details):

Step 1: *Clustering setup* - Define the optimal number and size of the agent clusters considering the network characteristics.

Step 2: *Cluster head setup* - (1) Assign one cluster-head per cluster, (2) Detect and update faulty links of the corresponding information, (3) Define the minimum number of agents for timeout-based control and inform the cluster head, (4) Define the rerouting paths for faulty links and the back-up agents.

Step 3: *Execution* - Run the network in real-time and evaluate the performance rate according to the timeout-based control through having the agents operate and interact with the cluster head at the minimum rate.

Step 4: *Timeout-based control* - Define the timeout value according to the minimum number of agents required in each cluster to achieve an acceptable performance.

Step 5: *Integration* - Integrate the outputs of agents in each cluster by the cluster head.

Step 6: *Reiteration* - Repeat the setup and operational cycles.

Step 7: *Analysis* - Analyze the network performance for further improvement and utilization.

2.1.6 Association and Dissociation (AD)

This principle addresses dynamic variations in the formation (topology), size, and operations of collaborative networks of agents. Examples include the growing number of virtual enterprises, the dynamic nature of enterprise alliances, self-organizing agent teams, sensor clusters, and modular systems. The AD principle analyzes the conditions and timing for individual agents (ω_a) or networks of agents (Ω) to *associate* with or *dissociate* from a collaborative network. The AD decisions are made at different levels in the entire network; by individual agents, sub-networks or clusters of agents, or among multiple networks (see Fig. 2.8). The analysis includes several phases, from creation and execution to dissolution and support. In general, the following questions are targeted by the AD principles:

(1) When and why should an agent ω_a join a network Ω?
(2) What are the rewards and costs for an individual agent ω_a to participate in a network Ω?
(3) Why would an agent ω_a opt to remain in a network Ω?
(4) For the above three decisions, what are the criteria to characterize and evaluate the performance of a network Ω?

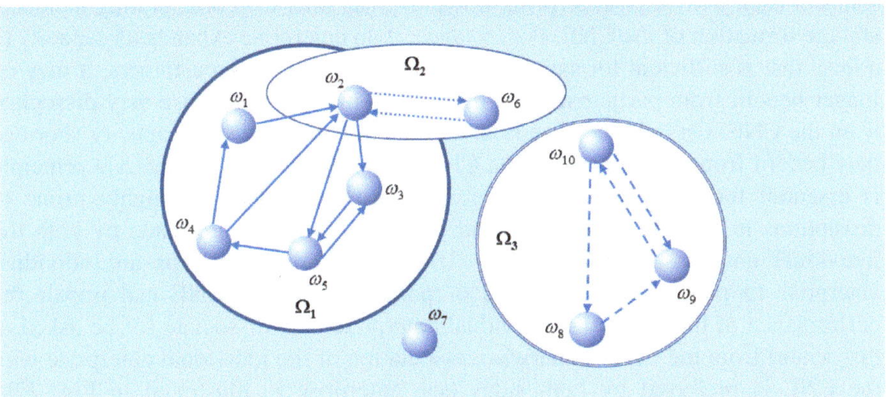

Fig. 2.8 Example of collaborative networks of agents in a competitive and dynamic environment; The overall network is composed of ten agents clustered in three collaborative networks. Each agent may belong to more than one collaborative network (e.g., ω_2). In this example, agent ω_7 may be associated to or dissociate from a collaborative network. The direction of the arrows may represent the direction of information or resource sharing.

Now let's go back to the robotic bag handling example and analyze the implications of the AD principle for that case. Recall that in our example, there are three robots in the system that may dynamically form teams of one, two or three robots, depending on the characteristics of the bags (e.g., weight, size, dimension) arriving on the conveyor. Each team that includes at least two robots (i.e., r_4-r_7) is a collaborative network. In this example, agents (robots) dynamically associate to a collaborative network (team) to process a specific task (bag handling), and after finishing the process, may remain in or dissociate from that team and associate to another team for processing the next task. This dynamic AD behavior imposes some constraints to the formation of collaborative networks over time. In case of having three robots, for instance, only one collaborative network (i.e., team of at least two robots) can be formed at a time. This phenomenon, in turn, influences the sequencing of the bags on the conveyor to avoid process infeasibility. Other implications of the AD principle are the failure of one (or more) robots (which temporarily dissociates from the whole network) or adding (associating) a fourth robot to the system, which non-polynomially increases the number of possible collaborative networks to form (i.e., six teams of two, four teams of three, and one team of four robots). Below, a similar AD behavior is further elaborated in the context of enterprise collaboration (see Fig. 2.9).

An important application of the AD principle is in enterprise collaboration, where enterprises collaborate within a collaborative network of enterprises (CNE) through demand and capacity sharing protocols. In a CNE, enterprises laterally collaborate with each other through sharing their information and resources, e.g., excess capacities or unfulfilled demand, such that mutual benefits (e.g., higher profit, service level, utilization, stability) are obtained. Dynamic changes in the status of each individual enterprise, however, may alter the collaborating members and the formation of the CNE. For instance, if an enterprise expands its capacity to a level that is sufficient for satisfying the demand of its own customers, it may no longer benefit from participating in the CNE. Thus, the enterprise may dissociate from the CNE. On the other hand, an individual enterprise with capacity shortage may benefit from associating with a CNE (Fig. 2.9). In general, the AD principle is essential for enterprises operating in environments that are highly prone to disruptions in capacity and/or demand. The AD decisions are made by both the individual enterprise and the CNE: If there is no benefit for an individual enterprise to participate in a CNE, or participation of a CNE can impair the performance of the CNE, the individual enterprise may dissociate (or be asked to dissociate) from the CNE. Otherwise, association of the individual enterprise with the CNE is preferred by both sides (see enterprise e_3 illustrated in Fig. 2.9). Service level and profit are the two foremost criteria two making the AD decisions in enterprise collaboration.

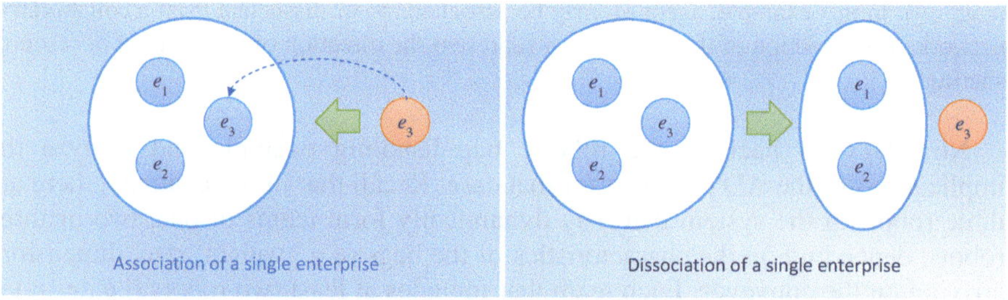

Fig. 2.9 Illustration of the AD decisions in a CNE.

Let Ω denote a collaborative network of N agents, where the demand (for any type of product or service) corresponding to each agent ω_a follows a general distribution. Each agent is responsible for fulfilling its own demand D_{ω_a} by its limited capacity K_{ω_a}. The service level of each agent ω_a is then formulated as follows:

$$SL_{\omega_a} = \Pr\left(D_{\omega_a} < K_{\omega_a}\right). \tag{2.9}$$

Agent ω_a can make AD decisions through comparing its service level with the service level of the collaborative network Ω, i.e., SL_{Ω}, formulated as follows:

$$SL_{\Omega} = \Pr\left(D_{\Omega} < K_{\Omega}\right), \tag{2.10}$$

where $D_{\Omega} = \sum_{a=1}^{N} D_{\omega_a}$ and $K_{\Omega} = \sum_{a=1}^{N} K_{\omega_a}$. That is, if

$$SL_{\omega_a} > SL_{\Omega}, \quad \forall \omega_a, \tag{2.11}$$

agent ω_a may decide to dissociate from the network (without considering the profit).

Now consider, without loss of generality, that the respective demand of all agents follows a Normal distribution and the capacities are fixed. Therefore, according to (2.11), the condition for an agent ω_a to associate with network Ω is defined as follows:

$$\int_{-\infty}^{K_{\omega_a}} f(D_{\omega_a})\, dD_{\omega_a} < \int_{-\infty}^{K_{\Omega}} g(D_{\Omega})\, dD_{\Omega}, \quad \forall \omega_a, \tag{2.12}$$

or

$$\Phi\left(\frac{K_{\omega_a} - \mu_{\omega_a}}{\sigma_{\omega_a}}\right) < \Phi\left(\frac{K_{\Omega} - \mu_{\Omega}}{\sigma_{\Omega}}\right), \quad \forall \omega_a, \tag{2.13}$$

where μ_{ω_a} and σ_{ω_a} respectively represent the mean and the standard deviation of D_{ω_a}, $\mu_{\Omega} = \sum_{a=1}^{N} \mu_{\omega_a}$, and $\sigma_{\Omega} = \sqrt{\sum_{a=1}^{N} \sigma_{\omega_a}^2}$. Finally, from the monotony of the CDF Φ we have

$$\frac{K_{\omega_a} - \mu_{\omega_a}}{\sigma_{\omega_a}} < \frac{K_{\Omega} - \mu_{\Omega}}{\sigma_{\Omega}}, \quad \forall \omega_a, \tag{2.14}$$

as the *association condition* for agent ω_a. Similarly, the association condition can be analyzed from the perspective of the collaborative network Ω, through comparing SL_Ω with $SL_{\Omega-\omega_a}$ (i.e., the service level after dissociating ω_a). That is,

$$SL_\Omega > SL_{\Omega-\omega_a}, \quad \forall\, \omega_a, \tag{2.15}$$

is the association condition for an agent ω_a, from the perspective of the collaborative network Ω. Calculation of (2.15) and the association conditions considering profit as the decision criterion can be derived too. The summary of notation used in this section is provided below.

N	Number of agents in Ω	D_{ω_a}	Demand at agent ω_a
K_{ω_a}	Capacity of agent ω_a	SL	Service level
μ	Mean of demand	σ	Standard deviation of demand

The AD decision model can be extended, modified, and applied to other collaborative systems analogous to the enterprise collaboration decisions described above, i.e., any collaborative network composed of a set of capacitated agents, which aims to process a set of tasks with certain demands at the desirable and possible service level.

2.1.7 Emergent Lines of Collaboration and Command (ELOCC)

The purpose of this principle is to overcome drastic changes facing networks, e.g., under emergency, and the volatility of formal and informal communications between the individual and clustered agents. Similarly, it attempts to find adjustments when networks evolve, undergoing gradual changes along time. Such conditions are among the utmost challenges of collaborative e-Work. In particular, under emergency conditions, all the designs following principles 2.1.1 to 2.1.6 above would fail, since established lines of command and collaboration may be disabled. The ELOCC principle seeks to enable e-Systems to 1) make effective decisions in complex and dynamic environments when they are being challenged or forced to change; 2) sustainably exchange information and create knowledge despite an emergency or evolution; and 3) control and optimize the *emergent* networks. Emergent networks refer to networks that *evolve* over time (see, e.g., Fig. 2.9) through adding or removing nodes (e.g., agents) and/or links (e.g., interaction, communication, collaboration). The term *emergent network* was coined by Barabási and Albert in 1999, who developed the first model (known as BA model) for formation of *scale-free* networks. In the BA model, nodes are dynamically added to the network, where the likelihood of their connections is a function of their preferences over the existing nodes in the network. The BA model was first applied to the formation of emerging Internet networks, where web pages (i.e., nodes) are dynamically added to or removed from the network, with high likelihood of being linked to central hubs (e.g., Google).

Consider a network at time zero with N_0 nodes. A new node added to the network at time τ is connected to $N_\tau \leq N_0$ nodes (assuming no other arrivals between the times zero and τ), with a probability proportional to the degrees of the existing nodes (i.e., the number of links to each existing node). The BA model calculates the probability of connection between the new node and node $i \in N_0$ as follows:

$$p_i = \frac{\eta_i}{\sum_{i \in N_0} \eta_i}, \quad \forall i \in N_0, \tag{2.16}$$

where η_i denotes the degree of node $i \in N_0$. Accordingly, the likelihood of connection with heavy nodes (with high degree) is very high, implying high preference from the new node over those nodes (see Fig. 2.10).

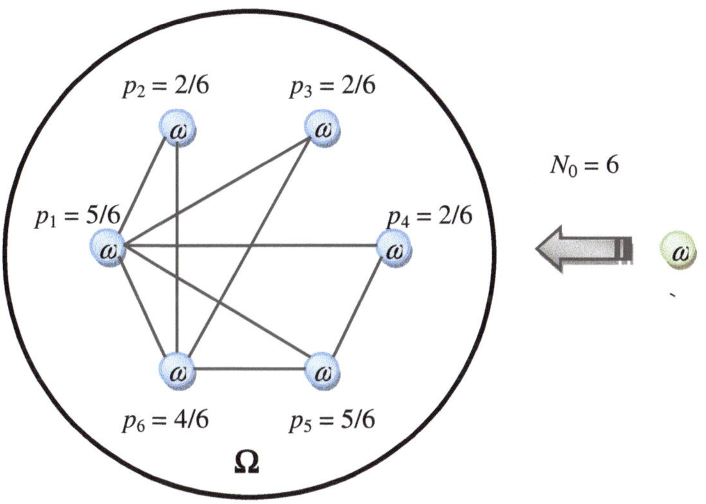

Fig. 2.10 Illustration of the BA model in an emergent network with six nodes.

The concept of emergence usually deals with incapacitated networks (e.g., social networks), where the collaborative agents can be matched to unlimited number of agents arriving into the system. The example of robotic bag handling system, on the other hand, is indeed a capacitated network; each robot/team of robots can handle only one bag at a time, and while handling it, it has zero capacity for another bag. It is natural that the task agents prefer the fastest and strongest robot in case robots are not the same (i.e., preferring the most popular node in the network; the BA model). However, the capacity constraints sometimes limit the choices of the arriving agents (task agents in this case) to join their preferred nodes (i.e., a robot or a team of robots) in the network.

Emergent networks are evolutionary interaction mechanisms, building upon well-established learning theories. Because of the changing situation, often under time pressure to respond, they have to be prepared and pre-programmed for *ad hoc* decisions, effective improvisation, on-the-spot creation of contacts, and best-matching protocols. The main challenge, beyond the basic models (e.g., BA), is to answer, at any given point in time, 1) with whom to interact and collaborate, 2) under which circumstances to begin interaction and collaboration, 3) when to interact and collaborate, and 4) how to begin interaction and collaboration considering certain criteria and constraints (e.g., time, cost, service level). These questions are initiated by the CRP-I principle, and now have to be re-addressed by the ELOCC principle. In some cases, re-evaluating principles 2.1.1 to 2.1.7 under the new, emergent or evolutionary conditions may be needed. In other cases, ELOCC solutions can be addressed by collaborative best matching decisions and protocols, as described in the next section.

2.1.8 Best Matching (BM)

Matching between the collaborative agents is a fundamental concern in the design and execution of collaborative e-Work networks. Often, as described for the AD and ELOCC principles, the challenge is how to match each individual agent to the best agent or set of agents in the network with whom to interact and collaborate (in general, and in particular cases; under normal conditions, and under emergency). The objective of the BM principle is to find the best match between two or more sets of agents, such that a set of objectives is satisfied. The simplest case is analogous to the original one-to-one assignment problem. In the two-sided scenario, i.e., matching the elements of two sets (e.g., robots and bags to unload), the problem to solve may be to match the elements of: (1) two static sets, where the sets are ready and the entities are fixed and known (e.g., three robots are ready, and ten bags are ready on a truck); (2) two dynamic sets, where the entities of both sets appear dynamically (moving robots arrive randomly and the bags appear randomly on a conveyor); and (3) one static set and one dynamic set: the entities of only one set appear dynamically.

To better understand potential BM problems in a collaborative network, consider the bag handling system where the robots are not the same in that (1) some bags that can be handled by a specific team of (one, two, or three) robots cannot be handled by another team with the same size, and (2) the robots have different speeds. In case of dynamic arrival of the bags, the following questions are of concern from the perspective of the BM principle: (1) Which robot is the best match for each arriving bag? (2) Which robot is the best match for another robot to interact and collaborate (for handling heavy bags)? Given different capabilities of the robots along with the dynamic behavior of the system, these fundamental questions must take into account both the current and the future states of the system. For instance, a specific robot/team of robots may be the best match for handling a specific bag considering only the current state of the system; but the system may need to reserve one (or more) robot for another bag that will arrive in the near future. (The same scenario may happen for dynamic team formation and collaboration between the robots).

The BM decisions are complex by nature, even in case all the agents and characteristics of the system are static and known. The simplest case is the two-sided one-to-one BM, also known as the original assignment problem, which is formulated as follows:

$$\begin{aligned}
&\min \sum_{i\in I}\sum_{j\in J}\beta(i,j)\chi_{ij},\\
&\text{s.t.}\\
&\sum_{i\in I}\chi_{ij}\le 1, \quad \forall j\in J,\\
&\sum_{j\in J}\chi_{ij}\le 1, \quad \forall i\in I,\\
&\chi_{ij}\ge 0, \quad \forall (i,j)\in I\times J,
\end{aligned} \tag{2.17}$$

where

$\beta(i,j)$ Best-fit index between entities i and j	χ_{ij} 1, if i and j are matched; 0, OW
I Set of entities i	J Set of entities j

A classic example of two-sided one-to-one BM is the matching between *bolts* and *washers* in *selective assembly*, assuming that even within the design tolerance, the mating parts may be selected to be best matched. The geometric best matching of two sets of bolts and washers is explained below.

(1) *Dimension of entities*:
   - $\delta(B_i)$: dimension of bolt B_i ($i=1,\ldots,N$)
   - $\delta(W_j)$: dimension of washer W_j ($j=1,\ldots,N$)

(2) *Best-fit index (BFI)*: represents the degree at which two entities match each other based on a set of e-Criteria (criteria defined for e-Work). In the bolt-washer BM problem illustrated, the relative *quality of match* can be considered as the BFI, assumed to be the minimum gap between the matched dimensions, calculated as follows:

- BFI for bolt B_i:

$$\beta_i(W_j) = \frac{\left|\delta(B_i) - \delta(W_j)\right|}{\delta(B_i)}, \quad \forall i, j = 1, \ldots, N, \tag{2.18}$$

- BFI for washer W_j:

$$\beta_j(B_i) = \frac{\left|\delta(W_j) - \delta(B_i)\right|}{\delta(W_j)}, \quad \forall i, j = 1, \ldots, N, \tag{2.19}$$

Then, the bolt B_i and the washer W_j with the minimum overall BFI are matched:

$$\beta(B_i, W_j) = \max\left\{\beta_j(B_i), \beta_i(W_j)\right\}, \quad \forall i, j = 1, \ldots, N, \tag{2.20}$$

(3) *Tie-breaking (TB) rules*: are applied for matches that yield identical BFI values.

- *TB*$_1$: If W_j is the "best" washer for both bolts B_i and B_k,
  (i) Compute the BFI for the next best match for B_i, denoted by W_{j+1}:
  $$x = \frac{\left|\delta(B_i) - \delta(W_{j+1})\right|}{\delta(B_i)}.$$
  (ii) Compute the BFI for the next best match for B_k, denoted by W_{j+1}:
  $$y = \frac{\left|\delta(B_k) - \delta(W_{j+1})\right|}{\delta(B_k)}.$$
  (iii) If $x < y$, match washer w_j with bolt B_k. Otherwise, match it with bolt B_i. If $x = y$, randomly match washers w_j and w_{j+1} with bolts B_i and B_k.
- *TB*$_2$: If washers w_j and w_{j+1} are the best matches for bolt B_i,
  (i) If neither washer matches another bolt, arbitrarily pick either washer W_j or W_{j+1}
  (ii) If either of the washers matches another bolt, pick the washer that does not have another match.

The number of matched sets may be more than two, or there may be dependent sequences of matches, where the BFIs are affected by former matches. For example, in selective assembly, some parts are assembled in parallel and then the outputs are assembled in the next steps (Fig. 2.11). There are several aspects for classification of BM problems:

(1) *Number of matches* per agent: one-to-one, one-to-many, and many-to-many.
(2) *Number of sets* being matched simultaneously (i.e., dimension of match).
(3) *Multi-level BM*, where the BM results depend on a sequence of matching in terms of BFI.
(4) *Static vs. dynamic sets*, as explained earlier.
(5) *Interdependent BFIs*, applied to one-to-many and many-to-many BM: the BFI of an agent from set *X* matched with an agent from set *Y* is influenced by the other agents from set *X* already matched to that agent from *Y* (Fig. 2.12). This characteristic is important to the AD and ELOCC principles.
(6) *Collaboration vs. competition*, where agents collaborate to optimize the overall BFI (collaborative BM), but at the same time compete with each other to optimize their own BFI (competitive BM; e.g., bidding).
(7) *Uncertainty*, where there is missing or incomplete information needed to calculate the BFI.

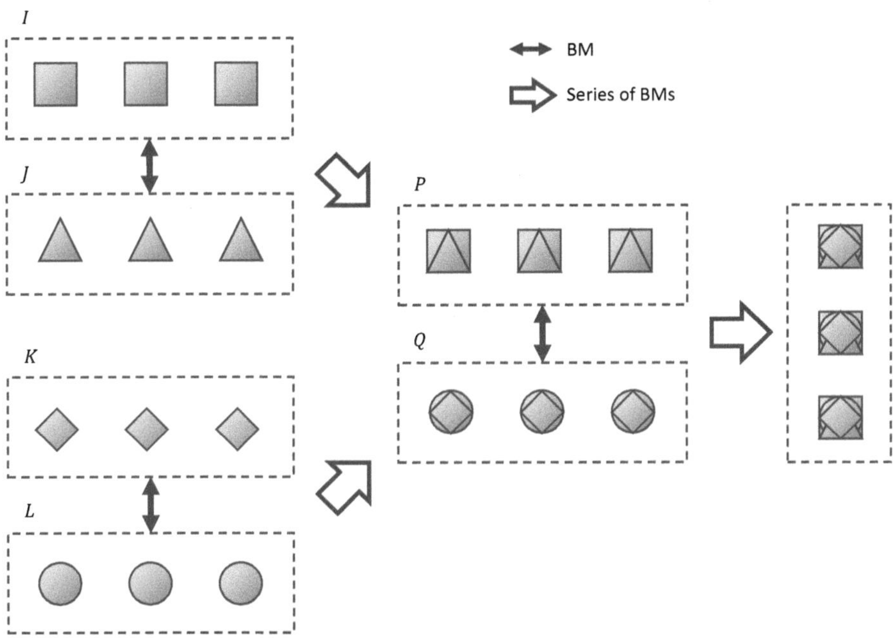

Fig. 2.11 Multi-level BM – The outcomes (sets *P* and *Q*) of matching the agents (e.g., sub-assembly parts) of earlier sets (sets *I*, *J*, *K*, and *L*) must be matched at the next level. Thus, the succeeding BM (between *P* and *Q*) is directly influenced by the outcomes of the preceding parallel BMs (between *I* and *J*, and *K* and *L*).

As explained earlier, the original BM problem is NP-hard, so is the extended BM problem dealing with multiple dimensions, dynamic and emergent networks of agents, and so on. Various exact and heuristic techniques have been developed over the years for solving different instances of this problem (see Table 2.3): scheduling, transportation and routing, telecommunication network design, production planning, facility location, allocation, supply network design and logistics and more. In Chapter 4, we will introduce two pioneering bio-inspired techniques that can be applied to tackle the inherent complexities of the BM principle through distribution of intelligence and autonomy among agents.

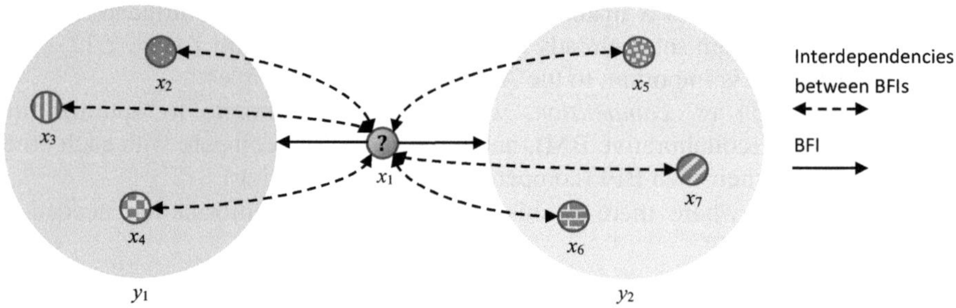

Fig. 2.12 BM with interdependent BFIs – In the above example (one-to-many matching), the preferences of agent $x_1 \in X$ for agents $y_1 \in Y$ and $y_2 \in Y$ may be influenced by other agents of set X already matched to y_1 and y_2. Agents x_1, x_2, for instance, may be altruistic, envious, or neutral about each other. That is, matching agent x_1 to agent y_1 may increase (altruism), decrease (envy), or not change (neutrality) toward the preference of x_2 for y_1 (and vice versa).

Table 2.3 Real-life examples of BM (two-sided).

Problem	Agents	Matched to
Project Management	Workers	Jobs
Load Balancing	Machines	Jobs
Aeromedical Routing	Flights	Patients
Vehicle Routing	Vehicles	Cities
Wireless Network Design	Base stations	Terminals
Storage Assignment	Locations	Items
Market Clearing	Asks	Bids
Stock Management	Demands	Stocks
Labor Force Scheduling	Labor	Sections
Production Planning	Agents	Tasks

2.1.9 Collaborative Visualization and Comprehension (CVC)

In most e-Work, human participants need to visualize data and information, and understand what is meant by what they visualize. Visual analytics is an emerging interdisciplinary field, which focuses on the integration of interactive visualization with analytic tools and techniques to deal with the rising complexities of e-Work systems. The aim is to integrate computer graphics, interaction, visualization, analytics, perception, and cognition domains to enhance and support the human-machine interactions, e.g., working based on multi-dimensional datasets or big data. Modern e-Business, e-Commerce, and e-Service enterprises are increasingly

looking for the ability to perceive the *interrelationships* between distributed and complex datasets, to collaboratively gain insight for planning, acting, and achieving their goals. These issues highlight the CVC principle in CCT as a field between collaboration and visualization, making such collaboration systems scalable. The CVC principle enables e-Systems to (1) collect and integrate insights of several distributed agents, (2) comprehend and investigate complex and incomplete data, for handling remote collaborative decision-making, and (3) provide common virtual environments for information visualization and data manipulation. CVC can be implemented in all scenarios of the classic *space-time matrix* (see Fig. 2.13), at three different levels (Isenberg et al., 2011):

(1) *Viewing*: static or dynamic visualization of data/information, without capability of manipulation (e.g., videoconferencing tools). The goal is to learn, discuss, interpret, analyze, or make decisions based on *pre-selected* data/information.
(2) *Interacting/Exploring*: co-located or distributed sharing of the *same interactive* visualization software, with the ability of selecting alternative views for exploration, analysis, and decision-making through communication means, e.g., e-mail, video/audio links (for distributed users). The goal is to explore various aspects of data/information and analyze alternative insights and interpretations.
(3) *Sharing/Creating*: contribution of users to creation and sharing of new data/information visualizations, beyond the properties of Levels (1) and (2). The goal is to increase awareness regarding certain issues.

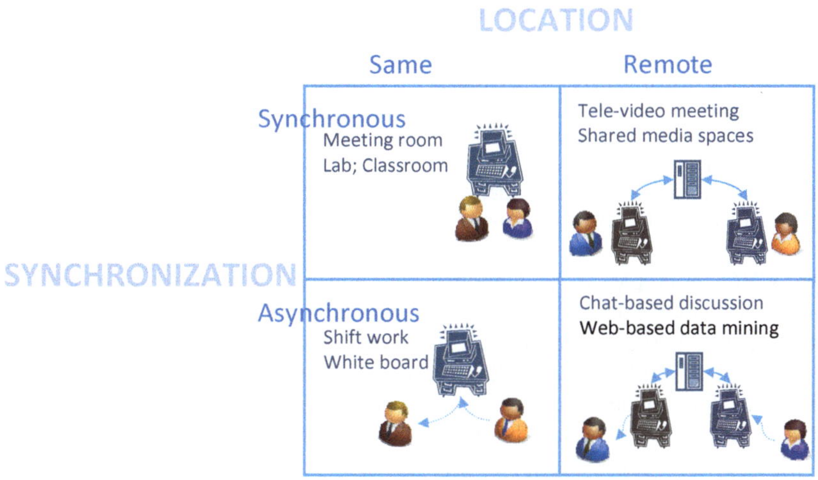

Fig. 2.13 CVC in four cases of the location-synchronization matrix.

Fig. 2.14 shows various concepts and applications of the CVC principle in different areas.

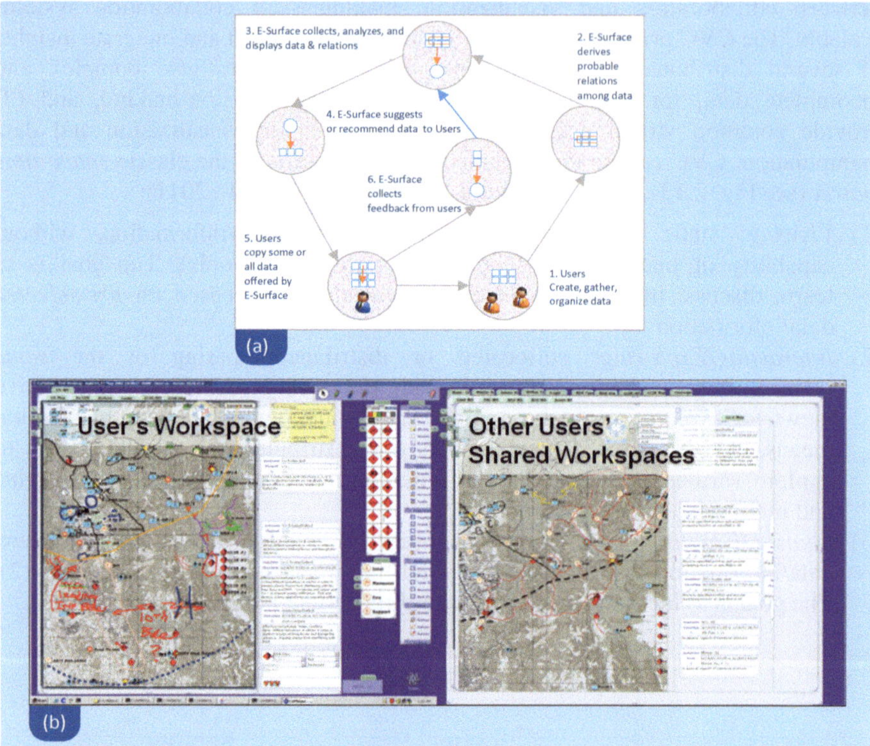

Fig. 2.14 CVC instances; (a) e-Surface model: An agent-supported computational environment for visual organization, analysis and comprehension of data and information; (b) Command Post of the Future: 2D-3D visualization of distributed information for improving command and control decisions, collaboration, and situation awareness in military applications (Courtesy of General Dynamics Corporation).

The CVC principle is a *human-in-the-loop* collaboration framework with automated collaboration support features that enables the exchange of insights between agents. It requires definition of the coordination network (Co-Net), tasks and participant models for collaborative insight (Co-Insight), and effective mechanisms for collaborative allocation of resources, tasks, participants, and interfaces (see Fig. 2.15). Accordingly, the CVC principle is tightly coupled with the BM principle for task-participant and participant-interface best matching and effective exchange of insights. The CVC principle relies on the TAP for collaborative requirement planning in the Co-Net (i.e., participants), allocation of resources (e.g., participants and interfaces) based on the tasks' profiles and the agents' expertise, and monitoring of time-out conditions in real-time (Ozsoy, 2011)

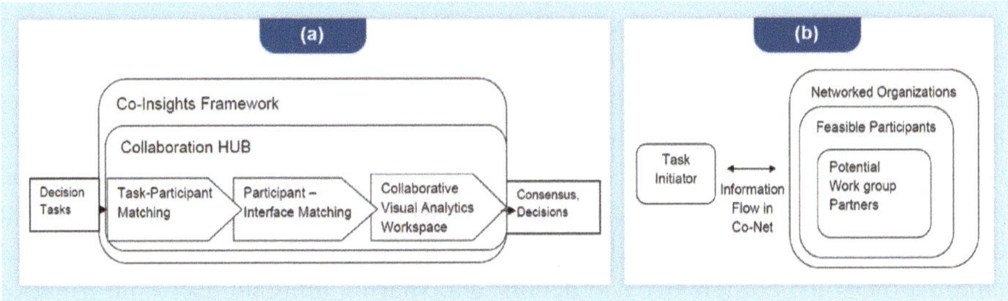

Fig. 2.15 (a) The Co-Insight framework: task-participant matching (with whom to collaborate for a specific decision problem), participant-interface matching (which tool to use for a specific analysis), within the collaborative visual analytics workspace; (b) The role of collaborators (i.e., participants) in Co-Net for Co-Insight.

2.1.10 Design e-Criteria and Evaluation e-Measures

As the conventional work models are influenced by the emerging e-Work models, so do the classic performance measures and criteria. As discussed in chapter 1, with the evolution of CIM and networked systems, several e-Criteria have emerged over the past decade for effective evaluation of e-Work systems, e.g., interoperability, integrability (includes connectivity and behavioral patterns), agility, scalability, reachability, viability, autonomy, dependability (includes availability, reliability, sustainability, integrity, and maintainability), and resilience [Nof and Huang, 1999; Huang and Nof, 2000a, b; Huang et al., 2002; Nof, 2007b]. Design e-Criteria, and evaluation e-Measures and guidelines, however, *must* address the classic criteria and measures, besides those added by certain e-Work requirements, such that failure can be avoided and success can be assured. The e-Criteria, e-Measures, and guidelines improve the e-Work design "*as it evolves*". Nevertheless, previously implemented e-Systems must be checked from time to time to avoid accumulation of changes in requirements and conditions over time, and assure compatibility of their design with real-time conditions and requirements. Examples of other emerging CCT e-Criteria for e-Work design are learning ability, collaboration ability, error and conflict detectability, error and conflict severity, and information significance.

2.2 Emerging CCT Design Principles

2.2.1 Bio-inspired Collaboration

Development of the agent-based collaboration and interaction models for e-Work is ever more inspired by the configuration and evolution of biological control systems, where each agent, in general, is considered as a member of a biological community with certain characteristics. Animal-to-animal interactions in various biological communities, e.g., ant and bee colonies, have been adapted to enhance the collaboration and interaction among the agents. For instance, the limited rationalizing capability of ants is enhanced by their ability to use pheromone as *hints* and therefore reduce information overloads. The idea is to reduce the overloads on individual agents (e.g., ants) in collaborative networks by local elimination of the complexity and details, and sharing minimal useful information and resources (e.g., pheromone). The objective is to minimize the communication efforts and the task overloads, while upholding (or even increasing) the performance level and the effectiveness of actions.

Bio-inspired applications include virtual insects (e.g., ants, bees) as mobile software agents, message-based interactions among agents and robots, and measures such as viability (survivability of each single insect) and autonomy (independence from the colony). Negotiation-based and bio-inspired techniques are the two pioneering techniques that continue to transform the multi-agent collaborative control (see Fig. 2.16).

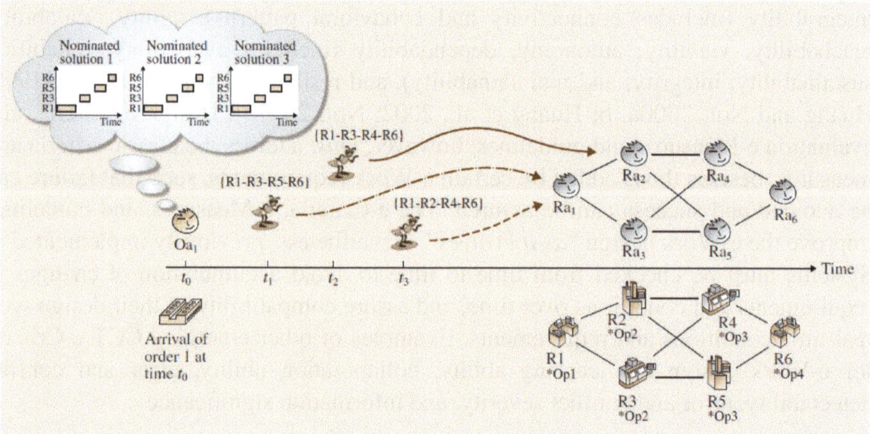

Fig. 2.16 The mechanism of ant agents in multi-agent manufacturing control (after Hadeli (2006)).

Inspired from the interactions and collaborations among synapses and spines for enabling survival and competitiveness in biological systems (Fig. 2.17), the human-made mechanisms of collaborative control, e.g., adaptability, dependability, discovery by data mining, decision support systems, and active middleware can be designed (Fig. 2.18). Bio-inspired collaborative control is still emerging and several discoveries on models, theories, and solutions are expected to satisfy the ever-growing needs of optimizing e-Work.

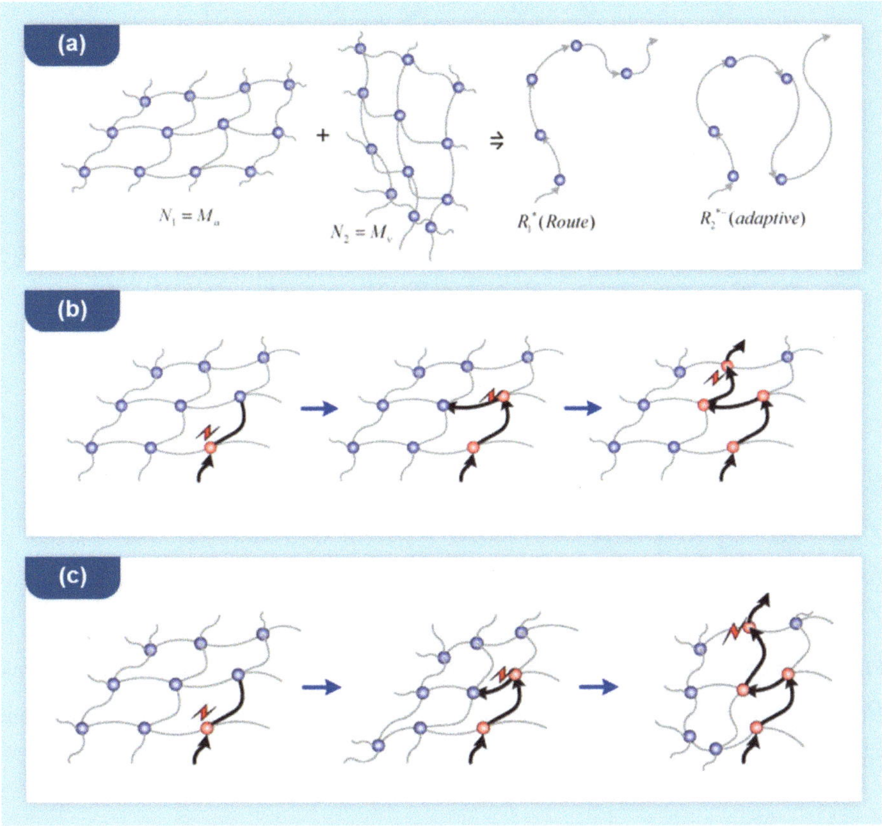

Fig. 2.17 (a) survival mechanisms of escape and prey hunting; calibrating by alignment of two topographic maps (left), one auditory (network N_1) and one visual (network N_2), the barn owl optimizes a route R_1^* and dynamically adapts to route R_2^* (rightmost); (b) cascading causality (propagation) consecutively leading to changes in certain affected arteries; (c) same, but changes are affected in both arteries and network structure.

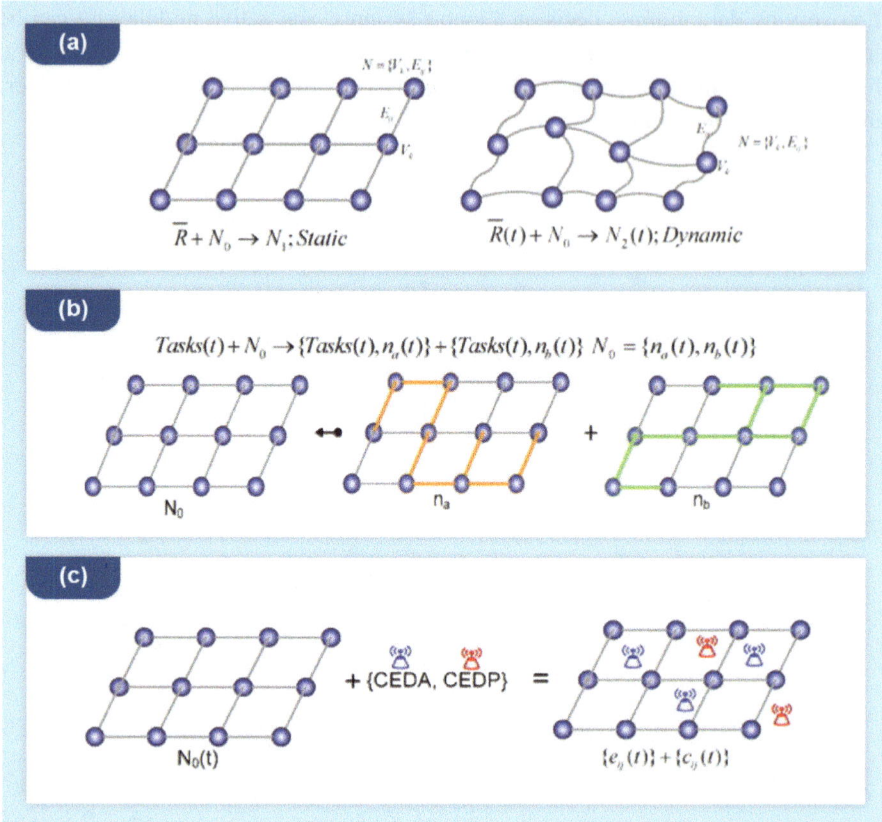

Fig. 18 Network models for (a) CRP on initial network N_0: static requirements R; dynamic $R(t)$; (b) Assignment of tasks to parallel sub-networks $\vec{N}_a$ and $\vec{N}_b$; (c) Conflict and error detection agents (CEDA) and protocols (CEDP) are assigned to Network $N_0(t)$.

2.2.2 Learning, Adaptation, and Evolutionary Collaboration

Learning and adaptive mechanisms are significant sources of inspiration from biological communities for automation and control of e-Work systems. The learning and adaptive behavior of species (including humans) have recently been adapted to develop learning methodologies that increase the autonomy of agents (e.g., intelligent robots) in driving their own development. Examples include adaptations and configurations in the topology of fish schools in face of danger (e.g., predators) and optimal path recognition through effective information sharing in ant/bee colonies. The learning and adaptation mechanisms for e-Work are mainly classified as:

(1) *Centralized*: is executed by the individual and autonomous agents without any interactions with the other agents. For instance, in market-based resource allocation, each resource agent locally adapts its behavior to gain more awards (e.g., utilization, bid awarding) compared to other peer agents.
(2) *Decentralized*: is executed under collaborative interactions mechanisms among several agents. Examples include local interactions in agent clusters and information diffusion across the entire collaborative network.

The key advantage of learning and adaptation by agents is the balanced and more effective allocation of tasks and resources through agent-based adaptation protocols. Learning and adaptation mechanisms in distributed networks of agents are addressed by different approaches. Cognitive networks, for instance, are an interdisciplinary class of communication networks, which utilize machine learning techniques, computer networks, and knowledge representation to learn (from the current network), plan and act according to the system's objectives (see Fig. 2.19). Other prominent approaches in learning and adaptation are neural networks, swarm intelligence, Bayesian networks and Bayesian learning. In Chapter 4, we will elaborate more on bio-inspired collaboration for optimization and control of e-Systems, inspired by natural evolution and swarm intelligence.

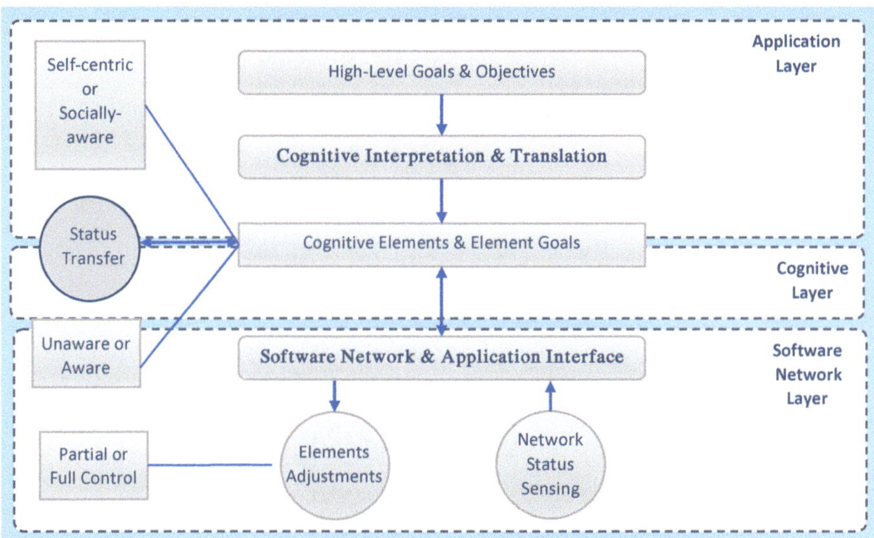

Fig. 2.19 Three-tier architecture of a cognitive network; Observations of the current situation are provided by the protocol layers and the optimal configuration and changes are decided by the cognitive process.

2.3 Collaboration Support Systems

Over recent decades, collaboration support systems (CSS) have drastically become an important area of research and development in e-Work, due to the ever-increasing need for powerful computing and communication supports. Recent and emerging advances in automation, CIM and ICT (e.g., robotics, virtual reality, ubiquitous computing, and augmented reality) are the main enablers of interaction and collaboration among systems of systems, beyond traditional time and space restrictions. CSSs are mainly concerned with

(1) *Hardware*: pervasive computing and telecommunication devices, wireless sensor networks, nano-electronics, cyber-physical systems hardware.
(2) *Software*: agent-based systems, integration and collaboration engineering models/protocols.

Collaboration tools, methodologies, and their applications in different domains have enhanced quality of service and reduced complexity of processes in collaborative e-Systems. However, inefficiency of interactions among hardware and software agents, information overloads, and ineffective collaboration are among problems that still exist. In spite of considerable progress in CCT engineering tools and design principles, the foundation to support and raise the sophistication and anticipated benefits of collaboration methods is still an ongoing area of research and requires extensive investigations (Ko and Nof, 2010). Fig. 2.20 presents a reference architecture for collaborative design (Huang et al., 2010), as a prototype system

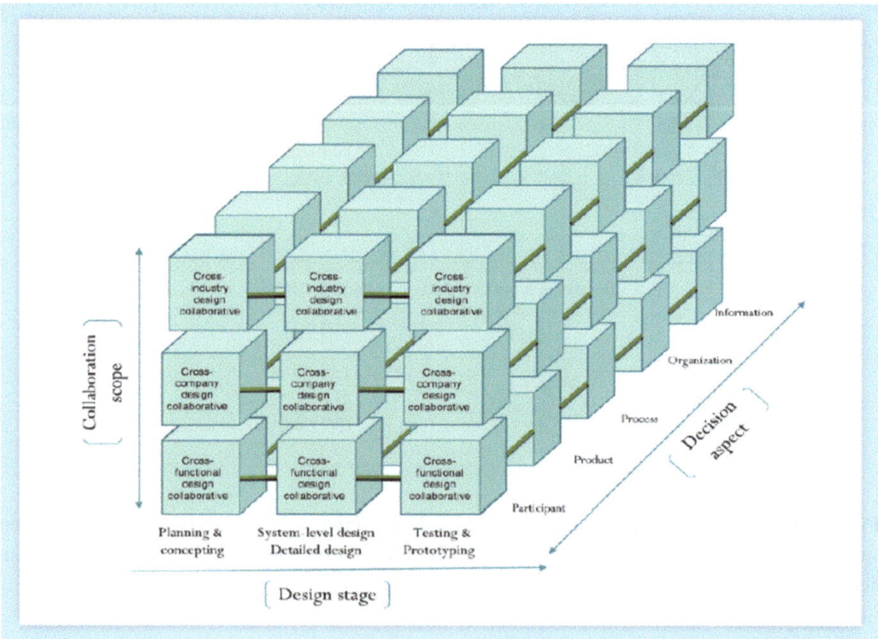

Fig. 2.20 Reference architecture for collaborative design; Supports decisions, designs, and collaboration; Represents collaboration status, details, and boundaries and support.

supporting decisions, design, and collaboration (three dimensions of collaborative design). In the following sections, three application examples and classifications of CSS are presented, and their relations to the CCT design principles are discussed.

2.3.1 Example 1: e-Surveillance Systems

Sensors, especially micro- and nano-sensors, are the most vulnerable type of agents to failure over time. Interoperability and allocation of resources are then two critical issues in design and implementation of sensor networks. An e-Surveillance system (de Freitas et al., 2010) is an example of CSS, which enables heterogeneity of sensors and smooth communication (interoperability), and supports coordination mechanisms (i.e., resource allocation) in heterogeneous sensor networks. Heterogeneity is defined according to three dimensions of (1) computer platform, (2) mobility, and (3) sensor capabilities. The intelligent surveillance system includes a pheromone-based coordination strategy, as a communication mechanism between the mobile and static nodes, for guiding the movements of UAV (Unmanned Automated Vehicle) swarms. The system is inspired by the foraging behavior of real ants (see Chapter 4 for details). The pheromone trails comprise information about the UAVs (e.g., in surveillance and patrolling applications), and are used for selection and assignment of UAVs to handle an alarm issued by a ground sensor node (Fig. 2.21). The intelligent surveillance system, enhanced with a customizable hardware platform and middleware, improves the flexibility, dynamicity, resource usage control, and data segregation.

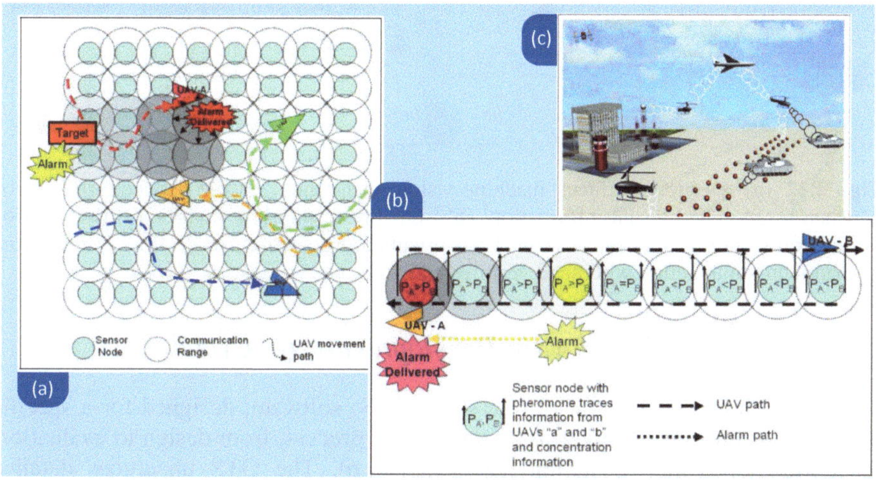

Fig. 21 (a) Detection of the target by a ground sensor, alarm generation and routing to the strongest UAV; (b) Choice of the best UAV based on the pheromone trails; (c) Case illustration: A surveillance application scenario.

2.3.2 Example 2: e-Operating Rooms

A significant portion of surgery cases in the US are nowadays performed in integrated and collaborative operating rooms enhanced with robotic surgery and telemedicine. The main purpose is to minimize the surgeons' focus shifts through minimizing (1) the distance between surgeon's head or gaze and target (i.e., focus spatial offset), and (2) the distance surgeon covers physically (movement spatial offset) (Wachs, 2010). An application of machine vision and computer graphics, as a CSS improving the performance of surgeons in collaborative intelligent operating rooms (i-OR), is shown in Fig. 2.22. The objectives are to (1) improve dynamic i-OR layout based on the behavioral cues of surgeons, (2) minimize unnecessary layout shifts, and (3) enable collaborative interaction in the i-OR to visualize sensitive information.

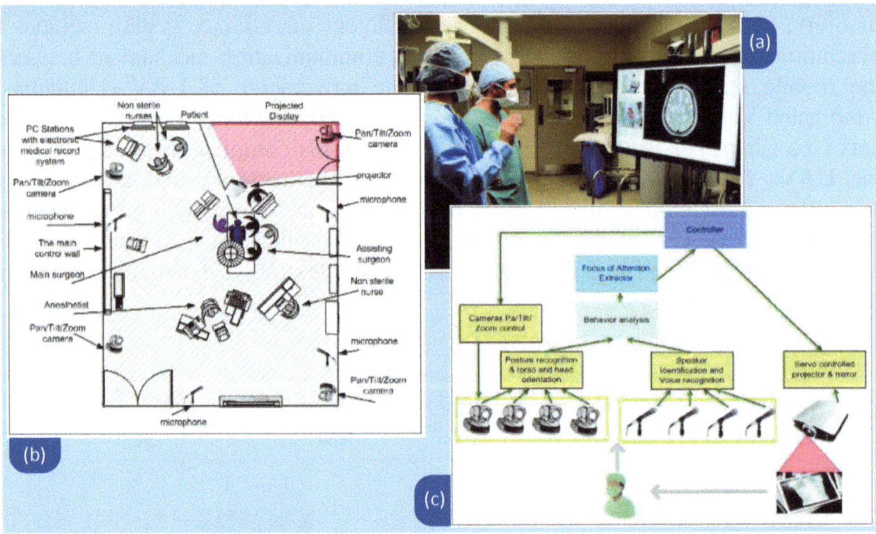

Fig. 2.22 (a) Sterile interface: machine-vision techniques enable isolated interaction between surgeon and visual information; (b) i-OR layout (concept image); (c) Recognition of surgeon's posture and orientation, head orientation detection, projection system setup and surgeon's interaction with hand gestures and voice commands.

2.3.3 Example 3: Software Development Life Cycle

The life cycle of commercial off-the-shelf (OTS) software, designed for a specific enterprise or the entire market, is a complicated process, from design to evaluation, integration, and implementation (see Chapter 6). The OTS undergoes detailed evaluations from various perspectives (e.g., robustness, significance, and reliability) by the organizations and stakeholders engaged in the transactions. A collaborative multi-tier framework for system development life cycle (SDLC) is presented in this section, as a CSS to enhance communication and collaboration among stakeholders involved in the OTS life cycle process, i.e., OTS provider, consumer, consultants, and integrators (see Fig. 2.23). The outcomes of implementation of the collaborative multi-tier SDLC in five enterprises from different vertical market segments are summarized into five major hypotheses (Cohen et al., 2010):

(3) The consumer's willingness to adopt a methodology is proportional to his/her willingness to adopt the OTS associated with that methodology.
(4) The level of uncertainty in the process is proportional the OTS provider's tendency to improvisation intensive development methods.
(5) The use of improvisation intensive development methods by an OTS provider shortens the market response time.
(6) Participation of human resources in selling the OTS to hesitant consumers increases the chance of successful implementation.
(7) Affiliation of an OTS provider with an established third party (e.g., a distinguished scientist) increases the chance of successful implementation.

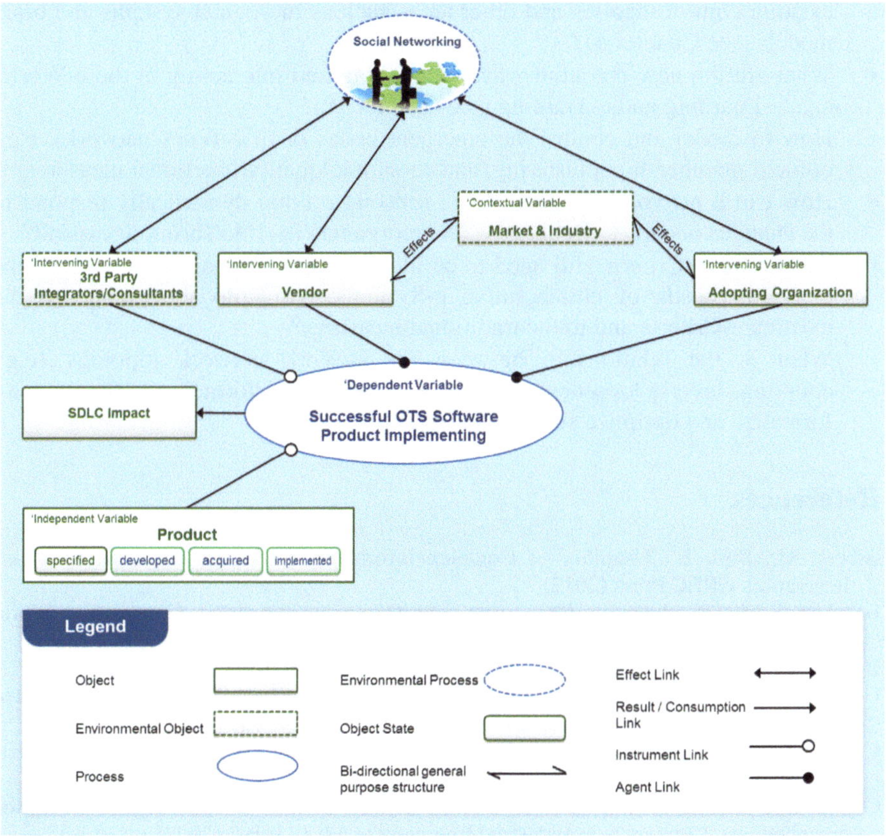

Fig. 2.23 The four states of OTS lifecycle (specified, developed, acquired, and implemented) in the collaborative multi-tier SDLC framework and interaction of stakeholders via non-systematic social networks.

2.4 Emerging Trends and Challenges

Successful design and implementation of next generation e-Systems depends highly on the understanding and effective deployment of collaborative e-Work principles, models, techniques, and guidelines. It is clear that the design of emerging e-Systems should mimic but not duplicate the traditional "non-e" principles, and must address the rising challenges formalized by the emerging e-Criteria. The future of collaborative e-Systems is promising significant improvements in our quality of life. It will require the exploration of different areas and the definition of further collaborative theories, principles, methodologies, and implementations, as described in this book. Proper network models must be developed and applied to describe the nature and design of intelligent teams of agents, from nano-sensors to extended enterprise networks. In view of that progress, the following interesting questions still need to be addressed, as open questions and challenges for better collaboration engineering (see also Chapter 11):

- How to define and enrich the foundation of CCT, and find synergies between existing control theories and other areas such as biological systems and brain models (see Chapter 4)?
- What are the new potential ways of working and interacting in the e-World, e.g., e-Learning and e-Training (see Chapter 10)?
- How to model and control the emergent behavior of e-Work networks, e.g., optimal membership, clustering, and economic/quality/functional measures?
- How can a network of agents (e.g., sensors, robots) dynamically respond to the changes occurring frequently, and reorganize itself to shrink or expand?
- What other e-Criteria still need to be identified and formalized to address the emerging needs of collaborative e-Systems? How do they relate to the existing e-Criteria and to the traditional measures?
- What is the relationship between the e-Work network topology (e.g., coupling, layers, looseness, redundancy) and its performance? How can we formalize and optimize such interactions?

References

Albers, M., Still, B.: Usability of Complex Information Systems: Evaluation of User Interaction. CREC Press (2012)

Baecker, R.M.: Readings in Groupware and Computer-Supported Cooperative Work. Morgan Kaufmann Publishers, San Francisco (1993)

Boy, G.A.: Orchestrating Human-Centered Design. Springer (2013)

Ceroni, J.A., Nof, S.Y.: A workflow model based on parallelism for distributed organizations. Journal of Intelligent Manufacturing 13, 439–461 (2002)

Chen, X.W., Nof, S.Y.: Error detection and prediction algorithms: Application in robotics. Journal of Intelligent Robotic Systems 48(2), 2–225 (2007)

Chituc, C.M., Nof, S.Y.: The Join/Leave/Remain (JLR) decision in collaborative networked organizations. Computers & Industrial Engineering 53, 173–195 (2007)

Cohen, S., De Haan, U., Dori, D.: A Software System Development Life Cycle Model for Improved Stakeholders' Communication and Collaboration. International Journal of Computers, Communications & Control 5(1), 20–41 (2010) de Freitas, E.P., Larsson, T., Heimfarth, T., Wagner, F.R., Allgayer, R.S., Pereira, C.E., Ferreira, A.M.: Coordinating Aerial Robots and Unattended Ground Sensors for Intelligent Surveillance Systems. International Journal of Computers, Communications & Control 5(1), 52–70 (2010)

Isenberg, P., Elmqvist, N., Scholtz, J., Cernea, D., Ma, K.L., Hagen, H.: Collaborative Visualization: Definition, Challenges, and Research Agenda. Information Visualization 10(4), 310–326 (2011)

Hadeli, K.: Bio-inspired multi-agent manufacturing control systems with social behavior. Ph.D. Thesis (Katholieke Universiteit Leuven, Leuven) (2006)

Huang, C.Y., Nof, S.Y.: Formation of Autonomous Agent Networks for Manufacturing Systems. International Journal of Production Research 38(3), 607–624 (2000a)

Huang, C.Y., Nof, S.Y.: Autonomy and Viability – Measures for Agent-Based Manufacturing Systems. International Journal of Production Research 38(17), 4129–4148 (2000b)

Huang, C.Y., Yang, T.T., Chen, W.L., Nof, S.Y.: Reference Architecture for Collaborative Design. International Journal of Computers, Communications & Control 5(1), 71–90 (2010)

Huang, C.Y., Ceroni, J.A., Nof, S.Y.: Agility of Networked Enterprises - Parallelism, Error Recovery and Conflict Resolution. Computers in Industry 42(2-3), 275–287 (2000)

Huang, C.Y., Nof, S.Y.: Evaluation of Agent-Based Manufacturing Systems Based on a Parallel Simulator. Computers and IE 43(3), 529–552 (2002)

Keel, P.E.: Collaborative visual analytics: Inferring from the spatial organization and collaborative use of information. In: IEEE Symposium on Visual Analytics Science and Technology, pp. 137–144 (2006)

Ko, H.S., Nof, S.Y.: Editorial – Special Issue on Collaboration Support Systems (CSS). International Journal of Computers, Communications & Control 5(1), 6–7 (2010)

Ma, K.-L., Wang, K.-L.: Social-Aware Collaborative Visualization for Large Scientific Projects. In: Proceedings of the IEEE International Symposium on Collaborative Technologies and Systems, Los Alamitos, CA, USA, pp. 190–195 (2008)

Nof, S.Y.: Robot Ergonomics: Optimizing Robot Work. In: Nof, S.Y. (ed.) Handbook of Industrial Robotics, 2nd edn., ch. 32, pp. 603–644. John Wiley & Sons, New York (1999)

Nof, S.Y.: Collaborative control theory for e-Work, e-Production, and e-Service. Annual Reviews in Control 31(2), 281–292 (2007a)

Nof, S.Y.: Availability, integrability, and dependability – what are the limits in production and logistics (Plenary). In: Proceedings of MCPL, Sibiu, Romania (September 2007b)

Nof, S.Y., Huang, C.Y.: Enterprise Agility: A View from the PRISM Lab. International Journal of Agile Management Systems 1(1), 51–59 (1999)

Ozsoy, E.: Co-Insights for Multi-Criteria Decision Making. M.S. Thesis (Purdue University, West Lafayette) (2011)

Rajan, V.N., Nof, S.Y.: Cooperation requirement planning (CRP) for multi-processors: optimal assignment and execution planning. Journal of Intelligent Robotic Systems 15, 419–435 (1996)

Thomas, R.W., Friend, D.H., DaSilva, L.A., MacKenzie, A.B.: Cognitive Networks: Adaptation and Learning to Achieve End-to-End Performance Objectives. IEEE Communications Magazine (2006)

Velásquez, J.D., Nof, S.Y.: A best-matching protocol for collaborative e-Work and e-Manufacturing. International Journal of Computer Integrated Manufacturing 21(8), 943–956 (2008)

Wachs, J.P.: Gaze, Posture and Gesture Recognition to Minimize Focus Shifts for Intelligent Operating Rooms in a Collaborative Support System. International Journal of Computers, Communications & Control 5(1), 106–124 (2010)

Review and Challenge Questions

1. Consider several examples of your collaborative work experiences. For each, describe the purpose of effort, the key participating resources, and (i) any of the collaboration procedures that was (were) pre-planned; (ii) any computational tools applied for this planning purpose; (iii) whether you could separate the planning as CRP-I and CRP-II.
2. Consider the following teams and their examples:
   a. Human – Human: Salesperson/Customer; Furniture designer/Client; Component machinist/assembler.
   b. Human – Software Agent: You/A search engine; Airplane designer/CAD system; Manager/ Scheduling App.
   c. Human – Robot: Nurse/Robot nurse; Farmer/Milking robot; You/Robot cleaner.
   d. Sensor – Robot: Camera/Robot crane; RFID reader/robot loader-unloader;
      Navigation recommender/automatic vehicle.

   Select one example from each of the four team types and describe 6-8 work steps for this team. Then develop the Collaboration Requirement Matrix (CRM) for each of the four team cases.
3. What is the purpose of the Collaboration Requirement Matrix (CRM)? Is there a difference in this purpose for the different team types? Explain why or why not for each type.
4. The rational for CRP-I and CRP-II is not unique to collaboration engineering. The first, CRP-I is "static planning" or pre-planning for the best known systems' (participants') conditions; the second is "dynamic planning", adaptive and responsive, possibly even predictive to variations in the systems as a function of time. Think and describe other areas that can benefit from this rational. Examples: Real estate acquisition; Production scheduling; Project management; Business planning; Customer services.
5. Select an example from Question 2 and develop for it the CRP architecture (similar to Fig. 2.1, but specific to your example).
6. Select two examples from Question 2 and (i) describe their Task Dependencies; (ii) For which CRP principles (protocols) are Task Dependencies useful?
7. Explain the purpose of the Task Integration Module (DPIEM-1) and illustrate this purpose and the logic of this module, based on an example from Question 2.
8. Repeat Question 7 for the Task parallelism optimization module (DPIEM-2), and for the Distributed task assignment module (DPIEM-3).
9. Consider the release of a new generation of a document processor. (i) How would you follow the KISS principle in the design of the new version? (ii) What would be some ways in this case to balance between the availability of new functional features, and the need to simplify any retraining?
10. List (i) several errors, (ii) several conflicts that may occur for a team from Question 2, and (iii) using Table 2.2, characterize their method (type).

11. Explain and illustrate several types of costs that are incurred in the eight EPCR functions (Fig. 2.6) for the errors and conflict you described in Question 10.
12. What is similar and what is different in comparing the Error Prevention and Conflict Resolution (EPCR) principle, and the Collaborative Fault Tolerance (CFT) principle? Illustrate your answer with one of the team examples.
13. Suggest several approaches that the three robots in the bag handling example can help each other in tolerating errors, following the CFT principle.
14. Apply the AD principle to the bag handling example, for the decisions and considerations of each participant (i) with robots; (ii) with human operators; (iii) with a mixed team.
15. Discuss examples of emergencies which would result in disabled lines of collaboration and command in an e-Work system. For each, suggest how the CRP, AD, and ELOCC principles would apply.
16. Explain and illustrate how the Best Matching, BM principle, is related to the Association and Dissociation, AD principle, and to the Emergent Lines of Collaboration and Command, ELOCC principle.
17. (i) Define a different BFI, Best-Fit Index, for several cases in Table 2.3. (ii) Define a dependent BFI, based on a series of two different BFIs (as illustrated in Fig. 2.11 and 2.12).
18. Suggest several definitions of e-Criteria and e-Measures for the three examples of CSS, Collaboration Support Systems in this chapter. Then develop a table showing for each CCT design principle how those e-Criteria and e-Measures can help evaluate alternative design options.
19. Consider three levels of matching: Best (BM), Adequate (AM), and Poor (PM). Illustrate each of them for the following cases, and explain the implications and impact on the resulting outcomes:
    a. Matching components in jet engine assembly
    b. Selecting suppliers and sub-contactors
    c. Planning routes for navigation and exploration
    d. Choosing a tool, a robot gripper, or a decision algorithm.
20. Explain the role of the Collaborative Visualization and Comprehension (CVC) design principle in each of the three CSS examples in this chapter.
21. Based on what you have learned so far in Chapters 1 and 2 about revolutionizing collaboration, and related emerging trends and challenges, identify a few weaknesses in e-Systems that are designed without attention to collaboration engineering as defined in this book.

Lab Questions

22. For the e-Work lab system you selected in Question 1.13, describe several CCT design principles that you think are followed in its design, or could be followed.
23. For the e-Work Pair you selected in Question 1.14, define one e-Criterion and one e-Measure, and apply them to refine your answer to 1.14.

Chapter 3
Rationalization

Abstract. Similar to any investment decision, implementation of e-Work solutions requires a comprehensive rationalization from various perspectives, especially economy, productivity and sustainability. Considering e-Work, in general terms, as a sort of technology encapsulation (Nof, 2003 and 2007a), this chapter focuses on the prioritization of e-Work implementation plans. Realization of e-Work solutions depends highly on the demonstration of its inherent benefits to individual, networked e-Systems, and the users. The strategic requirements of an e-System are considered as a starting point to identify proper e-Work processes and solutions. As discussed earlier in this book, in line with the emergence of e-Systems, various e-Criteria are continuously evolving, which must be incorporated in the rationalization process as key performance indicators (e.g., reconfigurability and reusability of software and processes). Therefore, a robust framework is indeed required to support rationalization of e-Work implementation projects based on multiple (and somehow conflicting) e-Criteria.

In this chapter, basic operational strategies and traditional economic rationalization procedures are first reviewed. It is shown that traditional procedures cannot sufficiently address some emerging aspects of e-Work and disregard strategic benefits of technology. Several scenarios for production of goods and services are then analyzed in the form of value chains, corporative level strategy, business units strategies, and functional strategic planning to determine the operational strategies. Process specification is performed through Value Stream Maps (VSM) for each enterprise to achieve a cross-company operational specification. e-Work alternatives can then be evaluated through defining and prioritizing the implementation plans. Technology evaluation models are introduced for evaluating of the results. Such models conform the implementation plans for different e-Work alternatives. The goal is to maximize the benefits of enterprises and enhance the specification of e-Work processes and technologies through prioritizing the enterprise plans for research and development. Finally, a comprehensive Multi-Criteria Decision-Making (MCDM) framework, based on Analytic Hierarchy Process (AHP) and Data Envelopment Analysis (DEA), is introduced for evaluation of e-Work alternatives based on multiple e-Criteria.

S.Y. Nof et al., *Revolutionizing Collaboration through e-Work, e-Business, and e-Service*,
Automation, Collaboration, & E-Services 2, DOI: 10.1007/978-3-662-45777-1_3

3.1 Operational Strategy

Operational strategy of an enterprise is commonly structured based on decision categories and competitive priorities, i.e., a set of goals for the business strategy based on market requirements. Selection of proper operational functions significantly affects the enterprise competitiveness in the market. Such selection decisions are typically grouped into decision categories (Hayes and Wheelwright, 1984), generating operational strategies. Hence, the effectiveness of an operational function depends highly on the consistency of the decision patterns in supporting the competitive priorities and the operational functions. In general, decision categories are classified as *structural* (i.e., long-term commitments or investments) and *infrastructural* (i.e., support functions in a short-term perspective) (Hayes and Wheelwright, 1984). A common set of decision categories includes *process technology*, *capacity*, *facilities*, and *vertical integration* as structural class, and *quality*, *organization*, and *manufacturing planning and control* (MPC) *system* as infrastructural class (Hayes and Wheelwright, 1984; Leong et al., 1990; Miltenburg, 1995). Process selection, i.e., choosing an operational process that supports the competitive priorities of an enterprise (Hill 2000), is a key element defined according to the relationships between the market requirements and the operational strategy. Market requirements are proven as key factors influencing the process selection decisions (Hayes and Wheelwright, 1979, 1984) for different products and services. The relationship between the product/service aspects and the process selection decisions depends mainly on the *congruence* between the market characteristics and the specification of the operational strategy (Bozarth and Berry, 1997; Schroeder et al., 1995; Sheu, 1994; Kim and Lee, 1993).

An e-Work implementation project must be evaluated as a new integrated system, based on the enterprise operational strategy, capable of providing financial, operational, and strategic benefits, taking into account specific market requirements. This type of projects entails some unique characteristics with clear economic values, which differentiate them from any other capital equipment project:

- Providing flexibility in production capability, enabling timely responses to market changes.
- Enabling users to rethink, define, and integrate systematic operational functions.
- Enabling reusability (e.g., reprogrammable solutions) and relatively longer lifecycles.
- Reducing requirements for services and related facilities.

The above differences lead to several operational benefits such as optimized flexibility, productivity, precision, quality of products/services, operational costs, etc. e-Work solutions enhance competitive strategies through time-based competition and mass customization of global markets. Extremely shortened product lifecycles demand agile automated systems, modularity of processes, and rapid deployment in alternative configurations. Such strategic aspects noticeably differentiate the rationalization of an e-Work project from other traditional systems. In fact, traditional economic evaluation methods cannot sufficiently address the benefits of capital reutilization in multiple projects, and therefore, must be modified to incorporate long-term strategic values of e-Work in the selection decisions, based on various e-Criteria.

3.2 Traditional Economic Rationalization

Economic rationalization enables comparison between the financial benefits of alternative uses of capital in an investment project. Evaluation is performed through

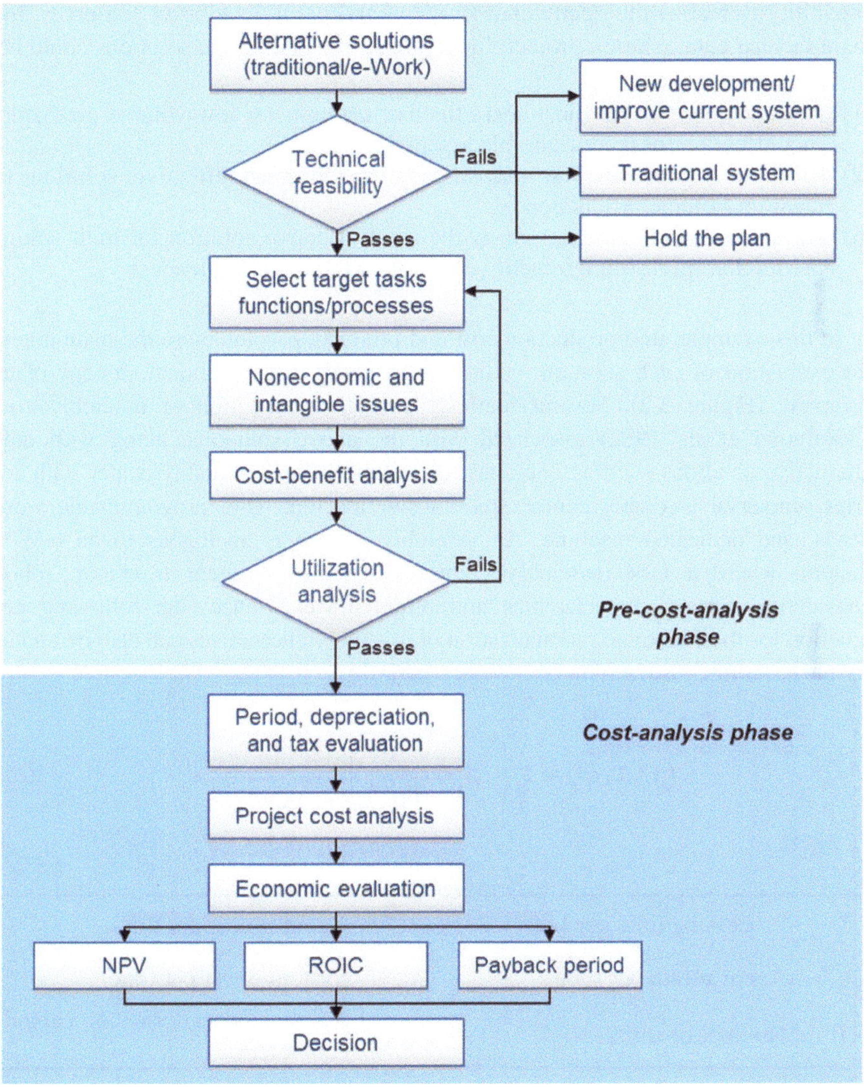

Fig. 3.1 Economic rationalization of e-Work projects (NPV: Net Present Value; ROIC: Return-On-Invested Capital).

measurement of cost plus operating expenses against cash-flow benefits estimated for each alternative. The general procedure of (traditional) economic rationalization is shown in Figure 3.1, which consists of two phases described below.

3.2.1 Pre-cost-Analysis

In this phase, the feasibility of alternative solutions for an e-Work implementation project is evaluated. The evaluation criteria (traditional) are (1) technical capability to realize the operational strategy, and (2) utilization of resources. In a manufacturing automation project, for example, the alternative solutions could be:

(1) *Manual labor*: is the most cost-effective solution for low-volume production (not effective for reconfigurable assembly).
(2) *Flexible (programmable) automation*: is the most effective solution for medium-volume production.
(3) *Hard (dedicated) automation*: is the most effective solution for high-volume production (performing repetitive operations in fixed sequences).

In this example, unit production cost and production volume are the main criteria for evaluation of each solution, taking into account the operational strategy of the company (Figure 3.2). Several methods exist for estimation of assembly costs (Boothroyd et al., 1982) associated with the above solutions, along with other alternative solutions such as one-operator assembly line, assembly center with two arms, universal assembly center, free-transfer machine with programmable work-heads, and dedicated machine. An assembly system is analogous to an e-Work system, where a task (assembly) is processed by an agent (operator, robot), considering certain characteristics and requirements. Hence, the following cost function for the economic rationalization of assembly alternatives can also be applied for the economic justification of e-Work solutions:

$$f(t,\tau,\omega) = \tau_t \left(c_\tau^\omega + \frac{r_\tau^\omega c_\tau^s}{\mathrm{T}C^\omega} \right), \tag{3.1}$$

where

τ_t	Processing time per task	c_τ^ω	Agent ω cost per time
r_τ^ω	Agent ω rate per time	c_τ^s	System s cost per time
T	Number of shifts	C^ω	Agent cost in terms of capital equivalent

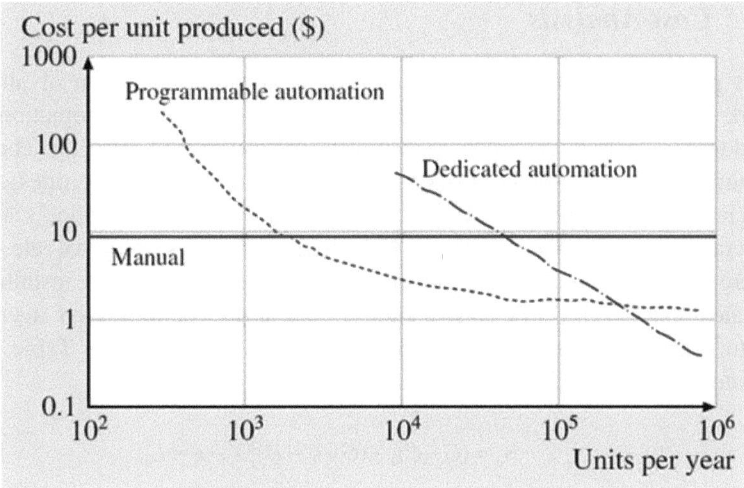

Fig. 3.2 Comparison of manufacturing automation solutions (production volume vs. cost).

In spite of cost effectiveness, an e-Work alternative may be infeasible to implement from non-economic perspectives (e-Criteria), such as:

(1) *Integrability*: Is it technically feasible to integrate data from distributed agents in the system?
(2) *Scalability*: Are processes and enterprise network capable of handling the growth and increasing load of tasks associated with each solution?
(3) *Dependability*: Can successful execution of tasks be ensured as a component of the total system?
(4) *Maintainability & integrity*: Is it possible to maintain safety, process continuity, and the designed quality level?
(5) *Reconfigurability & resilience*: How agile and capable is the solution in responding to alterations in information, processes, and system's requirements?
(6) *Enterprise-specific issues*: Are issues related to specific enterprise characteristics, policy, standards, social responsibility and morale, etc. properly addressed?

In addition, utilization of different e-Work solutions and other intangible benefits in long-term (e.g., flexibility, business opportunities and competitiveness, and error and conflict reduction) must be incorporated in the feasibility analysis. The mentioned e-Criteria, however, cannot be addressed properly by traditional economic rationalization methods. These considerations are further discussed and alternative approaches for comprehensive rationalization based multiple e-Criteria are introduced in Sections 4 and 5.

3.2.2 Cost-Analysis

In this phase, detailed cost analysis for economic rationalization of alternative e-Work solutions is performed. A numerical example from an automation project is used to demonstrate the mechanics of different methods (Table 3.1). The project is evaluated over a 6-year horizon. The evaluation period, tax rates, and tax depreciation are determined using the US Internal Revenue Service's Modified Accelerated Cost Recovery System (MACRS) for 5 years, with tax rate of 40%. The project cost (including equipment, software integration, and installation) is calculated using Eq. (3.1), aggregated for all tasks, periods, and agents (i.e., operators and/or robots). The yearly cash flows presented in Table 3.1 are calculated using the following formula:

$$N_\tau = (G-C)_\tau - (G-C-D)_\tau X - K + L_\tau, \tag{3.2}$$

where

G_τ	Gross income at period τ	C_τ	Total cost at period τ
D_τ	Tax depreciation at period τ	X	Tax rate
K	Project cost	L_τ	Salvage value at period τ

Table 3.1 Project data

End of τ	X	K & L	G_τ	C_τ	D_τ	N_τ
0	-	168,000 (K)	-	-	-	(168,000)
1	20.00	-	92,000	25,000	33,600	53,640
2	32.00	-	92,000	25,000	53,760	61,704
3	19.20	-	116,000	19,000	32,256	71,102
4	11.52	-	116,000	12,000	19,354	70,142
5	11.52	-	116,000	12,000	19,354	70,142
6	5.76	12,000 (L)	116,000	12,000	9,677	78,271

Three methods for economic evaluation of alternative solutions are discussed in the following sections.

NPV (Net Present Value)

Upon determining the cash flow (Table 3.1), the NPV can be calculated using the following equation:

$$\mathrm{NPV} = \sum_{\tau=0}^{\mathrm{T}} \frac{N_\tau}{(1+r)^\tau} = \sum_{\tau=0}^{\mathrm{T}} N_\tau (\mathrm{P/F}, r, \tau), \tag{3.3}$$

where r denotes the Minimum Acceptable Rate of Return (MARR), usually designated as $(\mathrm{P/F}, r, \tau)$. According to the cash flows shown in Table 3.1 and assuming r = 25%, the NPV for the alternative solution in the given example is

$$\mathrm{NPV} = -168{,}000 + 53{,}640(\mathrm{P/F}, 25, 1) + 61{,}706(\mathrm{P/F}, 25, 2) + \cdots + 78{,}271(\mathrm{P/F}, 25, 6) = \$18{,}431.$$

The alternative solution is economically acceptable, since the corresponding NPV is positive, which implies that the rate of return is greater than r.

ROIC (Return-On-Invested Capital)

The ROIC, also referred to as Internal Rate of Return (IRR), is equal to the interest rate at which NPV = 0. Mathematically, the ROIC is defined as

$$0 = \sum_{\tau=0}^{\mathrm{T}} \frac{N_\tau}{(1+i)^\tau} = \sum_{\tau=0}^{\mathrm{T}} N_\tau (\mathrm{P/F}, i, \tau), \tag{3.4}$$

where i denotes the ROIC. For the numerical example, the ROIC can be obtained from the following expression (through trial-and-error and linear interpolation):

$$0 = -168{,}000 + 53{,}640(\mathrm{P/F}, i, 1) + 61{,}706(\mathrm{P/F}, i, 2) + \cdots + 78{,}271(\mathrm{P/F}, i, 6) \quad \Rightarrow \quad i \approx 31\%.$$

The ROIC is now compared with the MARR, and since ROIC ≥ MARR the alternative solution is acceptable. Note that the definitions of cash flow and MARR are mutually dependent. Moreover, exclusion of debt interest from the cash-flow equation does not necessarily imply that the initial capital expenditure is not financed by some combination of debt and equity capital. Accordingly, when total cash flows are used, the (approximate) debt interest is included in the MARR definition as follows:

$$\mathrm{MARR} = k_e(1-b) + k_d(1-X)b, \tag{3.5}$$

where

k_e	Required return for equity capital	k_d	Required return for debt capital
b	Debt ratio of the pool of capital	X	Tax rate

In practice, it is common to adjust (increase) k_e and k_d to account for the risks and uncertainties of alternatives in economic conditions. In the automation project example the ROIC is influenced, at different levels, by different solution alternatives as shown in Figure 3.3.

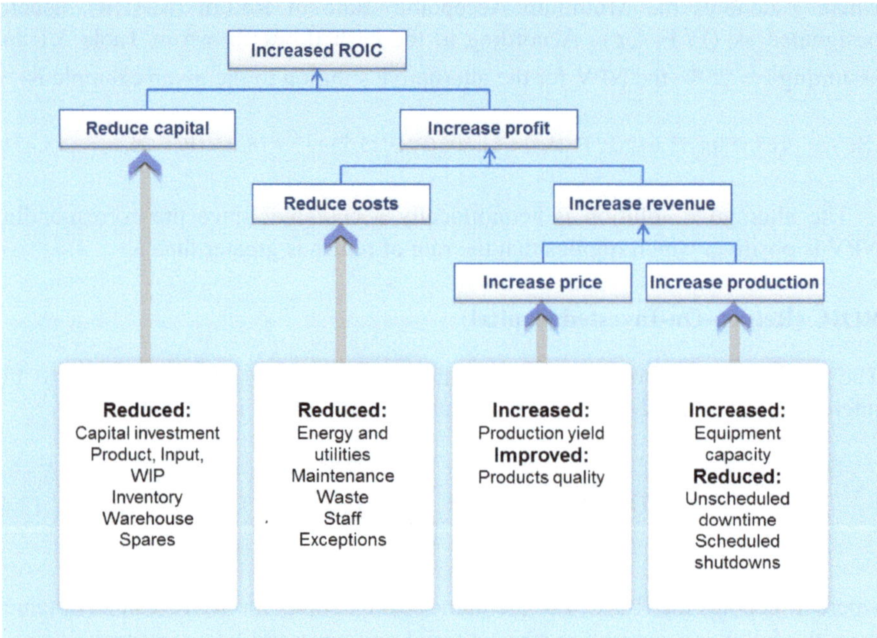

Fig. 3.3 Effects of automation projects on ROIC.

Payback Period

Another method that can be used for economic rationalization of e-Work solution alternatives is payback (or payout) period, and is defined as the number of periods (e.g., years) required by the cash inflows to balance the cash outflows. The payback period p is calculated as follows:

$$0 = \sum_{\tau=0}^{p} N_{\tau}. \tag{3.7}$$

In practice, an extended definition is often used that employs *discounting procedures*. According to the cash flow data given in Table 3.1, the payback equation for 2 years gives

$$-168,000+53,640+61,706=\$-52,656,$$

and for three years

$$-168,000+53,640+61,706+71,102=\$18,446.$$

Therefore, using linear interpolation, the payback period is obtained as follows:

$$p=2+\frac{52,656}{71,102}=2.74 \text{ Years}$$

3.3 Value Mapping and Technology-Based Rationalization

3.3.1 Value Stream Mapping (VSM)

VSM is a method for representing the processes carried out by manufacturers, suppliers, and distributors in a value chain for delivering products and services to end-customers. The method generates process maps that consider both products and information for signaling and controlling the flows among the processes. The goal is to categorize processes, according to their value-adding capabilities, into (1) value adding (i.e., those to potentiate) and (2) non-value adding (i.e., those targeted for reduction or elimination). VSM enables enterprises to drive modification of their processes and infrastructure such that leadtime, level of work-in-process (WIP) inventories, and/or any other criteria/e-Criteria (that are positively correlated to value in the chain) are optimized. VSM method is based on two types of maps:

(1) *Current state*: considers logistics and production processes, and their hierarchy in the value chain. Products and information flows are recorded in each process as indicators of the current state of production in the value chain. The process execution data, usually flowing backwards in the form of demand over the processes, is recorded for each process identified and represented in the map. All processes in the VSM are represented along a timeline in the map, which (1) temporarily represents each process and its relationships with other processes in the map (usually in the form of delays), and (2) differentiates between value-added and waste processes in the current state. Segmentation of processes is performed by comparing their processing times against the *takt time*, the ratio between the available capacity and the customer demand.

(2) *Future state*: represents the improvements in the current state. The future state map is generated based on the current state map by improving the value-adding processes and (possibly) eliminating non value-adding processes, following seven guidelines (Rother and Shook, 1999):

   (i) Produce to takt time
   (ii) Develop continuous flow
   (iii) Use downstream nodes in the chain to control production where continuous flow does not extend upstream

(iv) Schedule based on the pacemaker operation
(v) Produce different products at a uniform rate
(vi) Level the production load on the pacemaker process
(vii) Develop the capability to make "every part every time period"

Table 3.2 shows the advantages and limitations of the VSM method.

Table 3.2 Advantages and limitations of VSM method.

Advantages	Limitations
• Links manufacturing processes to supply networks, distribution channels and information flows • Integrates product and information flows • Links production control and scheduling functions to production scheduling and shop floor control using operating parameters for the manufacturing system, e.g., takt time • Enables integration of flow analysis methods, e.g., production flow analysis, business process reengineering, and process analysis and improvement • Provides descriptive information for the operation and storage icons in the standard flow process charts • Enables lean manufacturing design, e.g., dock-to-dock flow time for a product family • Provides a *blueprint* for strategic planning based on *lean thinking*	• Fails to map multiple non-identical products (in terms of routing and processes) • Fails to relate transportation and queuing delays, and changes in transfer batch sizes to operating parameters and criteria • Lacks an economic measure for value, e.g., profit, operating/inventory costs • Lacks the spatial structure of facility layout, and how that impacts inter-operation material handling delays • Lacks capabilities of handling low-volume high-variety manufacturing facilities • Fails to address allocation and utilization of important resources, inefficient product flows, handling complete BOMs, queuing delays, sequencing rules for multiple orders • Lacks the capability of rapid evaluation of *what-if* analyses for prioritizing multiple alternatives under time/budget constraints

3.3.2 Value Network Mapping (VNM)

A major limitation of the VSM method (Table 3.2) is high variety of products produced and services delivered in small batches involving lots of process routing to be followed by production. This issue makes the generation of the current state map complicated and exhausting. For instance, manual generation of a current state map for a small case with 20 products in 10 different work-centers is as cumbersome, time consuming, and repetitive as mapping multi-product flows while designing a facility layout. A good solution to this problem is to aggregate different product flows. This aggregation, however, makes the current state maps overlap, which in turn requires continuous revisions of the maps. VNM is a promising solution approach, which usually includes numerous value streams and processes. A VNM must be clear to read and use. The size of the map limits the number of process boxes defined and the value streams included in that specific map, which in turn necessitates connections among multiple maps. This gap can be addressed by *bubble diagrams*, which enable clutter-free drawing of the complete network of product and information flow (Figure 3.4).

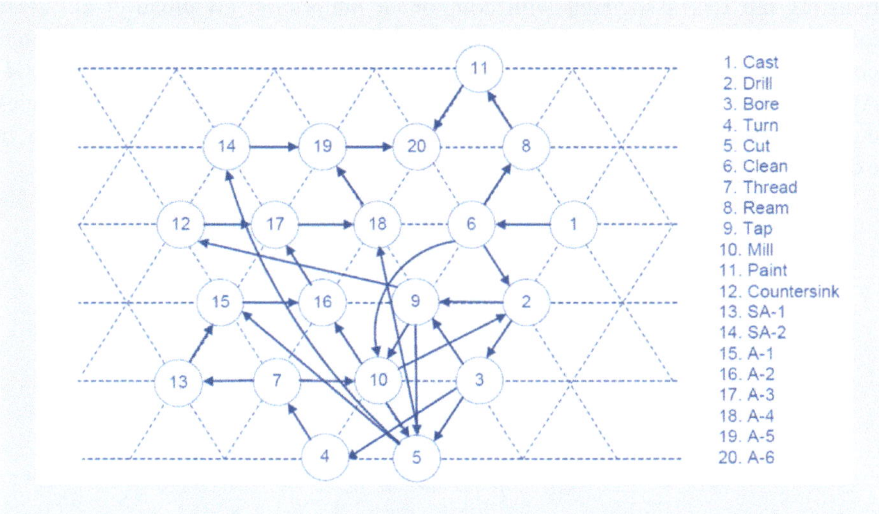

Fig. 3.4 A bubble diagram for the VNM of a gate-valve assembly process (Irani and Zhou, 2011); the process boxes represent the adjacent strongly connected pairs of work centers in the product flow network.

A typical make-to-order assembly facility produces a wide range of products that require different combinations of parts and sub-assemblies, whose routings feature different sets of work centers located in the same facility. Hence, the VNM method must represent, and possibly aggregate, multiple operations process charts for different products produced in the same facility. The notion of VNM method can be generalized to generate a prioritized list of operational tasks candidate to improve their value contribution. Then, the evaluation process presented in the next section can be applied to define whether an alternative e-Work solution can provide additional values to the enterprise or the entire CNE.

3.3.3 Technology-Based e-Work

To determine the level of correspondence between systems and technologies, it is necessary to evaluate technologies within the context in which they will be used in the system. All technologies work well within a specific context and under certain conditions. For example, Web services work well for asynchronous communication over the Internet. In a business environment, these conditions are very common. However, this may not be the case in a security tactical command and control environment, where high performance and availability requirements prevail. The process outlined in Figure 3.5 is a context-based technology evaluation process for determining the fitness of a technology within a specific context. An integral part of the process is to develop models, as a technique initially codified for building systems from commercial components. This technique prescribes the use of focused hands-on experimentation as a way of reducing the risk associated with component integration (Wallnau et al., 2001, Lewis 2005). The process is also largely based on the process for the evaluation of commercial off-the-shelf (OTS) software products (Comella-Dorda et al., 2004). Although the focus of this technique is on evaluation of software technologies, e.g., Web services, database systems, and architectural frameworks, it can be extended for analyzing other types of technology-based solutions.

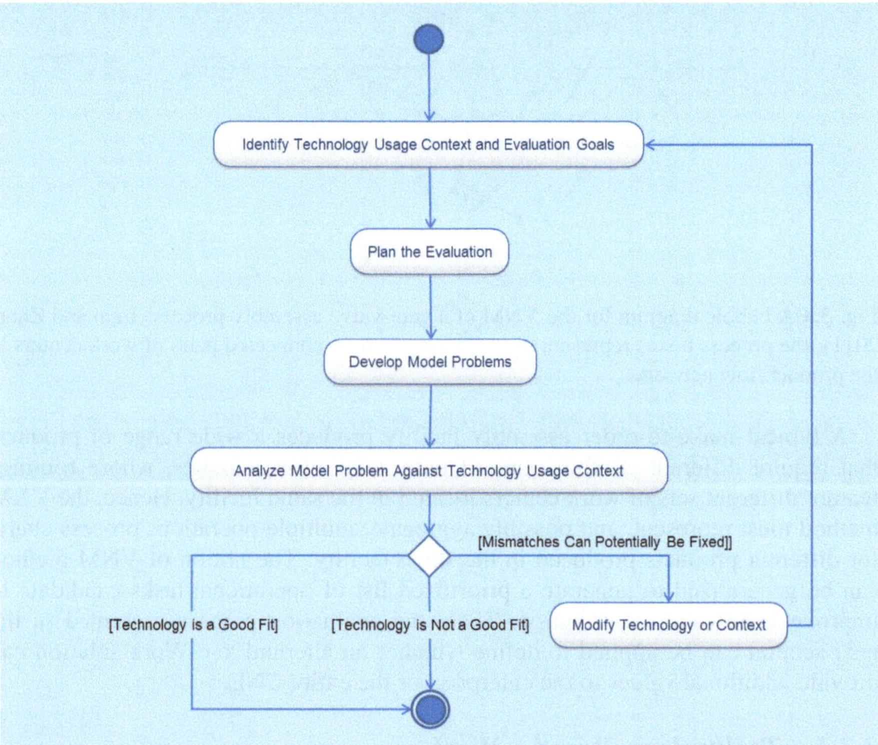

Fig. 3.5 Context-based technology evaluation process (source: Lewis and Wrage, 2005).

The systematic procedure shown in Figure 3.5 is further described below:

Step 1: *Identify technology usage context and evaluation goals*

- Determine and document the reasons for and the expectations from the evaluation, (e.g., interoperability, quality of service).
- Define the scope of evaluation (e.g., budget, time, alternative technologies available).

- Identify the existing constraints (e.g., legal requirements or organizational policies).

Step 2: *Plan the evaluation*

- Form and authorize an evaluation team (including technical and domain experts).
- Identify stakeholders (e.g., system architects, managers and administrative staff, change agents, sponsors), and their goals and concerns.
- Select a best-fit approach through evaluating all technologies in scope with respect to the constraints and benefits to stakeholders.
- Estimate efforts and schedule, and develop a preliminary plan and an overall schedule.

Step 3: *Develop model problems*

- Develop hypotheses, i.e., claim about sustaining or refuting technologies.
- Develop criteria (measurable) based on the expectations documented in Step 1.
- Design model solution (e.g., J2EE and .NET).
- Implement and evaluate model solution against criteria, i.e., implement, run, and observe the model solutions until obtaining enough information to sustain or refute the set of hypotheses.

Step 4: Analyze model problem results against technology

- Determine if technology is a good, a partially good, or not a good fit with the system based on the requirements.
- Modify either the context or the technology (if required) through reset technology expectations, relaxing constraints, and/or negotiating technology requirements with stakeholders (e.g., complementary technologies whose combinations satisfy the requirements).

3.4 Multicriteria Rationalization

Traditional rationalization techniques for justification of advanced technologies are mostly economical and short-term. The limited scope and short-term nature of traditional approaches cause difficulties in adopting systems and technologies with strategic implications and intangible benefits, represented in terms of various emerging e-Criteria. Examples include high capital costs and risks, difficulty of quantifying indirect and intangible benefits, inappropriate capital budgeting procedures, technological uncertainties, cultural and organizational issues, and most importantly, sustainability. In line with the emergence of e-Work technologies and transition from conventional systems to e-Systems, several solutions are continuously provided for resolving the emerging challenges of sustainability, concerning the three pillars of economy, environment, and society (see Table 3.3). The impact of implementing flexible e-Work systems crosses many organizational boundaries, where success or failure depends on the buy-in of all individuals involved in the system.

Table 3.3 Emerging challenges in sustainability and collaborative e-Work solutions.

Challenge	Description	e-Work solutions
Sustainable business	Product service systems models, mass customization,	Agent based systems, SOA
Sustainable supply chain	Multi-dimensional inventory management, interoperable production planning and control, sustainable design	Collaboration protocols, e.g., demand and capacity sharing, best matching, TAP
Sustainable product design	Sustainable product design, integrated product data management system	Applications of CIM, e.g., RFID
Maintenance and asset lifecycle management	Asset lifecycle management, mobile and remote maintenance	Application of MASs, distributed control systems, e.g., autonomous service oriented devices, wireless sensor networks
Reduces use of scarce resources	Recycling technology and material reuse, remanufacturing, sustainable packaging, resource recovery from alternative fuel and raw material	CIM, e.g., RFID for management of *end-of-life* systems
Energy efficient manufacturing	New energy sources, management and control of energy consumption	Active agents and protocols, distributed control systems, wireless sensor networks
New production processes	Adaptability, re-configurability, and "brown field" reengineering, eco-factory, management of hazardous substances in manufacturing, minimal manufacturing	CIM, automated control, MAS, and coordination protocols, e.g., CEDP, for improved adaptability and fault tolerance (CFT principle)
Advanced manufacturing technologies	Additive forming processes for manufacturing, mechanical micro-matching, high performance and quality-embedded manufacturing, emission reduction technology	Advances in CIM
ICT infrastructures	Model-based manufacturing, semantic and ontology-based manufacturing, high-accuracy modeling of companies, application service for embedded devices	Active middleware technology, design and development of SOAs

Many other intangible and non-quantifiable issues such as improved product quality, better customer service, improved response time, improved product consistency, reduction in inventories, improved safety, better employee morale, improved management and operation of processes, shorter cycle times and setups, and support for continuous improvement need to be incorporated in the rationalization process as well. Hence, in line with the technology-based evaluation methods discussed earlier, an integrated and all-inclusive multi-criteria rationalization framework capable of handling multiple, heterogeneous criteria (both qualitative and quantitative) is required. Our suggested framework (Figure 3.6) applies Analytic Hierarchy Process (AHP) for the quantifying of qualitative criteria, and Data Envelopment Analysis (DEA) for ranking different solution alternatives, as Decision-Making Units (DMU).

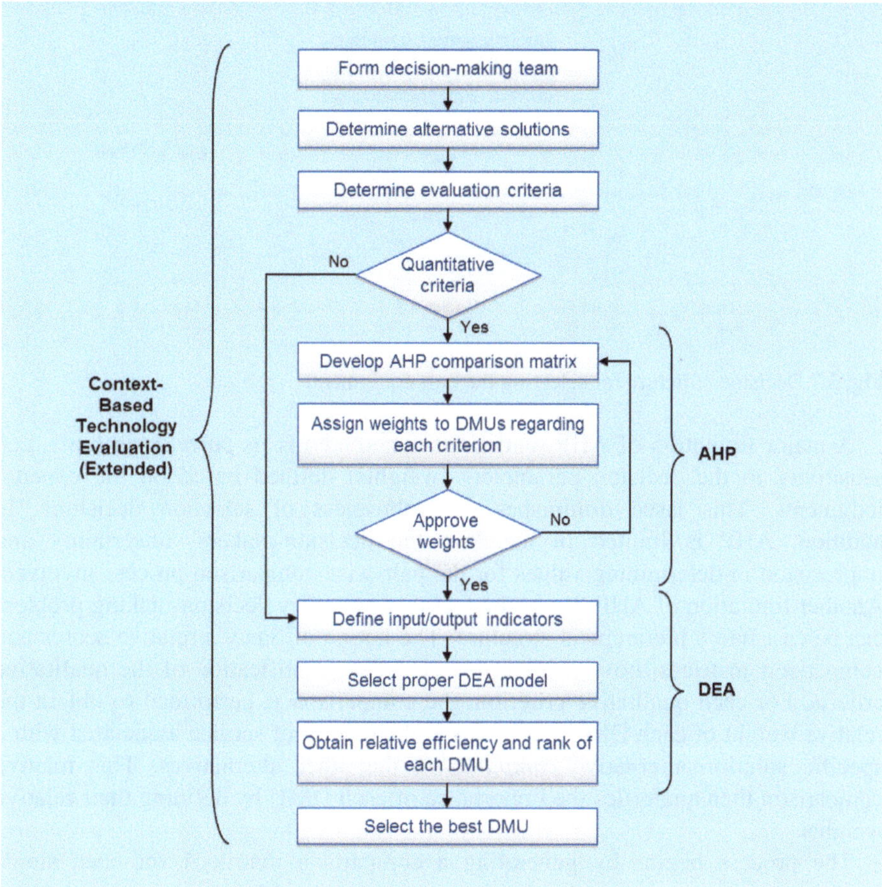

Fig. 3.6 The integrated AHP-DEA framework for context-based technology evaluation.

3.4.1 AHP

Most automated systems, including e-Work systems, nowadays comprise a collection of equipment properly integrated into an effective solution. This integration process makes the evaluation of alternative solutions more complex, due to the existence of many possible and variable combinations of configuration of available components. AHP is applied, as a promising decision support system to assist the selection process in such complex situations. AHP is a method that conveys the relative importance of a set of alternatives, based on a set of quantitative and qualitative criteria, following three principles: (1) the structure of the model, (2) comparative judgment of alternatives and criteria, and (3) synthesis of priorities. Figure 3.7 shows an application of AHP in selection of equipment out of five alternatives in a manufacturing system based on six criteria. Through assigning weights to the criteria and alternatives based on experts' judgments and using comparison matrices, the best alternative can be selected.

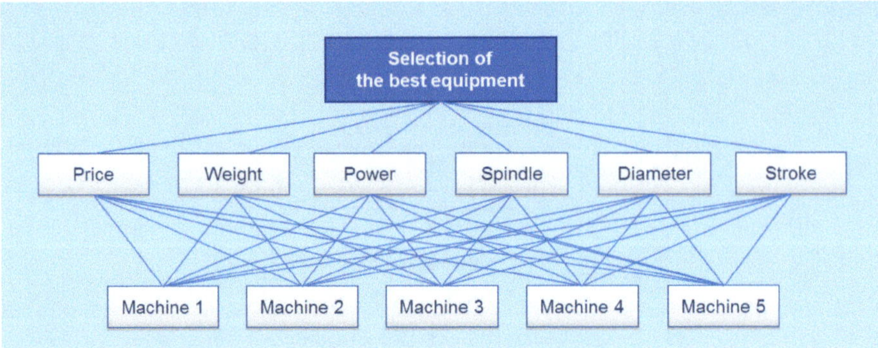

Fig. 3.7 Decision structure for selecting the best equipment.

A major limitation of AHP (and similar methods) is its *parametric basis*, i.e., sensitivity to the decision parameters (weights) defined based on the experts' judgments. This issue diminishes the robustness of selection decisions. In addition, AHP is limited in handling the decision-makers' uncertainty and imprecision in determining values for the pair-wise comparison process involved. Another limitation of AHP lies in the fact that not every decision-making problem can be cast into a hierarchical structure. The notion of Saaty's relative scores and comparison matrices, however, can be used for quantification of the qualitative criteria. For each qualitative criterion, the comparison is performed to obtain the relative weight of each DMU, e.g., how is the quality of service associated with a specific solution alternative *compared to* the other alternatives. This relative comparison then quantifies the importance of each DMU by defining their relative weights.

The process begins by generating a comparison matrix A for each single qualitative criterion. The comparison matrix M is $N \times N$, where N is the number of DMUs. Each entry m_{AB} of matrix M represents the relative importance of DMU A compared to DMU B, defined in terms of relative scores. If DMU A is better than DMU B in terms of a given criterion, $m_{AB} > 1$; otherwise, $m_{AB} < 1$, and for two DMUs with the same performance, $m_{AB} = 1$. An important consistency condition that should be satisfied is then

$$m_{AB} m_{BA} = 1, \quad \forall A, B = 1, \ldots, N. \tag{3.8}$$

The relative scores m_{AB} vary from 1 to 9 (see Table 3.4), which represent the *interpretation* of qualitative evaluations performed by the decision-makers on the relative comparisons between two DMUs (intermediate score can also be assigned).

Table 3.4 Saaty's relative scores.

m_{AB}	Interpretation
1	Performances of *A* and *B* are equal
3	*A* slightly outperforms *B*
5	*A* moderately outperforms *B*
7	*A* strongly outperforms *B*
9	*A* very strongly outperforms *B*

After building matrix *M*, it can be normalized through setting the sum of the column entries equal to 1:

$$\hat{m}_{AB} = \frac{m_{AB}}{\sum_{l=1}^{N} m_{lB}}, \quad \forall A, B = 1, \ldots, N. \tag{3.9}$$

Finally, the weight vector associated with each DMU can be calculated by averaging the row entries of the normalized matrix:

$$w_A = \frac{\sum_{l=1}^{N} \hat{a}_{Al}}{N}, \quad \forall A = 1, \ldots, N. \tag{3.10}$$

Some consistencies may occur during defining the weights for different DMUs. A consistency index is defined and calculated using the following formula

$$CCI = \frac{z - N}{N - 1}, \tag{3.11}$$

where z denotes the average of the elements of the vector whose A^{th} element is the ratio of the Ath element of the vector $M \cdot w$ to the corresponding element of the vector w. Absolute consistency occurs where $CCI = 0$. However, small inconsistencies could be tolerated, i.e., if

$$\frac{CCI}{RI} < 0.1, \tag{3.12}$$

where RI is the consistency index when the entries of A are random. For example,

$$A = \begin{bmatrix} 1 & 3 & 1/3 \\ 1/3 & 1 & 3 \\ 3 & 1/3 & 1 \end{bmatrix} \Rightarrow CI/RI = 1.5 \Rightarrow \text{Inconsistent}$$

$$A = \begin{bmatrix} 1 & 3 & 3 \\ 1/3 & 1 & 3 \\ 1/3 & 1/3 & 1 \end{bmatrix} \Rightarrow CI/RI = 0.12 \Rightarrow \text{Slightly inconsistent}$$

$$A = \begin{bmatrix} 1 & 3 & 5 \\ 1/3 & 1 & 3 \\ 1/5 & 1/3 & 1 \end{bmatrix} \Rightarrow CI/RI = 0.03 \Rightarrow \text{Consistent}$$

3.4.2 DEA

DEA is a *non-parametric* MCDM method for evaluation of relative efficiency of DMUs, e.g., e-Work alternatives, based on multiple criteria. DEA has been extensively applied for evaluation and benchmarking of schools and universities, hospitals and clinics, bank branches, airports, distributed production/ manufacturing enterprises, utility/power sectors (e.g., electricity, oil, water), quality control, risk evaluation, location-/allocation decisions, facilities layout design, and, in general, any system which produces certain products or services.

DEA is one of the most popular MCDM methods for performance evaluation of DMUs, due to several unique characteristics that are also useful in e-Work rationalization, such as:

- Handling multiple criteria.
- Non-parametric nature and requiring few a priori assumptions.
- Calculating relative efficiency of DMUs against a peer or combination of peers.
- No restriction on the measurement units of criteria.

Before discussing details of DEA, some basic definitions and assumptions are needed:

(1) *DMU*: refers to an alternative which *produces* some products or services.
(2) *Indicator*: is the decision criterion, classified into two classes of (1) inputs (*undesirable* or *bad* indicators) and (2) outputs (*desirable* or *good* indicators), taking into account the definition of the criterion and requirements of the problem. For instance, re-configurability is an output (or desirable, good) indicator, while leadtime is an input (or undesirable, bad) indicator.
(3) *Efficiency*: is the ratio of the weighted sum of outputs to the weighted sum of inputs.
(4) *Relative efficiency*: defines the efficiency of each DMU compared to the *best* DMU.

(5) *Return to Scale*: refers to the behavior of rate of increase in the output to the subsequent increase in the inputs, which can be constant, increasing, or decreasing.
(6) *Input-/output-oriented model*: refers to improvement of efficiency through proportional reduction of inputs (input orientation) or proportional augmentation of outputs (output orientation).

A simple example is developed for graphical description of the mechanics of DEA. Consider a comparison between five DMUs (different alternatives) based on three indicators: one input and two outputs (both quantified using AHP) as shown in Table 3.5.

Table 3.5 Input data.

DMU	Investment Cost (input)	Reconfigurability (output)*	Scalability (output)*
A	$10M	0.20	0.90
B	$10M	0.50	0.65
C	$10M	0.30	0.70
D	$10M	0.60	0.40
E	$10M	0.80	0.30

* Weights obtained using AHP and normalized between zero and one.

Without loss of generality, we assume constant return-to-scale. Hence, since the input indicator is the same, it can be eliminated from the calculations. Then, considering reconfigurability and scalability as the two main indicators, the problem can be visualized as shown in Figure 3.8. Now assume that we are allowed to have virtual DMUs that are convex combinations of actual DMUs, i.e., all points on the line segment connecting DMUs *A* and *B* are virtual DMUs. Since the segment *AB* lies beyond the line segments *AC* and *BC*, it is obvious that a DMUs *A* and *B* and all virtual DMUs corresponding to the line segment *AB* are capable of creating more outputs (i.e., reconfigurability and scalability) compared to DMU *C* and the virtual DMUs on the line segments *AC* and *BC* (see Figure 3.8). The line above all possible line segments is called *efficient frontier*. In this example, the efficiency of DMU *C* is defined through finding the fraction of the inputs that the virtual DMU V_C would need to produce the same outputs as DMU *B*. The relative efficiency of DMU *C* is calculated by looking at the line from the origin, *O*, to V_C. The efficiency of DMU *C* is then equal to OC/OV_C, (see Figure 3.8). DMUs *A* and *B* are efficient, since they are on the efficiency frontier, and thus, their relative efficiency scores are equal to 1.0. Similarly, this procedure must be applied to obtain the relative efficiency scores of DMUs *D* and *E*.

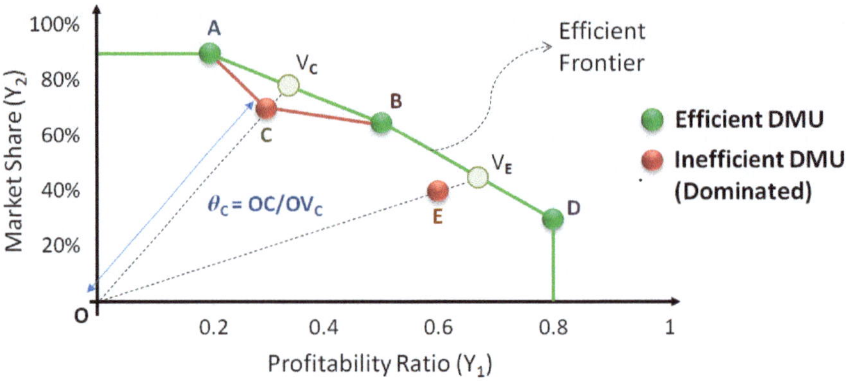

Fig. 3.8 Graphical representation of DEA for the given example.

Given I inputs and O outputs, the CCR (named for the developers of DEA, Charnes, Cooper, and Rhodes) model yields (1) the relative weight for each indicator, and 2) the relative efficiency score of DMU k out of N DMUs, as follows:

$$\max \frac{\sum_{o=1}^{O} u_o y_{ok}}{\sum_{i=1}^{I} v_i x_{ik}}$$

$$\text{s.t.} \quad \frac{\sum_{o=1}^{O} u_o y_{oj}}{\sum_{i=1}^{I} v_i x_{ij}} \leq 1, \quad \forall j \in N, \tag{3.13}$$

$$u_o, v_i \geq 0, \quad \forall i \in I,\ o \in O,$$

where

u_o	Weight of output o	y_{oj}	Output o for DMU j
v_i	Weight of input i	x_{ij}	Input i for DMU j

The fractional program (3.13) can be converted to an *input-oriented* linear program as follows (see Charnes et al., 1978):

$$\begin{aligned}
\max \ & \sum_{o=1}^{O} u_o y_{ok} \\
\text{s.t.} \ & \sum_{i=1}^{I} v_i x_{ik} = 1, \\
& \sum_{o=1}^{O} u_o y_{oj} - \sum_{i=1}^{I} v_i x_{ij} \leq 0, \quad \forall j \in N, \\
& u_o, v_i \geq 0, \qquad \forall i \in I,\ o \in O,
\end{aligned} \tag{3.14}$$

which can also be rewritten as the following dual program:

$$\begin{aligned}
\min \ & \theta_k \\
\text{s.t.} \ & \sum_{j=1}^{N} \lambda_j x_{ij} - \theta_k x_{ik} \leq 0, \quad \forall i \in I, \\
& \sum_{j=1}^{N} \lambda_j y_{oj} - y_{ok} \geq 0, \quad \forall o \in O, \\
& \lambda_j \geq 0, \quad \forall j \in N,
\end{aligned} \tag{3.15}$$

where θ_k denotes the relative efficiency score of DMU $k \in N$ (see Banker et al., 1984).

The DEA models above need to be solved for each DMU, separately, to obtain its relative efficiency score. The best alternative(s) can then be identified based on the relative efficiency of all DMUs. Interestingly, the only parameters required for solving the DEA models are the input and output indicators (i.e., decision criteria/e-Criteria). This non-parametric property makes DEA a unique technique compared to other approaches, such as Principal Component Analysis, Numerical Taxonomy, and Technique for Order of Preference by Similarity to Ideal Solution. There are several types of DEA models in literature, addressing different aspects of multicriteria evaluation of alternatives (see [Cook et al., 2013] for more details).

3.4.3 e-Work Portfolio Selection: A Case Study

Considering the set of e-Work solutions being developed for a given budget period at an organization as a portfolio, the selection of the suitable set of e-Work alternatives is of utmost importance for the future of the organization. The selection process is nevertheless complex and highly dependent on the available budget, complexity of e-Work solutions and their interactions (for example the ability of one e-Work solution for solving multiple needs in the organization). Also, the selection process must consider the needs to be solved by the e-Work solutions and their hierarchical relative position in the organization, i.e., e-Work solutions at operational restricted level should be adequately compared with solutions involving support of top control or management activities. Among the extensive research for evaluation technology development projects portfolios, we selected the method proposed by Eilat et al. (2006) for static evaluation of research and developments projects portfolio, based on Balanced Scorecard (BSC) (Kaplan and Norton, 1992, 1996a, b) and DEA.

BSC has the objective of generating a balanced representation of the organization performance. To accomplish this objective, the method classifies organizational performance measures into four groups (the "cards"): financial, market, internal growth, and innovation. A number of performance measures need to be defined at each group, from which a comprehensive evaluation of the organization's performance should be obtained. Key advantages of BSC are minimization of information overload, congregation of dissimilar performance measures, and protection from sub-optimization by considering simultaneously all relevant performance measures.

The e-Work portfolio evaluation methodology consists of seven steps, starting from resource allocation process for the organization strategic areas where the candidate e-Work solutions will be developed (Step 1). Once resources are strategically distributed, the e-Work solutions are modeled and evaluated as DMUs with the DEA-BSC model (Step 2). In the next step, e-Work projects are classified into two classes: candidate or reject projects. Candidate projects are also evaluated in terms of variability of the risk, efficiency, and balance of their expected outputs (Step 3). A branch and bound model is then used for generating the alternative e-Work portfolios to proceed to their evaluation (Step 4). Projects considered in each portfolio are evaluated to generate a combined measurement of the portfolio benefit, outcome, and resource interactions (Step 5). Evaluation of alternative portfolios is performed using the DEA-BSC model (Step 6). The final step involves a sensitivity analysis in order to test the decision for a given e-Work portfolio.

In order to show the essential characteristics of the methodology, its application to a hypothetical case is presented. The case considers a hierarchical organizational structure with divisions organized for the key technological development areas. Specific details on the methodology are available in Eilat et al. (2006).

Step 1: The e-Work projects are defined in each division with the corresponding resource and budget assignment for achieving the organization's strategic objectives. The set of e-Work projects in the division is assumed as 10 from which the e-Work portfolio is to be defined. Project data includes measures in 4 BSC cards: economic contribution (better quality, higher productivity or lower costs), scientific contribution (advancement of the scientific knowledge body), social contribution (job creation or improved working conditions), and resource requirement (personnel, material, equipment, etc.). The general procedure to determine the BSC cards may involve utilization of consensus building methods such as Delphi. For each e-Work project one output measure is defined for the economic (measured in monetary terms), scientific (measured in an arbitrary scale from 0 to 100), and social (measured in an arbitrary scale from 0 to 100) contribution cards, while 2 input measures are defined for the resource requirement card (work content and material costs). A success probability is estimated for each e-Work project. Interaction among e-Work projects is considered in terms of their requirements in resources, value and outcomes and modeled into interaction matrices $\mathbf{U}_1$ and $\mathbf{U}_2$, the value interaction matrices $\mathbf{V}_1$, $\mathbf{V}_2$, and $\mathbf{V}_3$, and the outcome interaction matrix $\mathbf{P}$. Data for the e-Work projects is shown in Tables 3.6-a and 3.6-b.

Table 3.6a e-Work projects data.

e-Work Project	Labor	Material cost	Economic Output	Scientific Output	Social Output	Success probability
1	9	7	134	26	34	0.6
2	9	15	2636	77	81	0.3
3	97	4	1054	60	17	0.6
4	11	6	116	9	17	0.7
5	46	18	1115	77	34	0.7
6	54	6	365	81	21	0.8
7	42	17	667	81	17	0.9
8	16	3	235	13	9	0.5
9	84	13	837	30	77	0.9
10	63	12	439	60	81	0.5

Table 3.6b e-Work projects interactions.

Labor		Material cost		Economic		Scientific		Social		Outcome	
$\mathbf{U}_1(2,1)$	-1	$\mathbf{U}_2(4,3)$	-1	$\mathbf{V}_1(4,1)$	370	$\mathbf{V}_2(4,2)$	5	$\mathbf{V}_3(4,2)$	15	$\mathbf{P}(1,5)$	0.2
$\mathbf{U}_1(3,2)$	-4	$\mathbf{U}_2(5,2)$	-4	$\mathbf{V}_1(6,2)$	940	$\mathbf{V}_2(5,1)$	10	$\mathbf{V}_3(6,5)$	40	$\mathbf{P}(2,3)$	0.5
$\mathbf{U}_1(4,1)$	-2	$\mathbf{U}_2(7,5)$	-1	$\mathbf{V}_1(7,3)$	200					$\mathbf{P}(8,3)$	0.1
$\mathbf{U}_1(8,3)$	-1									$\mathbf{P}(10,9)$	0.3
$\mathbf{U}_1(9,7)$	-15										

Table 3.7 e-Work projects attractiveness rating.

e-Work Project	Attractiveness	Efficiency	Balance	Risk
2	1.0000	1.0000	1.0000	51
6	1.0000	1.0000	1.0000	64
9	0.8133	1.0000	0.8133	51
5	0.8101	0.8433	0.9606	87
7	0.8051	1.0000	0.8051	27
1	0.6996	1.0000	0.6996	23
3	0.6540	0.9687	0.6751	230
10	0.6362	0.7769	0.8189	192
4	0.5381	0.6325	0.8508	22
8	0.5102	0.5224	0.9766	50

Step 2: With the data of Tables 3.6-a and 3.6-b, the DEA-BSC model is applied to the e-Work projects. The application of the DEA-BSC model yields the attractiveness of each project, depicted in Table 3.7.

Step 3: In order to generate the e-Work project portfolios efficiency, balance, and risk indices are computed for each project. The objective is to limit the variability of the projects indices to include in the portfolio. Projects input and output overlaps and interactions are also considered for the portfolio generation. The results of such calculations are shown in Table 3.7.

Step 4: Using the branch and bound procedure, the portfolios are generated and their BSC cards obtained. The result is shown in Table 3.8 (see Eilat et al., 2006 for details). Portfolio generation is based on the following logic. Let $G = \{1, 2, ..., n\}$ and Q_k respectively denote the particular groups of candidate e-Work projects, and the group of e-Work projects in portfolio k taken from G ($Q_k \subset G$). Also, let vector z_k be a particular selection of e-Work projects in portfolio k, where $z_{jk} = 1$, if e-Work projects j is included in portfolio k, and $z_{jk} = 0$, otherwise. The generation procedure focuses on input requirements and availabilities. Let x_{ij} and Ri respectively denote the amount of input i required for e-Work project j, and the total availability of the same input. The amount of input i allocated to portfolio k is denoted by $\hat{x}_{ik}$ is calculated by applying an accumulation function on the input values of the e-Work projects in the portfolio. We consider the following simple input-accumulation function:

$$\hat{x}_{ik} = \sum_{j=1}^{n} x_{ij} z_{jk} .$$

A portfolio k with the set $Q_k \subset G$ of e-Work projects is said to be maximal with respect to group G and resource availabilities R_i, if the two conditions next hold:

a) The portfolio is feasible, i.e., $\hat{x}_{ik} \le R_i$, $\forall i$.
b) Inclusion of any additional e-Work projects in this portfolio leads to violations of the resource constraints, i.e., $\forall j \notin Q_j, \exists i: \hat{x}_{il} > R_i$, where $Q_l = Q_k \cup \{j\}$.

Generation of alternative maximal portfolios from G is accomplished through a branch and bound procedure that operates as explained next. Starting with an empty portfolio at node 0 of the tree, branching to n portfolios is performed by adding e-Work projects to the portfolio, one by one, following the order of their indices. Node i at the first level of the branching process represents the problem of generating a maximal portfolio that includes e-Work project i included and e-Work projects with lower indices not present. This rule, when applied recursively to the generated sub-problems, creates the complete tree for the combination of all possible e-Work projects into different portfolios. A bound is set for each of two situations: (a) after the e-Work project with the highest index n is included, or (b) at least one of the resource constraints is violated. No further development of the portfolio is needed after a bound point is reached. In vector z_k, the relative position of value 1 indicates the inclusion of the specific e-Work project according to the order (1,2,4,5,6,7,8,9,10) after e-Work project 3 was dropped.

Table 3.8 e-Work projects portfolios.

Portfolio	z_k	Labor	Material cost	Economic	Scientific	Social	Attractiveness
1	111100100	104	54	2416	128	106	1.0000
2	111110000	148	57	2847	197	144	1.0000
3	000011011	270	56	2349	249	195	1.0000
4	000110011	290	57	2561	220	227	1.0000
5	010110010	227	57	3304	191	175	1.0000
6	110110100	156	54	2682	196	131	0.9943
7	010010111	266	58	2936	198	206	0.9839
8	101110010	237	58	2578	200	197	0.9764
9	100100011	237	58	2344	173	217	0.9761
10	100001011	217	57	2101	191	199	0.9699
⋮	⋮	⋮	⋮	⋮	⋮	⋮	⋮
15	110100010	173	58	2862	144	165	0.9422
20	101011010	217	57	2282	218	157	0.9303
25	110011100	151	57	2438	214	91	0.9041
30	101110001	212	57	1949	204	164	0.8860
35	010101100	133	58	2693	176	75	0.8759
40	011100010	177	57	2831	123	150	0.8657
45	000101110	206	59	2649	181	132	0.8428
⋮	⋮	⋮	⋮	⋮	⋮	⋮	⋮
47	111010101	187	58	2241	172	138	0.8200
48	001011101	218	52	1542	211	105	0.8107
49	001110101	223	53	1754	182	137	0.8020
50	010100101	158	53	2245	126	105	0.7697
51	011100001	152	56	2203	126	117	0.7653
52	011001001	147	59	1991	156	107	0.7651
53	101100101	168	54	1744	135	127	0.7644
54	000101101	196	58	2021	184	99	0.7542
55	010001101	153	56	2033	155	95	0.7468
56	101001101	163	53	1449	153	109	0.7242

Step 5: The e-Work portfolio accumulated inputs and outputs are calculated. At this point, available budget and labor must be considered as constraints for the portfolio, as well as the projects interaction within the portfolio and their possible savings in resource consumption and output contribution.

Step 6: The DEA-BSC model is applied to generated portfolios considering their accumulated measures. This step produces the attractiveness rating of each portfolio as shown in Table 3.8. The portfolio ranking can now be used to select the most desirable combination of e-Work projects. Further sensitivity analysis may be applied to ensure the robustness of the result.

3.5 Challenges, Advantages and Benefits for Rationalization

Before concluding this chapter, let us review significant challenges, benefits and advantages of collaborative e-Work, e-Business and e-Service already discussed in Chapters 1 and 2. A summary of the main challenges addressed by collaboration engineering and augmentations, which are challenging also the rationalization process is as follows:

1. The principles of the work system have changed.
2. Advancing technology increases productivity, yet complexities and challenges also increase the need to:
   a. Optimize operations despite information and task overloads.
   b. Overcome interactive work complexity.
   c. Design and operate with and despite higher inter-dependence.
   d. Design for integrity and trust; safety and security.
   e. Seamless coordination, cooperation, and synchronization.
   f. Overcome communication and interference challenges and failures.
   g. Overcome the mismatch obstacles -- inconsistent versions, cultural differences.
   h. Eliminate or minimize repeated training requirements and associated costs.

The collaboration engineering tools and augmentation techniques that are described in this book are viewed as valuable enablers and major contributors for addressing these challenges.

At the end of Chapter 1, a list of eight common, tangible and measurable e-Criteria which are useful for design and rationalization is provided: *Integrability; Agility; Scalability; Reachability; Viability; Autonomy; Dependability; and Resilience.* While these e-Criteria for performance outcomes may also depend on other design factors, they are all impacted by the CCT design principles discussed in Chapter 2. For example:

1. Cost cuts and time reduction of business and service processes, by designing e-Work parallelism (EWP) which leads to increased capacity, and also increases agility and viability.
2. Error elimination and early detection and recovery, by collaborative fault-tolerance (CFT) and by error prevention and conflict resolution (EPCR) which improve quality of products and services; safety and security; enable continuity of business and services, and also increase dependability and resilience.

Similarly, other CCT design principles also influence the outcomes and performance, as explained and illustrated throughout the book (and the references, e.g., Velasquez and Nof, 2009, Table 88.2). Many of the relatively better outcomes resulting from augmenting collaborative processes, and optimizing collaboration activities and interactions, have been found and proven in scientific research and in industry practice (see examples in Figure 3.9). These benefits and advantages can serve as a useful source for rationalization efforts.

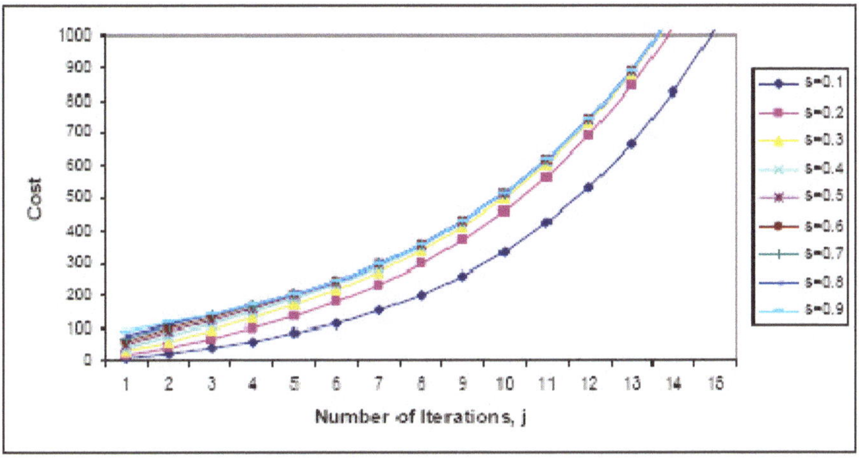

Fig. 3.9 Benefits and advantages of collaboration engineering and augmentation design with CCT, as found and validated in previous work (examples):

a) Cost of error recovery and conflict resolution efforts in e-Work, e-Business and e-Service, in repeated attempts (iterations). S in this example is the probability of success in each attempt. (Source: Velasquez and Nof, 2009)

(a-i) below: Cost of these efforts when human involvement is 20% or more, and engineering tools for augmenting collaboration are applied for the rest (<80%). Observation: The cost is exponential and unbounded.

(a-ii) below: Cost of the same efforts when human involvement in these efforts is less than 20% and engineering tools for augmenting collaboration are designed for over 80% of these efforts. Observation: The costs are bounded (they reach a limit), so the system does not get out of control.

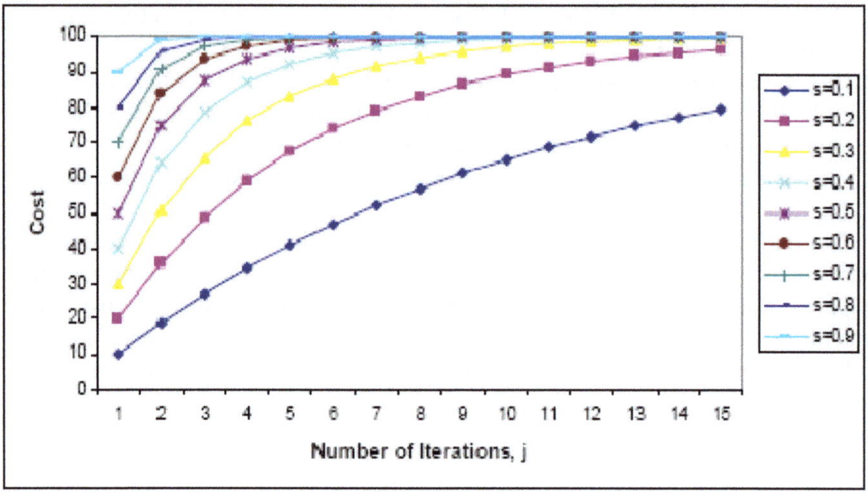

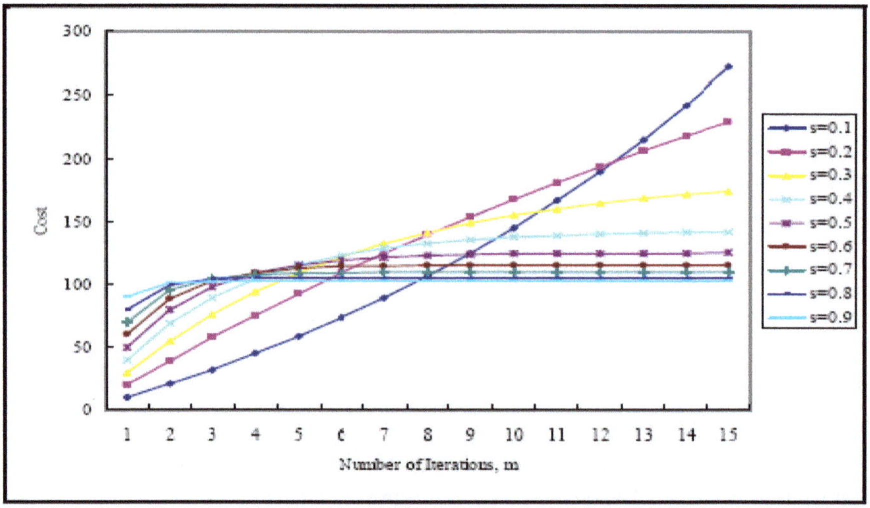

(a-iii) above: Border line near 80:20 percent of these efforts (engineering tools for augmenting collaboration: Human involvement). Observation: Some costs are exponential, some are bounded.

Figure 3.9 (cont.) b) Economic benefits of applying a CCT Best Matching Protocol (Source: Velasquez and Nof, 2009)

(b-i) below: The Best Fit Index indicating relative quality level.

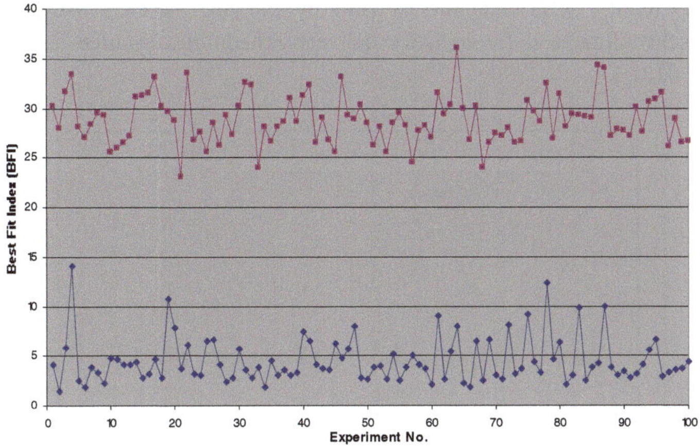

Top graph: Overall better fit when applying the CCT Best Matching protocol in producing batches of 100 parts AB, each part built of one component type A, and one other component, type B. Lower graph: Lower overall fit with random matching of the component pairs. (Components' attribute by which they are matched are assumed to be uniformly distributed.) The part's resulting fitting information is then used (together with other factors) to determine which supplier would best match a given customer, as shown in b (ii).

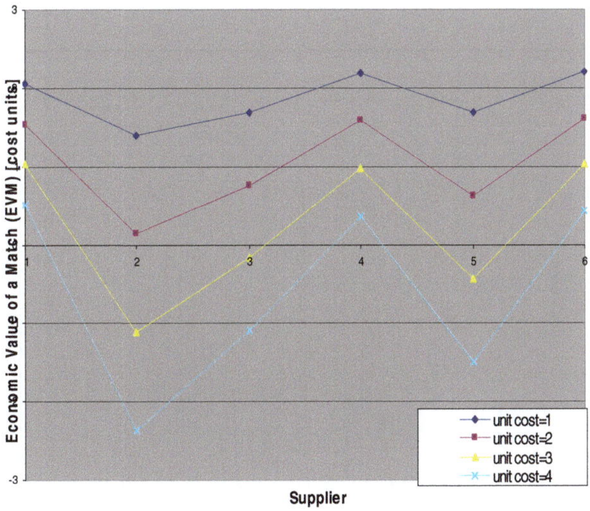

(b-ii) above: Relative economic value of a match to identify which supplier out of six to match to a given customer according to the cost to produce and match parts.

Figure 3.9 (cont.) c) Sustainability benefits by applying the CCT S-DSP Protocol for production scheduling, and for logistics delivery scheduling. (Source and assumptions: Seok et al., 2012)

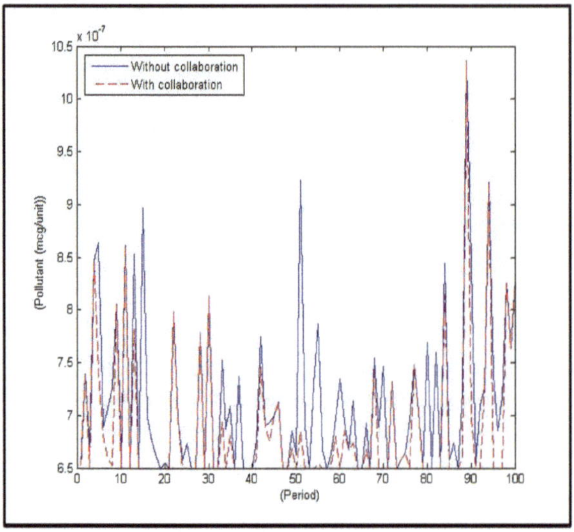

(c-i) above: Pollutant per product (resulting from production process) decrease by 0.5% to 27. 5%, with 6.8% average decrease.

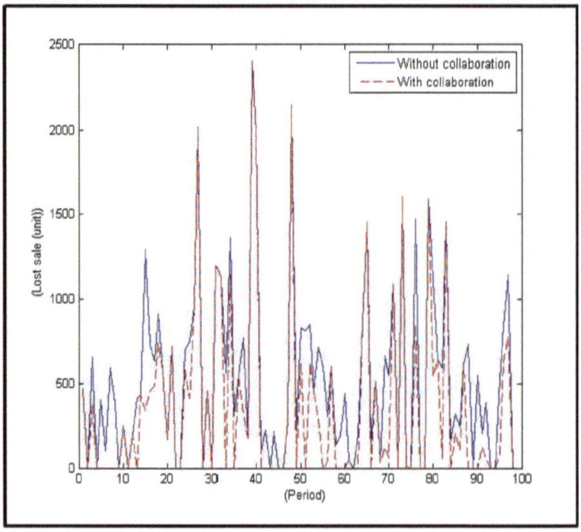

(c-ii) above: Lost sales of manufacturers decrease on average by 72.5%

(c-iii) below: Idle capacity of manufacturers decreases on average by 79.7%

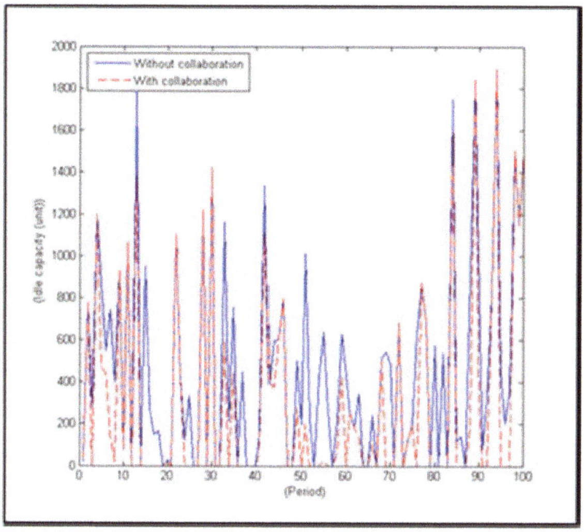

(c-iv) below: Additional profit of manufacturers due to collaborative production scheduling (always non-negative).

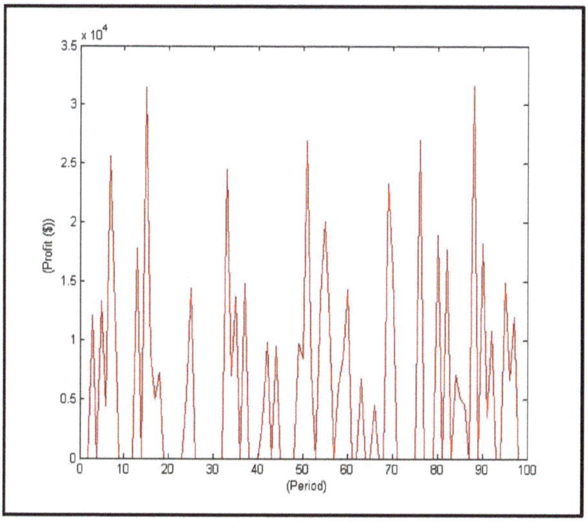

(c-v) below: Air pollutants caused by logistics delivery process decrease (left); Employee overall workload decreases (right) both by between 9.7% and 20.6%, with 12.8% average improvement.

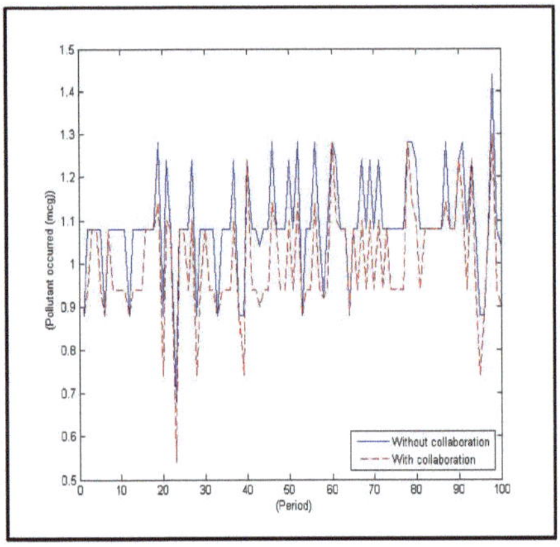

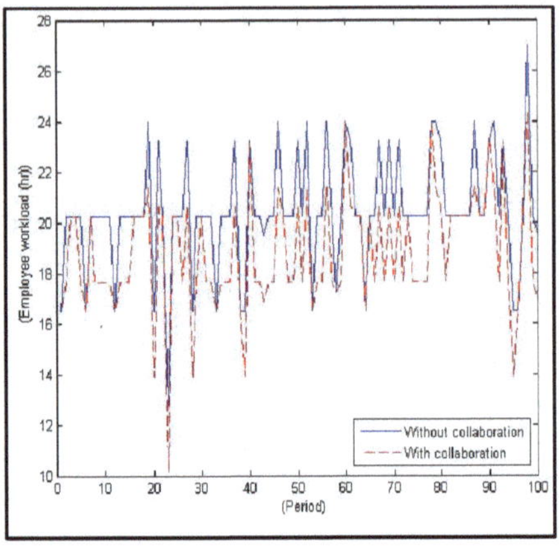

(c-vi) below: Utilization of logistics delivery trucks increases on average by 19.3%.

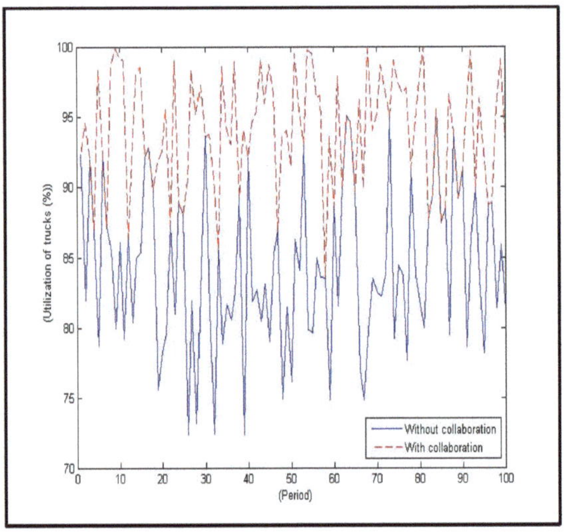

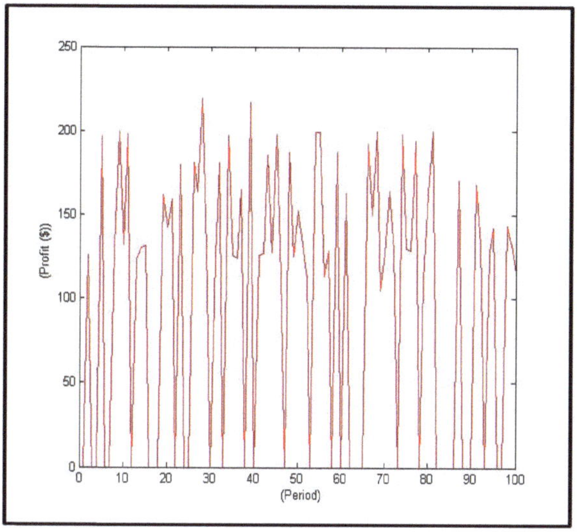

(c-vii) above: Additional profits of manufacturers due to collaborative scheduling of logistics delivery increase (always non-negative).

Figure 3.9 (cont.) d) Agility, Availability, and Dependability in multi-agent systems. (Source: Nof, 2007b)

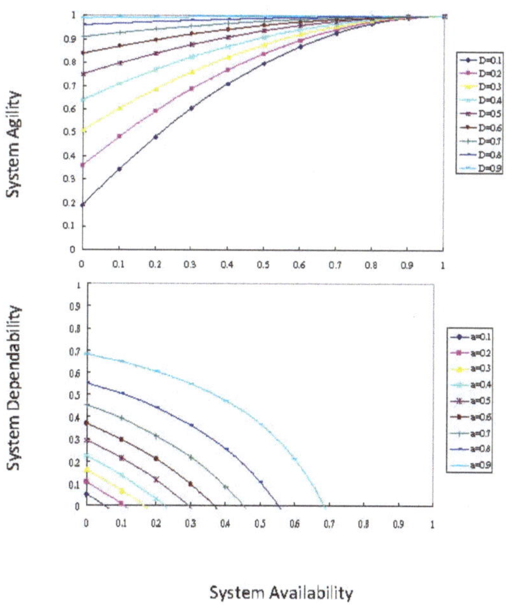

Agility (a) and Dependability (D) are inter-dependent with systems' Availability.

Finally, it should be noted that in many cases, no outcomes can be possible at all without engineered, augmented collaboration. For example, all supply chains and supply networks, where suppliers and customer enterprises must apply effective e-Work in their collaboration, could not effectively survive without it. And careful design rationalization would help guide the relatively more successful outcomes.

Another example is by Aragon (2009), who writes about collaborative e-Work in a large-scale astrophysics project: "In July 2005, at the commencement of the Sunfall project, the SNfactory had discovered no supernovae. As of November 2008, the SNfactory has discovered over 1000 supernovae....By focusing on improvements in automation, Sunfall has led to more efficient data processing and labor savings for project scientists, enabling them to concentrate on physics research .... rather than on repetitive tasks."

3.6 Emerging Trends and Challenges

By their nature of technology encapsulation, e-Work solutions will always be linked to the company operations strategy. From that perspective, and similarly to the analysis performed to incorporate new processes into e-Systems, e-Work solutions need to be evaluated in terms of classic performance measurements such as costs, savings, payout periods, etc. However, as the complexity and scope of the e-Work solution increase, the classic performance evaluation process starts to present important shortcomings, necessitating the application of more sophisticated evaluation methods. For example, consider the difficulties in analyzing the net present value or the internal return rate of a collaborative platform for software development. Nevertheless, the future generations of e-Work solutions will keep depending on financial metrics for their development. But additional considerations are needed. In that sense, utilization of less structured methods such as AHP by Saaty (2008) or DEA by Charnes et al. (1978) in combination with strategic information from the BSC method by Porter (1985) have benefits: They can capture the less structured and wide coverage nature of the tools and projects being proposed in the present design to adapt the selection process in a useful, rational manner.

In the future, we expect more comprehensive evaluation models for rationalization of collaborative e-Work, e-Business, and e-Service systems. Such models will include aspects left out by currently available methods, transforming and evolving towards fully integrated evaluations of e-Work solutions. These integrated evaluation models will be able to include at once strategic, tactical, and operational aspects of the e-Work solutions in their relationship with the organization, the business and service. So far, however, no feasible methods have been developed to cope with such demanding task as considering all multi-level, multi-scale aspects an e-Work project may involve when all three levels of its requirements and performance are simultaneously considered.

References

Aragon, C.B.: Collaborative analytics for astrophysics explorations. In: Nof, S.Y. (ed.) Springer Handbook of Automation, vol. 93, pp. 1645–1670. Springer (2009)

Banker, R.D., Charnes, A., Cooper, W.W.: Some models for estimating technical and scale inefficiencies in data envelopment analysis. Management Science 30(9), 1078–1092 (1984)

Boothroyd, G., Poli, C., Murch, L.E.: Automatic Assembly. Marcel Dekker, New York (1982)

Bozarth, C.C., Berry, W.L.: Measuring the congruence between market requirements and manufacturing: a methodology and illustration. Decision Sciences 28, 121–150 (1997)

Ceroni, J.: Economic Justification of Automation Systems. In: Nof, S.Y. (ed.) Springer Handbook of Automation, ch. 40, pp. 699–713 (2009)

Charnes, A., Cooper, W.W., Rhodes, E.: Measuring the efficiency of decision making units. European Journal of Operational Research 2, 429–444 (1978)

Comella-Dorda, S., Dean, J., Lewis, G., Morris, E., Oberndorf, P., Harper, E.: A Process for COTS Software Product Evaluation (CMU/SEI-2003-TR-017) Pittsburgh, PA., Software Engineering Institute, Carnegie Mellon University (July 2004)

Cook, W.D., Tone, K., Zhu, J.: Data Envelopment Analysis: Prior to Choosing a Model. Omega 44, 1–4 (2013)

Eilat, H., Golany, B., Shtub, A.: Constructing and evaluating balanced portfolios of R&D projects with interactions: A DEA based methodology. European Journal of Operational Research 172, 1018–1039 (2006)

Hayes, R.H., Wheelwright, S.C.: Restoring our Competitive Edge – Competing Through Manufacturing. Wiley, New York (1984)

Hill, T.: Manufacturing Strategy: Text and Cases, 2nd edn. Palgrave, Hampshire (2000)

Irani, S.A., Zhou, J.: Value Stream Mapping of a Complete Product. Research memo. Department of Industrial, Welding and Systems Engineering, The Ohio State University (2011)

JBoss Application Server (2005), `http://www.jboss.org/products/jbossas`

Kaplan, R.S., Norton, D.P.: The Balanced Scorecard. Measures that Drive Performance. Harvard Business Review, 71–78 (January-February 1992)

Kaplan, R.S., Norton, D.P.: Using the Balanced Scorecard as a Strategic Management System. Harvard Business Review, 75–85 (January-February 1996a)

Kaplan, R.S., Norton, D.P.: Linking the Balanced Scorecard to Strategy. California Management Review, 53–79 (Fall 1996b)

Kim, Y., Lee, J.: Manufacturing strategy and production systems: an integrated framework. Journal of Operations Management 11, 3–15 (1993)

Leong, G.K., Snyder, D.L., Ward, P.T.: Research in the process and content of manufacturing strategy. Omega 18, 109–122 (1990)

Lewis, G.A.: An Approach to Analysis and Design for COTS-Based Systems. In: Franch, X., Port, D. (eds.) ICCBSS 2005. LNCS, vol. 3412, pp. 236–247. Springer, Heidelberg (2005)

Lewis, G.A., Wrage, L.: Approaches to Constructive Interoperability, Carnegie Mellon Software Engineering Institute CMU/SEI-2004-TR-020, ESC-TR-2004-020

Miltenburg, J.: Manufacturing Strategy - How to Formulate and Implement a Winning Plan. Productivity Press, Portland (1995)

Nof, S.Y.: Design of effective e-Work: Review of models, tools and emerging challenges. Production Planning and Control, Special Issue on e-Work: Models and Cases 15(8), 681 (2003)
Nof, S.Y.: Collaborative control theory for e-Work, e-Production, and e-Service. Annual Reviews in Control 31, 281–292 (2007a)
Nof, S.Y.: Availability, integrability, and dependability – what are the limits in production and logistics. In: Proceedings of MCPL, Sibiu, Romania (September 2007b)
Oracle Corporation. Oracle Database (2005), http://www.oracle.com/database/
Porter, M.: Competitive Advantage. Free Press, New York (1985)
Rother, M., Shook, J.: Learning to See. Version 1.2. Lean Enterprise Institute, Brookline (1999)
Saaty, T.: Decision Making with the Analytic Hierarchy Process. International Journal of Service Sciences 1(1), 83–98 (2008)
Schroeder, D.M., Congden, S.W., Gopinath, C.: Linking competitive strategy and manufacturing process technology. Journal of Management Studies 32, 163–189 (1995)
Seok, H.S., Nof, S.Y., Filip, F.G.: Sustainability Decision Support System based on Collaborative Control Theory. Annual Reviews in Control 36(1), 85–100 (2012)
Sheu, C.: Linking market factors to manufacturing design. Omega – International Journal of Management Science 22, 269–282 (1994)
Velásquez, J.D., Nof, S.Y.: Collaborative e-Work, e-Production, and e-Service. In: Nof, S.Y. (ed.) Springer Handbook of Automation, ch. 88, pp. 1549–1576 (2009)
Wallnau, K., Hissam, S., Seacord, R.: Building Systems from Commercial Components. Addison-Wesley, New York (2001)

Review and Challenge Questions

1. (i) Characterize e-Work solutions in the fields of: (a) medicine; (b) space exploration; (c) education; and (d) defense. (ii) For each, explain one or two main rationalization criteria.
2. State clearly the need for evaluating e-Work solutions following the investment logic for an organization.
3. (i) List relevant evaluation criteria that are possible to apply in the evaluation of e-Work based systems. (ii) Illustrate some of these evaluation criteria specifically for (a) an e-Work company; (b) an e-Service company.
4. Elaborate on the difficulties of mixing quantitative and qualitative criteria in the evaluation and rationalization process of e-Work solutions. (ii) Illustrate your answer in reference to Question 1.
5. Describe the organization scenario where traditional economical rationalization would be sufficient for selecting the best e-Work solutions to be developed (for this organization).
6. Relate the analysis of value added activities to the development of e-Work solutions.
7. Suggest one improvement to the branch and bound procedure used for generating the e-Work projects portfolios as presented in the chapter.
8. Analyze the advantages and disadvantages of using techniques such as Analytic Hierarchical Process (AHP) for selecting e-Work projects.

9. In relation to Question 8, explain how Data Envelopment Analysis (DEA) solves the shortcomings of AHP.
10. Explain (i) the benefit, and (ii) the limitation in combining the Balance Score Card (BSC) analysis and the DEA methods.
11. (i) Illustrate some examples of benefits and e-Criteria that have been applied for evaluating collaboration engineering projects. (ii) Suggest other examples of e-Business and e-Service that can be evaluated by those benefits and e-Criteria (from your answer in (i)).

Lab Questions

12. Consider an e-Work investment project with five alternatives A, B, C, D, and E. The best alternative must be selected with respect to three criteria: (1) Investment cost; (2) System maintainability; (3) System expandability. The investment cost is the same for all projects and equal to $20M. The last two criteria, however, are qualitative and their relative weights are defined using Saaty's scores as follows:

System Maintainability	A	B	C	D	E
A	1	1/5	1/3	1/7	1/9
B	5	1	3	1	1/5
C	3	1/3	1	1/3	1/7
D	7	1	3	1	1/3
E	9	5	7	3	1

System Expandability	A	B	C	D	E
A	1	3	7	5	9
B	1/3	1	1/5	3	7
C	1/7	5	1	3	5
D	1/5	1/3	1/3	1	7
E	1/9	1/7	1/5	1/7	1

(a) Are the relative weight matrices above consistent? If not, are the respective consistencies tolerable?

(b) Define, using AHP, the relative importance of the two qualitative criteria (i.e., system maintainability and expandability) for the five e-Work project alternatives.

(c) Now suppose the relative importance of the three criteria are defined using Saaty's scores as follows. Identify the best e-Work project alternative with respect to these criteria.

Criterion	Investment cost	System Maintainability	System Expandability
Investment cost	1	3	7
System Maintainability	1/3	1	1/5
System Expandability	1/7	5	1

13. Consider the e-Work investment project described in the previous question.

(a) Formulate the CCR Model (3.14) based on the quantitative values obtained for each criterion. How do you interpret the CCR model's objective function and constraints? (Note: You do not have to solve the model)

(b) Select, using DEA, the best e-Work project with respect to the three criteria: Investment cost; System maintainability; System expandability (Hint: You can solve the DEA graphically).

Chapter 4
Optimization and Control

Abstract. Considering the foundations, tools, and emerging discoveries of collaborative e-Work, as discussed in Chapters 1 and 2, it is realized that optimization and control are focused primarily on the core elements of e-Systems; agents, protocols, and workflows. In this chapter, we will show that these elements compose a solid framework for optimization and control of collaboration in emerging distributed e-Work systems. In order to be able to efficiently pass the benefits on to more complex constructs such as autonomous agents systems, production units configuration, highly reactive control protocols, and so on, these elements must be optimized as well. In order to show the evidence of the latest developments in optimization and control involving agent, protocol, and workflow theories, this chapter reviews the state-of-the-art techniques for achieving optimal design and operational control, and collaboration engineering. This chapter covers the incentives to construct autonomous agent-based systems, the key e-Criteria emerging from the transformation from traditional centralized work systems to decentralized e-Work systems, and several real-life applications of agent-based systems. Basic agent-based optimization and control architectures are reviewed along with pioneering bio-inspired mechanisms based on swarm intelligence and natural evolution, and their impact on the intelligence and autonomy of agents. Several techniques for protocol and workflow optimization are also discussed.

4.1 Emerging Impacts of Agents

With the transformation from traditional monolithic systems to decentralized e-Work networks, optimization and control of e-Systems are facing new challenges. The emerging requirements for timely response to the growing complexity and dynamicity of e-Systems call for higher intelligence and autonomy of agents for smarter interaction and collaboration (Rosenschein and Ephrati, 1993). Classic monolithic and hierarchical systems are no longer sufficient for addressing such requirements, and often lead to situations where the entire system collapses by a single error or conflict at one point of the system hierarchy

S.Y. Nof et al., *Revolutionizing Collaboration through e-Work, e-Business, and e-Service*,
Automation, Collaboration, & E-Services 2, DOI: 10.1007/978-3-662-45777-1_4

(Colombo et al., 2006). Various e-Criteria have been defined to formalize the emerging needs of e-Systems at different levels, and with different contextual relationships (Figure 4.1): *adaptability*, *flexibility*, *reconfigurability*, *changeability*, *robustness*, and *scalability* (Putnik et al., 2013). The prerequisites for addressing these requirements, however, mean to enable and delegate intelligence and autonomy to the lowest-level individual units collaborating within the e-System, namely the agents.

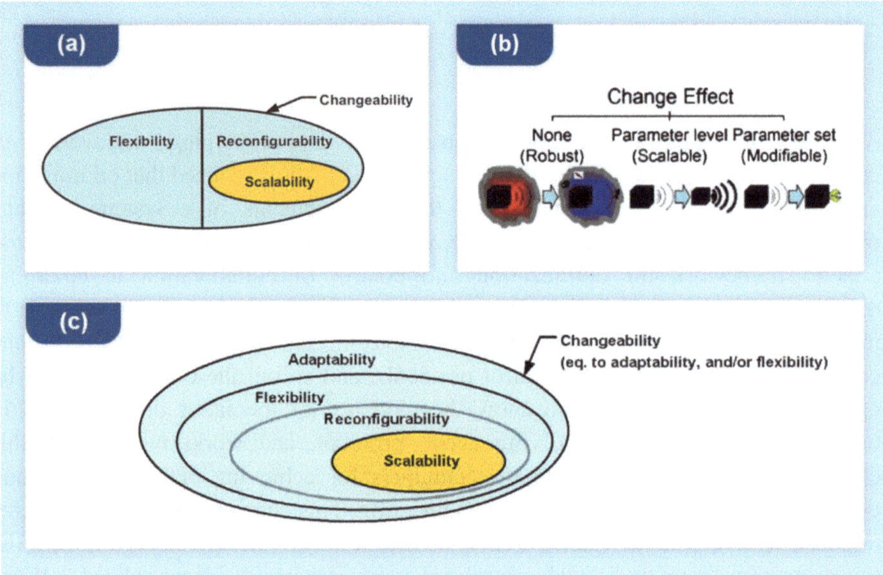

Fig. 4.1 (a) Flexibility is interpreted as the ability of a system to change its behavior without changing its configuration, and reconfigurability is interpreted as the ability to change the behavior of a system by changing its configuration (both implying a changeable system). Scalability implies changing configuration by adding/removing the configuration elements, which is regarded as a subset of reconfigurability (Wiendahl et al., 2007). (b) The changeability of a system can be classified into three categories of effects (Ross et al., 2008): robustness (i.e., the ability to remain constant in parameters), scalability (i.e., the ability to change the level of parameters), and modifiability (i.e., the ability to change the membership of the parameter set) (c) Conversion flexibility is similar to reconfigurability, and scalability is a subset of both flexibility and reconfigurability, which are in turn subsets of adaptability and changeability (Wiendahl et al., 2007). With permission from Elsevier B.V.

Agent-based systems, the first e-Dimension of collaborative e-Work, enable promising methodic, procedural, and algorithmic functionalities through *localization* of information, resources, and responsibilities, to handle the complexity and dynamicity of emerging e-Systems. An agent-based system is a loosely-coupled network of homogenous and/or heterogeneous agents with specific attributes, which interact with each other (either cooperatively or competitively) through certain mechanisms to solve problems that are beyond their individual capabilities. An agent is a computational element characterized by the following four features (Wooldridge and Jennings, 1995; Huang and Nof, 2000a, b; Nof, 2000):

1. *Independence* – action within an environment without the direct control of any other element (human beings or other devices).
2. *Sociability* – interaction with other elements of an environment through a communication language, such that their own goals are achieved.
3. *Re-activeness* – responding, in a precise manner, to the signals received from the environment.
4. *Pro-activeness* – taking the initiative in order to achieve their goals (goal-oriented behavior).

To pursue their individual goals, agents need to interact with each other and with their environment under specific coordination mechanisms. Their interaction paradigms are either *cooperation-based* or *competition-based*. Following the cooperation-based paradigm, agents cooperate to achieve a set of common goals, where the cooperation can be distributed (agents are endowed with self-organizing rules for resource sharing and goal pursuing) or centralized (a mediator agent is assigned with the task of regulating and supervise agents' behaviors). In the competition-based paradigm, agents are self-interested, compete under certain negotiation mechanisms, and no mediator is involved. The unique features and various architectures of agent-based systems provide several interesting possibilities for optimization and control of complex, modular, and dynamic systems. Applications of agents-based systems in manufacturing systems range from enterprise integration and collaboration to process planning and scheduling, shop floor control, and holonic manufacturing (Shen et al., 2000) (see Tables 4.1 to 4.4). Agent-based systems have also been widely applied in other domains such as healthcare, telecommunication, transportation, supply network design, air traffic control, and entertainment. Due to their importance in the design of e-Work, we will specifically discuss software agents in Chapter 5. Below are some miscellaneous application examples of agents:

- *Patient monitoring*: Agents are deployed to enhance collaboration in Surgical Intensive Care Unit, i.e., effective cooperation and information sharing among heterogeneous teams of experts: (1) Perception/action agents – control the interaction between the monitoring system and the world, mapping raw sensor inputs into usable symbolic forms, and translating action requests into raw effector control commands; (2) Reasoning agents – organize decision-making processes; (3) Control agents – responsible for top-level control of the system (Jennings et al., 1998). The motivation is based on the limitations on the number of expensive high-level agents (e.g., experts, doctors), and the expertise of inexpensive low-level (e.g., nurses).

- *Emergency response planning*: Agents represent patients, resources (e.g., emergency responder, ambulance), and care locations (e.g., hospitals), collaborating with each other to minimize the emergency response time and negative impacts of a disaster (Narzisi et al., 2006).
- *Conflict-resolution in telecommunication networks*: Negotiating agents are deployed to tackle feature (e.g., call forwarding and caller-ID functionalities) interaction problems in telecommunication networks. As the size of the network (i.e., the number of features) increases, the complexity of interactions among features and the rate of inconsistencies drastically increase (Jennings et al., 1998). Agent-based systems have been recently applied for resolving such conflicts through network control, transmission and switching, service management, and network management.
- *Transshipment of imported goods*: Three types of agents are deployed for (a) storage allocation (for temporary periods); (b) deployment scheduling (allocation of drivers to vehicles); (c) coordinating the local sub-plans and optimizing the global plan (Fischer and Gehring, 2004).

Table 4.1 Examples of projects on agent-based enterprise integration.

Project	Domain	Features	Reference
ABMA	Enterprise integration	Middleware architecture	Budenske et al. (1998)
ATP	Agile manufacturing	Plug-and-play framework	NIST (1998)
CIIMPLEX	Enterprise integration	Service (Facilitator and Gateway) agents	Peng et al. (1998)
GMPP	Enterprise integration	Order fulfillment process using GMPP (General Message-Passing Platform)	Yu and Huang (2001)
IA framework	Enterprise integration	Intelligent Agents	Pan and Tenenbaum (1991)
JACKAL	Enterprise planning and execution	JACKAL; Java-based multi-agent development platform	Cost et al. (1999)
KRAFT	knowledge Transformation	Mediators as knowledge brokers	Gray et al. (1998)
LekiNET	Integrated scheduling	Internet scheduling agents; Vickrey auction for agent coordination	Yen and Wu (2003)
MAKE-IT	Information Integration	XML and MSMQ for Structural communication modeling; CLIPS (if-then rule-based tasks) system for knowledge modeling	Sacile et al. (2000)
MetaMorph I	Intelligent manufacturing	Mediator-centric architecture; dynamic clustering and cloning; learning	Maturana et al. (1999)
POMAESS	E-maintenance	Remote maintenance processes; Case-based reasoning	Yu et al. (2003)

- *Multiple destination routing*: Multiple homogenous agents are deployed, which traverse different routes in an intelligent and parallel manner, constructing an adaptive search algorithm, i.e., providing information to posterior agents (Elliman and Youssef, 2004).
- *Air traffic control*: Agents represent aircrafts and air traffic control systems in operation. Upon arrival of an aircraft in the airport, an agent is assigned to it instantiating its information and corresponding goals, e.g., target runway, landing time, etc (Jennings et al., 1998).
- *Train coupling and sharing*: The system elements (e.g., travel units) are modeled as autonomous agents, negotiating under certain protocols to optimize their own utility (Böcker et al., 2001).
- *Interactive theaters*: "Believable" agents (i.e., software programs that provide an illusion of life and permit the audience's suspension of disbelief) enable the user to play out a role similar to those played by movie actors interacting with artificial characters that have the behavioral characteristics of real people (Jennings et al., 1998).

Table 4.2 Examples of projects on agent-based enterprise collaboration.

Project	Domain	Features	Reference
ADE	Supply network	Delegation-based event handling	Mehra and Nissen (1998)
AGENT-OPT	Supply network	Case-based model modification scheme	Chan and Lee (2004)
Co-OPERATE	Supply network	Integrated legacy transaction systems; XML for interoperation between heterogeneous systems	Azevedo et al. (2002)
DISPOWEB	Supply network	Autonomous production planning and logistics	Frey et al. (2003)
ISCM	Supply network	Agent Building Shell (ABS); Coordination Language (COOL); Functional agents	Fox et al. (1993)
MaBE	Virtual enterprise	Holonic manufacturing and logistics service planning; Nested Contract Net	Karageorgos et al. (2003)
MADEsmart	Collaborative engineering	Wrapper agents for legacy system encapsulation	Jha et al. (1998)
MAGNET	Supply network	Genetic market architecture; contract negotiation with temporal and precedence constraints	Babanov et al. (2003)
MASCOT	Supply network	Blackboard architecture; Mixed-initiative decision support	Sadeh et al. (2003)
MetaMorph II	Intelligent manufacturing	Hybrid architecture; mediators as subsystem coordinators and interfaces to the main system	Shen et al. (2000)
NIIIP-SMART	Integration-collaboration	Standards-oriented configurable object model; product date exchange and business integration of MESs and EISs	Barry et al. (1998)
SCADAS	Supply network	Mobile agents for information exchange and privacy	Gupta et al. (2001)
VIRTEC	Virtual enterprise	Virtual enterprise broker, virtual industry cluster, and virtual enterprise	Bremer and Molina (1999)

- *Resource allocation*: Agents represent firms in a virtual market-based supply networks, interacting and negotiating with each other over limited inventories of different products (Kaihara, 2003).
- *Supplier selection*: Agents represent production facilities and orders in a dynamic and networked production system. Agents are regarded as knowledge-bases, which collaboratively estimate the market demand, and negotiate with each other under Contract Net Protocol for finding the most suitable supplier according to multiple criteria (Ahn and Lee, 2004).
- *Order fulfillment*: In a collaborative supply network, each agent represents a collaborative partner (embedded in its information system), has a local view of the status of other participants, and negotiates with other agents under peer-wise negotiation mechanisms to maximize its own order fulfillment rate (Lin and Lin, 2006).
- *Supply network configuration*: Intelligent agents model functionalities and behaviors of resources they represent, seek for their own goals taking part in tasks (each set of tasks represent a single customer order), e.g., supply of parts, and assembly and transportation operations, and simultaneously collaborate with each other to achieve the common goal of satisfying customer orders (Akanle and Zhang, 2008).

Table 4.3 Examples of projects on agent-based process planning and scheduling.

Project	Domain	Features	Reference
ABACUS	Scheduling	Functional agents; Agent design via BDI approach	McEleney et al. (1998)
ADDYMS	Scheduling	Physical agents; Dynamic local resource scheduling	Butler and Ohtsubo (1992)
AMC	Scheduling	Physical agents representing parts and machines	Goldsmith and Interrrante (1998)
CAMPS	Planning-scheduling	Repair- and constraint-based mechanisms	Miyashita (1998)
CORTES	Scheduling	Micro-opportunistic technique	Sycara et al. (1991)
CSIMAS	Scheduling integration	Agent model; Composition model; Cooperation model	Wang et al. (2003)
DAS	Scheduling	Hierarchical architecture with agents representing resources, resource groups, and scheduling process	Burke and Prosser (1994)
HOLOS	Scheduling	Generic agent architecture; five levels of Integration	Rabelo (2003)
LMS	Scheduling	Functional agents; Voting protocol for communication	Fordyce and Sullivan (1994)
MAPP	Process planning	Combined sequential-blackboard architecture	Hayes (1998)
ProPlanT	Production planning	Tri-base acquaintance model; Expert knowledge	Pechoucek et al. (2000)
Sensible Agents	Scheduling	CORBA objects communicating through ILU object environment	Barber et al. (1998)

Table 4.4 Examples of projects on agent-based shop-floor control.

Project	Domain	Features	Reference
AARIA	Scheduling control	Agents represent physical entities, processes and operations	Parunak et al. (1998)
I-Control	Manufacturing control	Partial Dynamic Control Hierarchy	Brennan et al. (1997)
IFCF	Scheduling control	Physical agents; Market-like control model	Lin and Solberg (1992)
FABMAS	Manufacturing control	PROSA reference architecture	Mönch et al. (2003)
MASCADA	Scheduling control	Proactive disturbance handling; Hot pluggable agents	Bruckner et al. (1998)
MASCOT	Scheduling control	Shared ontology and a base set of realistic Modules	Parunak (1993)
Reagere	Scheduling control	Blackboard architecture	Berry and Kumura (1998)
YAMS	Scheduling control	Market-driven contract Net; Forward and backward scheduling	Baker (1991)

4.2 Basic Architectures

4.2.1 Agent-Based Optimization

Due to their unique properties in representation and simulation of distributed and complex systems, agents have been widely applied for solving a wide spectrum of optimization problems with modular domains, i.e., problems whose sub-domains and sub-problems can be clearly identified. An optimization problem, in general, involves searching for the best solution out of a feasible solution set, such that a given set of criteria is optimized (Figure 4.2). The problem may involve a single or multiple objectives, treated in different ways, e.g., weighted, lexicographic, min-max, goal programming, physical programming methods (Marler and Arora, 2004). An optimization algorithm may be exact, if it finds the optimal solution in a reasonable amount of time, or a polynomial-time heuristic, if it finds an acceptable solution but not necessarily optimal, following the rationale of sacrificing optimality for computational efficiency. Centralized optimization mechanisms may be more powerful and useful for optimization of e-Systems. Nevertheless, our focus here is to highlight the impact of agent-based systems and their unique contributions to optimization of large-scale and complex collaborative processes.

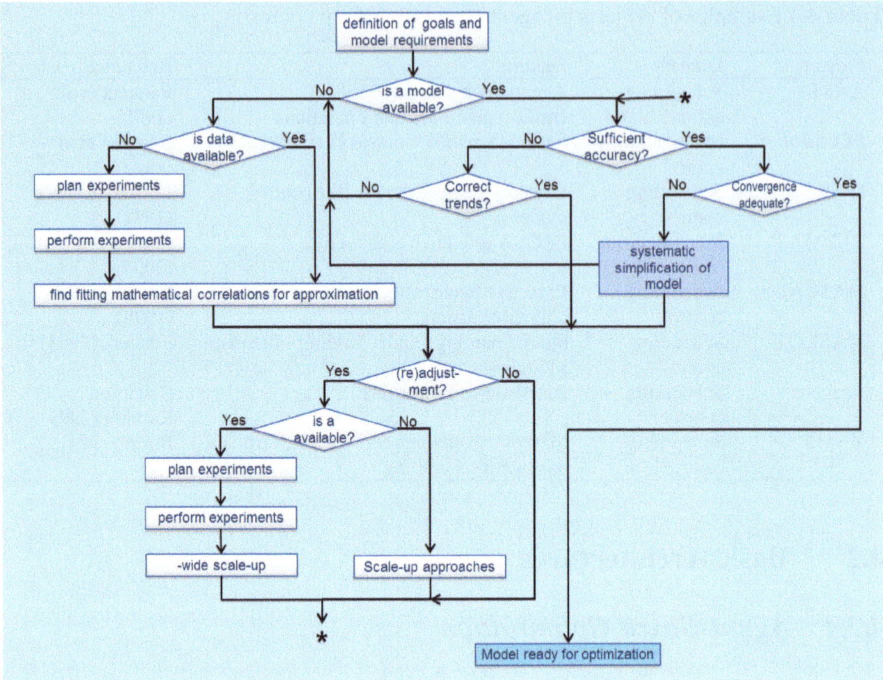

Fig. 4.2 A general workflow for an optimization problem (after Esche et al., 2013; with permission from Elsevier B.V.).

Agent-based optimization is typically applied to problems that are very difficult or impossible to be solved by centralized monolithic systems, due to their distributed, complex, heterogeneous, and frequently changing domains. Such problems can be effectively solved by networks of collaborating agents, thanks to some of their inherent characteristics:

- *Size and modularity* – Agent-based optimization partitions the problem into several local problems or *sub-problems*, which enables better handling of large-scale optimization.
- *Timescale/changeability* – Since agents are able to continuously monitor the state of their environment, and usually do not make complex decisions, they can provide higher re-activeness degrees, and thus better solutions to problems characterized by a high *dynamicity*.
- *Computational time/complexity* – Agent-based optimization provides considerable advantages in terms of computational time, because of the capability of decomposing large-scale problems into sub-problems.

Nevertheless, the computational benefits of agent-based systems may be diminished due to the high frequency of interactions and communications, which in turn requires extra resources such as time and facilities, as opposed to centralized optimization methods. Moreover, due to their distributed nature, individual agents do not have a holistic view of the entire system, and thus, the global optimality of the decision they take is not guaranteed. These facts further highlight the importance of effective interaction among agents enhanced by collaboration engineering; improving their intelligence, autonomy, and ability to work as a team. Agent-based optimization typically follows one (or a hybrid) of the following architectures: Autonomous; Mediator (Barbati et al., 2012). Production scheduling is used as a proper case here for describing these architectures:

(1) *Autonomous architecture* (Figure 4.3): Agents represent element of the manufacturing system such as jobs, resources, parts, and machines, and have the ability to generate their local schedules, react to local changes, and interact with each other to generate global optimal and robust schedules. The core process is based on *negotiations* between job and resource agents, under specific protocols (e.g., contract-net protocol, task administration protocol), to assign tasks to the machine agents. Part agents enter the system with different currency, evaluate the bids submitted by various resource agents, and select the ones that optimize their objectives. Job agents try to minimize the price paid, but the resource agents' goal is to maximize the price charged. In case of failure of a resource agent, the corresponding job agent(s) is informed to restart its negotiations with other resource agents. Resource agents are independently responsible for scheduling of operations along with error/conflict diagnosis and prognosis.

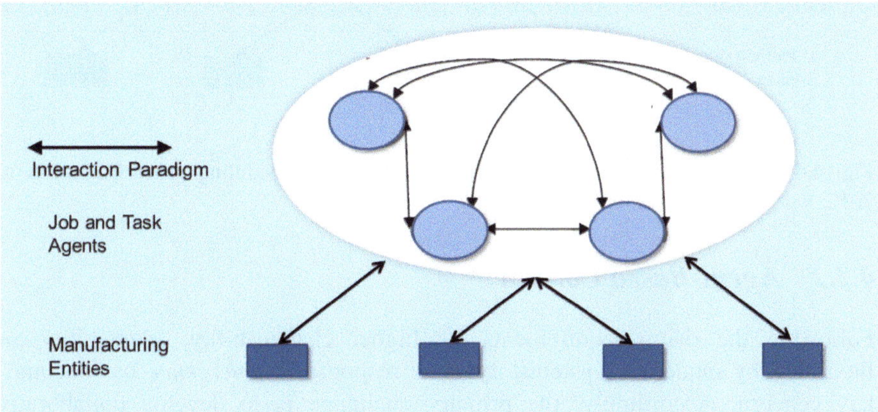

Fig. 4.3 Autonomous agent-based architecture for dynamic scheduling (after Barbati et al., 2012; with permission from Elsevier B.V.).

(2) *Mediator architecture* (Figure 4.4): This architecture aims to deal with the deficiency of autonomous architectures in finding the globally optimal solution, due to incapability of agents to comprehend the overall system

status. The architecture is composed of a set of autonomous agents interacting and negotiating with each other under direct control of a set of mediator agents responsible for global dynamic scheduling. The process is as follows: while the local agents autonomously make their own decisions, they request *advice* from the mediator agents, such that potential conflicts are resolved and the quality of solution is improved. The local and mediator agents have complementary characteristics; local agents have detailed and up-to-date local information about their local environment, while mediator agents have an overview of the entire multi-agent system. The mediator architecture involves relatively simpler computations compared to autonomous architecture, and is therefore more attractive for practical applications.

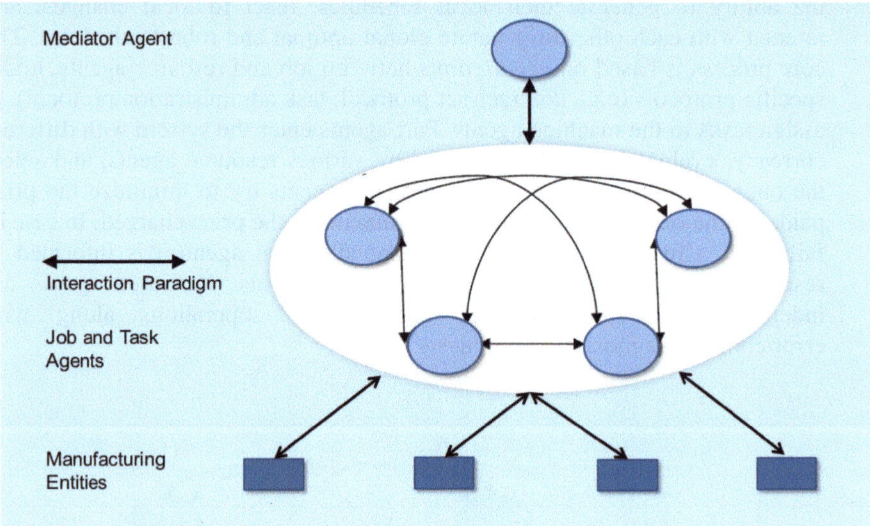

Fig. 4.4 Mediator agent-based architecture for dynamic scheduling (after Barbati et al., 2012; with permission from Elsevier B.V.).

4.2.2 Agent-Based Control

Following the rising requirements for higher changeability, adaptability and flexibility by modern e-Systems, dynamic response to *emergence* has become a key concern. Accordingly, the primary challenge is to develop collaborative control mechanisms (Nof, 2000) that facilitate intelligence and autonomy of agents for higher reconfigurability, robustness, and scalability of the e-System, which cannot be perfectly achieved by classic monolithic and hierarchical control architectures. Due to the inherent process complexities, production, service, and manufacturing control systems were traditionally designed on a centralized or hierarchical basis, comprising several components at different levels (strategic to tactical): planning, scheduling, execution and machine/device control (Figure 4.5).

As discussed in Section 1, such classic approaches are promising for achieving globally optimal systems; but their *rigid* structure limits their capabilities in addressing the above emerging requirements. Thanks to their unique features (i.e., independence, sociability, re-activeness, and pro-activeness) agent-based and *holonic* control systems are increasingly transforming classic control approaches to more modular, autonomous, scalable, reusable, and decentralized systems.

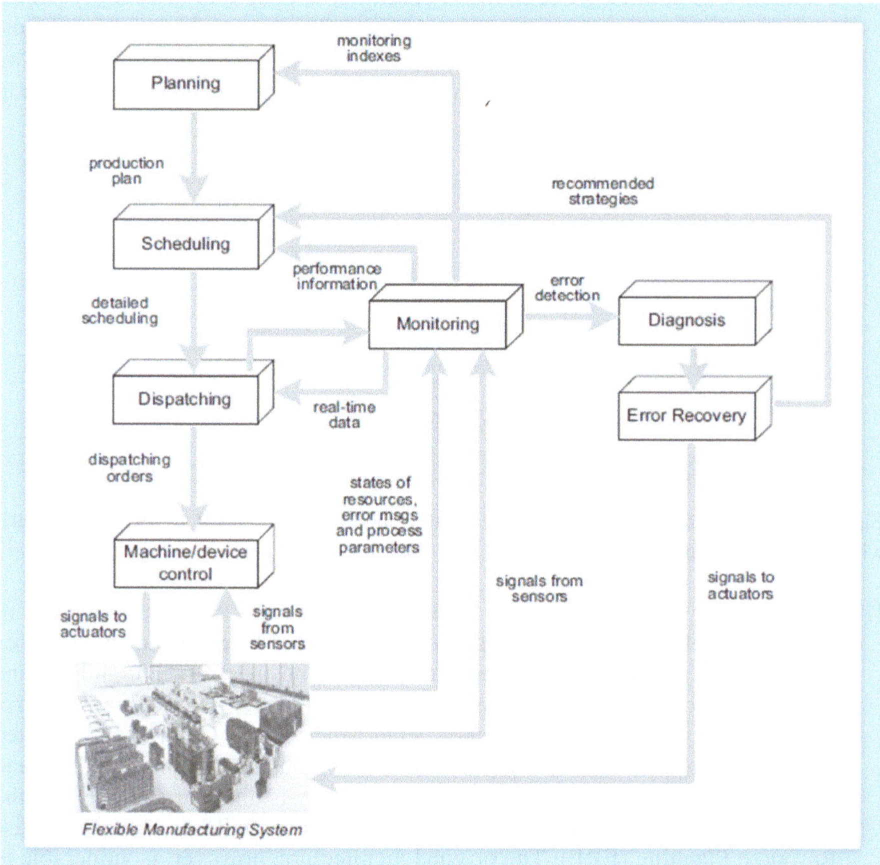

Fig. 4.5 Classic control architecture for production, service, and manufacturing (after Colombo, 2006); Long-term production plan is passed to the scheduling component, which is concerned with the assignment of operations to resources. The dispatching component decides utilization of resources considering their availability and the status of production/service system. The monitoring component provides plan execution information to the upper level components, and also the diagnosis component (if required). The machine/device control component involves initiation, coordination, and monitoring of different machine functions.

Agent-based control architectures are generally classified into the following three categories:

(1) *Centralized*: Characterized by a *central* decision node handling all planning and information processing tasks (Figure 4.6), centralized architecture is advantageous from the optimization perspective; but is substandard in terms of responsiveness, fault tolerance, and expansibility, especially in large-scale e-Systems. In addition, a centralized controller is unable to effectively handle system complexities, uncertainties, information incompleteness, real-time constraints, and so on. This drawback further highlights the importance of control decentralization, i.e., assignment of distributed decisional activities to local controllers.

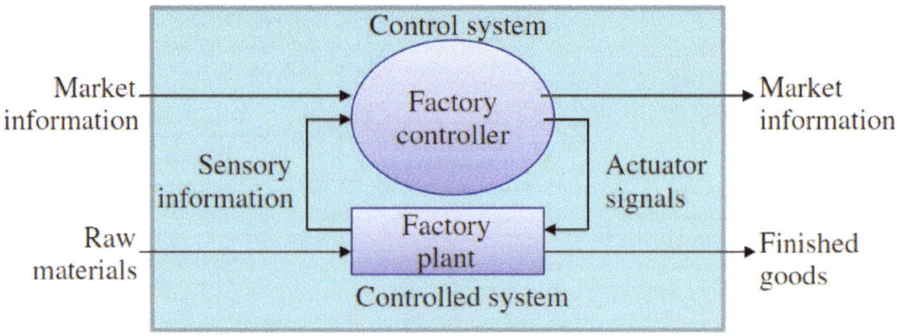

Fig. 4.6 Centralized production, service, and manufacturing control (after Baker, 1991; with permission from Elsevier B.V.).

(1) *Hierarchical*: Characterized by several control levels to make decisions of different levels (strategic; operational; tactical) in a hierarchical manner (see Figure 4.7), this architecture enables more robustness, predictability, and efficiency, compared to the centralized architecture. Hierarchical control architectures appeared by the emergence of Computer-Integrated Manufacturing (CIM) paradigms to enhance global optimality, agility, and pro-activeness of the system. This architecture is the cornerstone for the development of Manufacturing Resource Planning (MRP-II) and Enterprise Resource Planning (ERP). The original architecture, however, is vulnerable to disturbances, and has been modified by its next generation architectures. The modified hierarchical architectures improve the responsiveness of the original architecture to disturbances through enabling interactions between modules at the same hierarchy, leading to higher expansibility as well.

(2) *Heterarchical*: Disregarding client-server structures with fixed relationships, heterarchical architectures lead to higher responsiveness to disturbances, better expansibility, modifiability, and scalability of the system. To better understand the meaning of *heterarchy*, consider a directed *influence graph*, where nodes and arcs respectively represent decisional entities and master-slave interactions (master: decisional entity). If each node can be considered

as both a master and a slave, no hierarchy can be identified, and thus the graph is considered to be strongly connected, representing a heterarchy. Generally, hierarchy and heterarchy can be regarded as *vertical* and *horizontal* distribution of control, respectively (Figure 4.7). A critical limitation of heterarchical control architecture is lack of global optimality, since the decisional entities are autonomous and local, without a global view of the system.

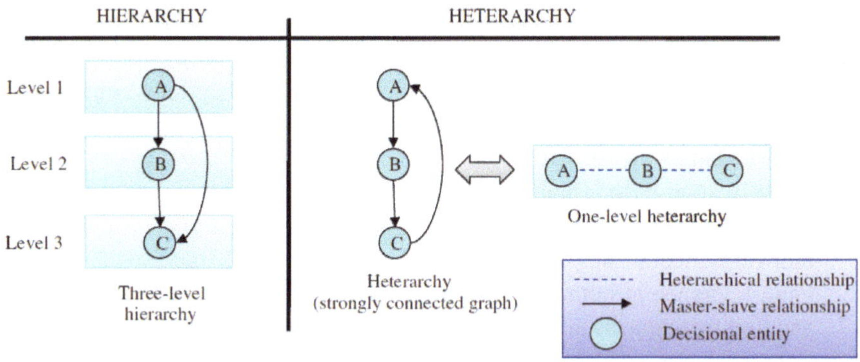

Fig. 4.7 Hierarchical vs. heterarchical control; formalization using graph theory (after Trentesaux, 2009; with permission from Elsevier B.V.).

Figure 4.8 summarizes different distributed control mechanisms, from centralized to decentralized, considering the two fundamental distributed design considerations described above, i.e., hierarchical and heterarchical relationships.

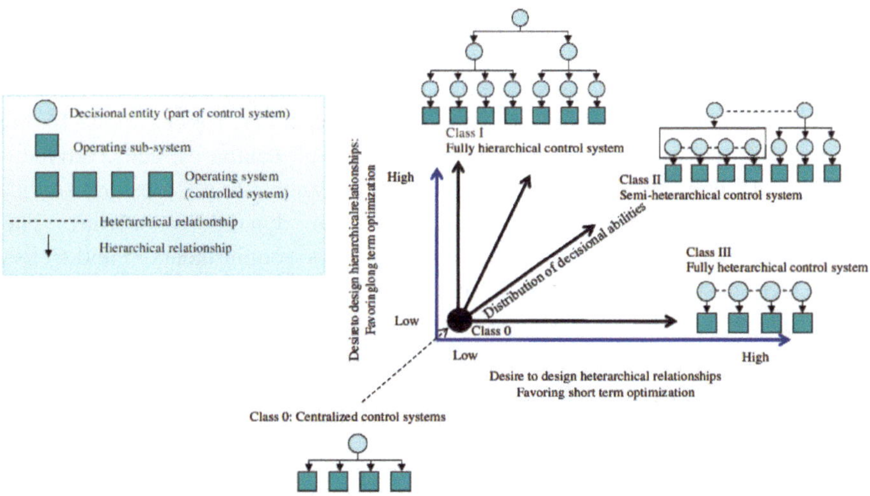

Fig. 4.8 Distribution of decisional capabilities in different control architectures (after Trentesaux, 2009; with permission from Elsevier B.V.).

Implementation of emerging heterarchical and hybrid control architectures (a.k.a. semi-hierarchical) requires economically justifiable systems with high level of agility and responsiveness, capable of satisfying the following six fundamental requirements (Fox, 1993; Shen et al., 2000):

1. High level of integration among heterogeneous hardware and software systems.
2. Reconfigurable, scalable, and evolvable architectures capable of accommodating new subsystems (hardware or software) or dismantling old subsystems (e.g., plug-and-play systems).
3. Effective interaction and collaboration among subsystems enabling co-existing and co-operation of entities with different hierarchies, without central supervision.
4. Agile response to changes and unanticipated disturbances rooting from internal or external sources, and thus lower long-term instability (e.g., bullwhip effect in supply networks).
5. Fault tolerance at both macro (entire system) and micro (subsystems) levels, and timely diagnosis and prognosis of errors and conflicts.
6. Enhanced product traceability throughout the entire lifecycle (i.e., design, manufacturing, inventory control, logistics, and future design).

Thanks to advances in CIM and AI, agent-based systems have brought up interesting possibilities for intelligent control of collaborative e-Systems. Agents are capable of encapsulating activities and wrapping legacy software systems in heterarchical architectures. Examples include agents for order processing, product design, engineering analysis, process planning, production planning and scheduling, simulation, and execution. These features properly address the first three requirements out of the above list. Agents can also be deployed to represent physical resources individually (e.g., products, machines, robots, tools, fixtures, AGVs, and operators) or at an aggregate level (e.g., databases, cells, production lines, and shop floors), and negotiation partners (e.g., design services, physical/virtual plants, financial institutions, or dynamic consortia). These characteristics can be fairly mapped into the last two requirements (5 and 6) listed above. For instance, requirement 6 can be satisfied by the use of intelligent agents enhanced with high-tech CIM technologies for identification and tracking. A common example is RFID (Radio-Frequency Identification), which utilizes wireless non-contact systems using radio-frequency fields for transferring data related to its corresponding product, with the purpose of enhancing automatic identification and tracking. RFID has a variety of applications in different fields/industries such as healthcare, privacy and security, safety, defense, airport, and primarily service, production, manufacturing, logistics, and supply network control (see Chapter 8).

Various classes of agents exist, ranging from fully reactive operating on a stimulus-response basis to intelligent deliberative agents capable of proactive reasoning and adaptive learning in volatile and dynamic environments. Agents organize themselves into heterarchical architectures with no more need for client-server structures with fixed relations (i.e., clients can be servers, and vice versa). The incomplete and local information/knowledge agents, however, necessitates effective interaction and collaboration protocols based on proper agent communication languages, mechanisms, and anthologies (see Chapters 1 and 2). As an example, consider a simple agent-based control system associated with a collaborative production/service system with three types of agents (Figure 4.9):

1) *Part agent*: responsible for the supervision of the operations executed on the part (installed on an Industrial Personal Computer).
2) *Machine agents*: responsible for numerical control of three CNC (Computer Numerical Control) machines.
3) *Transport agent*: running on a PCL (Programmable Logic Controller) and responsible for the control of an AGV (Automated Guided Vehicle).

Upon arrival of a part, part agent queries eligible (in terms of skills) and available machine agent(s) to perform a drilling operation on the part. Then, machine agents respond to the request independently: Machine Agent #1 reports its unavailability, as Machine #1 is out of service; Machine Agent #2 also reports its unavailability, as Machine #2 is overloaded; and Machine Agent #3 responds

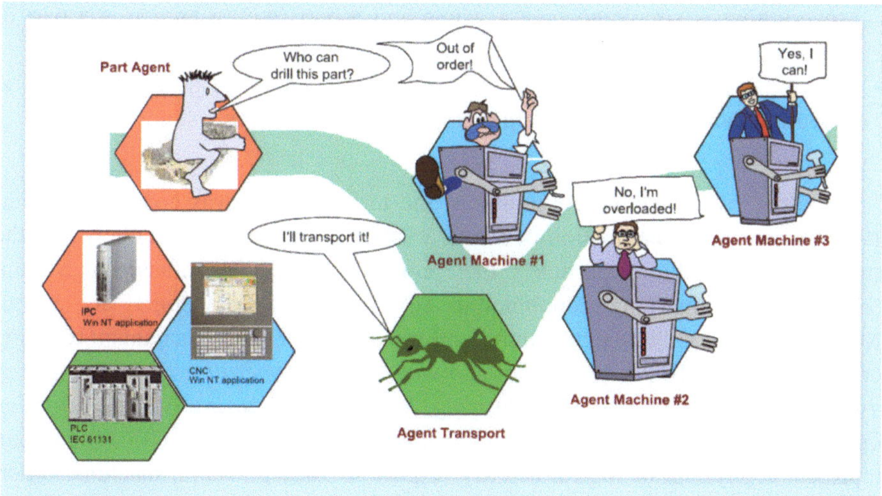

Fig. 4.9 Agent-based control of a production system with three parallel machines (adapted from Colombo et al., 2006).

about its availability. Following this negotiation, part agent assigns the drilling operation to Machine #3, and then starts negotiating with the transport agent. Note that the control system presented in this example is independent of the number of machines, and is insensitive to introduction of new machines, or removal of the existing machines. We will discuss more about interactions and collaborations between intelligent and autonomous agents in Section 4 below.

4.3 Emerging Bio-inspired Solutions

Social organization of agents is a key to guarantee timely resolution of conflicts and maintain the coherence of the entire e-System, which in turn, requires high levels of intelligence and autonomy for effective interaction and collaboration. Here is a simple example to explain why. Consider a system with 100 agents, each with 10 different behaviors, i.e., 1000 behaviors in general. Classic techniques require programming and control of a behavior space of 10^{100}, a number much larger than the total number of elementary particles in the universe (Van Dyke Parunak, 1997). A study by the philosopher Arthur Koestler in 1967 on the evolution of biological and social systems revealed that:

- Evolution of biological and social systems is based on some *stable intermediate forms* that are *self-reliant.*
- The "wholes" and "parts" of living organisms cannot be differentiated; almost everything is simultaneously a whole and a part (e.g., a human is a part of society, while comprising several basic units, body organs and cells).
- Living organisms are entirely self-supporting, while no auto-sufficient and non-interacting entities exist in their social organizations.

Followed by the above observations, the term "holon" was coined (Koestler, 1967), which is a combination of the Greek "holos" (meaning "whole"), and the suffix "on", (meaning "particle", like proton). Holonic Manufacturing System (HMS) is an emerging collaborative control paradigm, inspired by Arthur Koestler's work on biological control and social organization mechanism of living organisms. In HMS, a holon is "*an autonomous and cooperative building block of a manufacturing system for transforming, transporting, storing and/or validating information and physical objects*" (Van Leeuwen and Norrie, 1997), which can represent a physical or logical activity (Figure 4.10), e.g., robot, machine, order, flexible manufacturing system, or even a human operator.

A holon can be a part of another holon, where *holarchy* refers to a system of holons, organized in a hierarchical structure, collaborating to achieve a common goal, by sharing their information, knowledge, and responsibilities. Holons preserve their own autonomy and can dynamically join more than one holarchy. HMS can be applied into various areas of management and manufacturing, from order management to design, production, inventory control, marketing, agile manufacturing, and so on. Following this paradigm, a service or manufacturing system can benefit from the advantages of both hierarchical (high and predictable performance) and heterarchical (high reactivity against disturbances) architectures. Holonic manufacturing paradigm can be implemented using agent-based technology due to its inherent modularity, decentralization, reuse, and complex structure features (Figure 4.11). Real-time control, however, requires interconnection with physical devices in order to read data from sensors and to send action commands to actuators. Table 4.5 summarizes common HMS architectures.

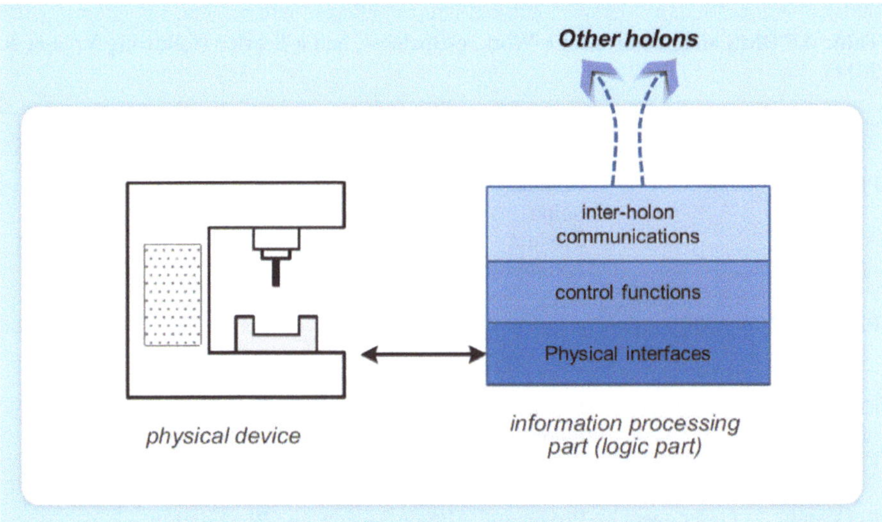

Fig. 4.10 Structure of a physical holon (after Leitão, 2009; with permission from Elsevier B.V.); A holon has information about itself and the environment, and (when representing a physical device) two information and physical processing elements.

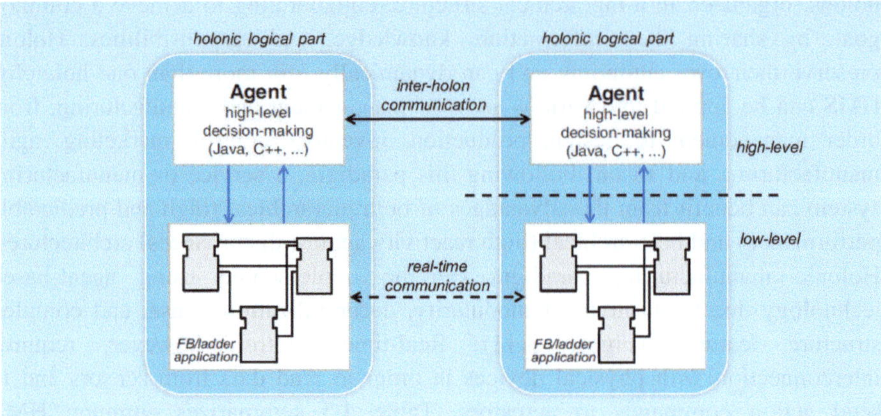

Fig. 4.11 Decision-making levels in the logical part of a holon (adapted from Marik and McFarlane, 2005).

Table 4.5 HMS architectures for e-Work, e-Business, and e-Service (following Vrba et al., 2011)

Architecture	Order	Product/ Service type	Product/Service instance	Resource/ Machine	Supervisor
PROSA[1]	-	Product holon	Order holon	Resource holon	Staff holon
ADACOR[2]	-	Product holon	Task holon	Operational holon	Supervisor holon
PABADIS[3]	Mfg. order agent	-	Product agent	Residential agent	Plant mgmt. agent
HCBA[4]	Coordination	Product holon	WIP agent	Resource holon	-
KASA[5]	Order agent	-	Supply agent	Machine agent	Contact agent
Rockwell	Order agent	Production plan agent	Product agent	Workstation/ facility agent	-

[1] Product-Resource-Order-Staff Architecture [2] ADAptive holonic COntrol aRchitecture
[3] Plant Automation Based on DIstributed Systems [4] Holonic Component-Based Architecture
[5] Knowledge-based Automation System Architecture

Despite many similarities, agents and holons are considered as different from several aspects[1]. Agents root in Computer Science and AI, while holons are basically derived from CIM, focusing more on the flexible manufacturing concept. Holon is a concept, while agent is both a concept and a technology, enabling implementation of HMSs. Unlike agents, a holon can be decomposed into several lower-level holons. Besides, an agent may or may not be able to satisfy real-time constraints, while holons must meet real-time constraints to ensure satisfactory system operation. Indeed, a holon is a reactive agent, capable of addressing real-time constraints and communicating with physical devices. An agent or multi-agents can also be designed and implemented for these purposes.

Going back to the observations of Koestler, it can be concluded that the evolution and survival of all living organisms have totally relied on their intelligence, as the main enabler of autonomy, and ability to interact and collaborate. Over the last 50 years, tremendous amount of work has been done on the development of AI tools and techniques to design intelligent agents that *emulate* the intelligence of natural organisms. AI is defined as the ability of agents, from nano-sensors, software agents to robots, to perceive, learn, reason, plan, communicate, and act, without the need for human intervention, i.e., agents with "minds". In the following sections, two fundamental bio-inspired AI techniques, anchored in swarm intelligence and natural evolution, are briefly introduced, and their application in the design and control of agent-based systems are reviewed using several real-life e-Work examples.

4.3.1 Swarm Intelligence: Ant Systems

Swarm intelligence is a type of intelligence built upon the collective behavior of a swarm, or community of decentralized and self-reliant organisms, which may not have a high level of intelligence individually. That is, individual elements of the swarm are not exposed by themselves to the complexity and dynamicity of their environment; they operate following a simple set of rules. What makes the global (swarm) behavior highly intelligent is the "collaboration" among individuals. There are several examples of swarm intelligence in the nature (Table 4.5): Colonies of ants and bees, bird flocking, fish schooling, etc. Although this is an emerging area of research, plenty of studies have been performed over recent decades on the inspiration from natural swarm behaviors in constructing agent-based systems. We will focus on ant systems, a revolutionary swarm intelligence technique inspired by real ants' behavior in seeking paths between their nest and a source of food, and will show how to imitate this behavior to increase the usefulness of artificial agents.

---

[1] For a comprehensive comparison between agent-based systems and HMS, see Giret and Botti (2004).

Table 4.6 Examples of natural swarm behaviors (adapted from Parunak, 1997).

Organism	Task	Description	Responsibilities	Behavior
Ants	Path planning	Ants find the shortest path between their nest and a food source	1) Avoid obstacles 2) Wander randomly considering pheromone trails 3) Deposit pheromone while carrying food back to the nest 4) Pick/drop food	Only food carrying ants deposit pheromone, as an interactive behavior, leading to a *reinforcement learning* mechanism for finding food and the shortest path.
Ants	Brood sorting	Ants continuously sort various things in their nest by kind, e.g., larvae, eggs, food	1) Wander around the nest 2) Sense and memorize local objects 3) Probabilistically pick up objects from surrounding and drop in the right place based on observations	With a similar behavior to path planning, ants continuously examine all objects in the nest, leading to higher concentration of similar things over time.
Termites	Nest building	Termites construct multi-story mounds to store food, and protect brood and population	1) Metabolize bodily waste, containing pheromone 2) Wander randomly considering pheromone trails 3) Probabilistically deposit waste considering local pheromone level	Reinforced waste deposit leads to transformation of scattered pheromone deposits to columns. Termites of each column are attracted to the other, creating an *arch* structure, and ultimately a *floor*. The process repeats to create the next floor
Wasps	Task allocation	Wasps assign three different roles: A single chief; Foragers; Nurses	1) Mature wasps engage in face-offs. 2) Broods demand reduces after receiving food. 3) Foragers randomly forage when they are nearby broods.	Confrontation dynamically shifts power among wasps, as a sort of interaction. More/less powerful wasps: Foragers/nurses. Brood's demand stimulates the foragers to forage.
Birds/ fishes	Flocking/ Schooling	Flocks/ schools stay together, coordinate turns, avoid collisions	1) Keep a minimum distance from the closest object. 2) Match velocity to the closest object. 3) Stay close to the center.	Birds and fishes have cooperative, self-reliant behaviors. Any change in the status of one object is propagated to the whole flock; global coordination.
Wolves	Group hunting	Wolves cooperate with each other to kill a big prey, e.g., moose	1) Surround the prey 2) Keep moving to the neighboring cell closest to the prey 3) Be faster in moves than the prey	Similar to flocking and schooling behaviors, each object influences and is influenced by the entire system. Wolves only interact with each other through this repulsion-attraction mechanism.

4.3.2 Theory

Ant systems were introduced by Marco Dorigo and his colleagues in 1990's, as a technique for systems engineering and optimization inspired by the foraging behavior of real ants. Ants wander randomly around the nest searching for food sources. Upon finding a food source, an ant evaluates its quality and quantity, carries a part of it back to the nest, and deposits a pheromone trail while carrying the food. The amount of pheromone deposited may be proportional to the quantity and quality of the food being carried. The pheromone trail acts as an indirect animal-animal interaction mechanism, a sign-based *stigmergy* that signals to the next ants the food source and alternative path to it. This *autocatalytic* behavior creates a positive feedback loop that over time enables an ant to detect the shortest path between the nest and the food source (Dorigo et al., 1996). The logic behind this behavior is that pheromone accumulation rate on a path is inversely proportional to the path length, resulting in a reinforcement learning mechanism for finding the shortest path over time (see Figure 4.12). The original algorithm was applied to the *traveling salesman problem*, but in general, it can be applied to any design and optimization problem *reducible* to a routing problem.

We use the original symmetric traveling salesman problem as a case to illustrate the procedure of ant systems. The problem is as follows: Given a list of cities and pair-wise distances, what is the shortest possible route that visits each city exactly once and returns to the origin city? Ant systems imitate the foraging behavior of real ants as follows. The procedure starts with A ants, starting their tours from an arbitrary node (i.e., city) in the network. In each trip from one node to another, ants update the pheromone trail of the corresponding link, which is called *local pheromone update*. The idea of local pheromone update imitates trail evaporation in real ants. That is, the deposited pheromone starts to evaporate, resulting in lower attractiveness of the path over time. This phenomenon is more noticeable in longer paths compared to shorter paths, as it takes more time for an ant to travel down the path. At the end of their tours, the best ant, i.e., the one with the shortest tour length, updates the pheromone trails of its tour, while the remaining tours remain unchanged. This procedure is called *global pheromone update*, which imitates the property of *differential pheromone trail accumulation* in real ants, and is based on reinforcement learning, i.e., giving higher reinforcement to the best solution. This information is *probabilistically* used by the next set of ants to identify the shorter paths following their preference for higher pheromone trail level. That is, with certain probabilities, the next ants either exploits the experience of the previous ants (i.e., pheromone trail) or apply a biased exploration using a probability distribution that is a function of both pheromone trail and pair-wise distances their respective positions and target nodes.

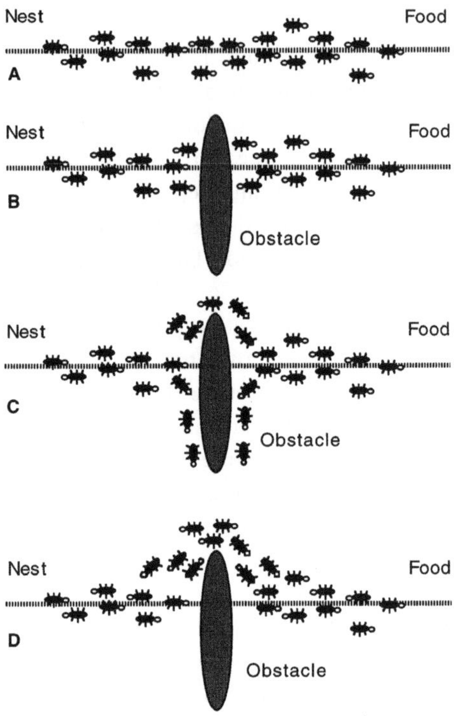

Fig. 4.12 Reinforcement learning mechanism of ant systems (A) Ants following the pheromone trail between the nest and the food source; (B) An obstacle destroys the trail, the first ants randomly turn left or right with equal probability; (C) Pheromone is accumulated faster on the shorter path; (D) All ants eventually follow the shorter path (after Dorigo and Gambardella, 1997; with permission from Elsevier B.V.).

The algorithmic definition of the ant systems procedure for the original symmetric traveling salesman problem is as follows (Dorigo and Gambardella, 1997). α_a, β_a, and γ_a, are auxiliary parameters defined for each ant a, c_a is the current position of ant a in the network, w_a is the working memory of ant a, and ρ_{ij} denotes the pheromone level of path i-j. The artificial ants consider in this technique have two extra features: They have memories (i.e., w_a) and are able to identify the distances of other nodes in the network (i.e., δ_{ij}). This is a population-based technique, which repeats by generation of new sets of ants over and over until the shortest path is detected. The population-based mechanism enables the next generations of ants to exploit the experiences of the previous generations and modify the path continuously. Three practical application examples of ant systems in optimization and control of e-Systems are discussed in the next section.

A Simple Ant Algorithm

Require: Network N; Pair-wise distances δ_{ij}, $i, j \in N$

repeat
    Initialize A ants
    for all $a \in A$ **do**
        Define $0 \le \alpha_a \le 1$, β_a, $0 \le \gamma_a \le 1$, $c_a = (0,0)$
        Set $w_a = \{c_a\}$
        repeat
            Select the next node to visit j
            Randomly select $\gamma \in [0, 1]$
            if $\gamma \le \gamma_a$ **then**
                $j = \arg \max_{k \notin w_a} \left\{ \rho_{c_a,k} \cdot \delta_{c_a,k}^{-\beta_a} \right\}$
            else
                Define probability of selecting j
                if $j \notin w_a$ **then**
                    $\Pr_j = \rho_{c_a,j} \cdot \delta_{c_a,j}^{-\beta_a} \Big/ \sum_{k \notin w_a} \rho_{c_a,k} \cdot \delta_{c_a,k}^{-\beta_a}$
                else
                    $\Pr_j = 0$
                end if
                Randomly select j according to its probability
            end if
            Update local pheromone trail
            $\rho_{c_a,j} \leftarrow (1 - \alpha_a) \cdot \rho_{c_a,j} + \alpha_a \cdot \rho_0$
            Update working memory
            $w_a \leftarrow w_a \cup \{j\}$
        until $|w_a| = |N|$
    end for
    Update global pheromone trail
    $a^* = \arg \min_{a \in A} \left\{ \sum_{i \in w_a} \sum_{j \in w_a} \delta_{ij} \right\}$
    $\rho_{ij} \leftarrow (1 - \alpha_{a^*}) \cdot \rho_{ij} + \alpha_{a^*} \cdot \sum_{i \in w_{a^*}} \sum_{j \in w_{a^*}} (\delta_{ij}/2), \quad \forall i, j \in w_{a^*}$
until termination condition is satisfied
output shortest tour

Example 1: Integrated Process Planning and Scheduling

As discussed earlier, the foraging behavior of real ants can be imitated for solving any optimization problem that can be recast as a routing problem. Graph-based Integrated Process Planning and Scheduling (IPPS) is a problem of selecting and scheduling processes following a set of predefined precedence relations between the operations. The set of feasible machines for each operation is recorded in an AND/OR graph (Ho and Moodie, 1996). In order to generate the optimal IPPS, the problem must be mapped to the characteristics of ant systems. The network is composed of a number of nodes corresponding to different operations. The information stored in each node includes the set of processing operations and their corresponding information such as processing times and set of feasible machines. Each ant generates a solution by moving through the network, taking into account the precedence relations between the nodes. Figure 4.13 shows a disjunctive graph of ten operations on two parts.

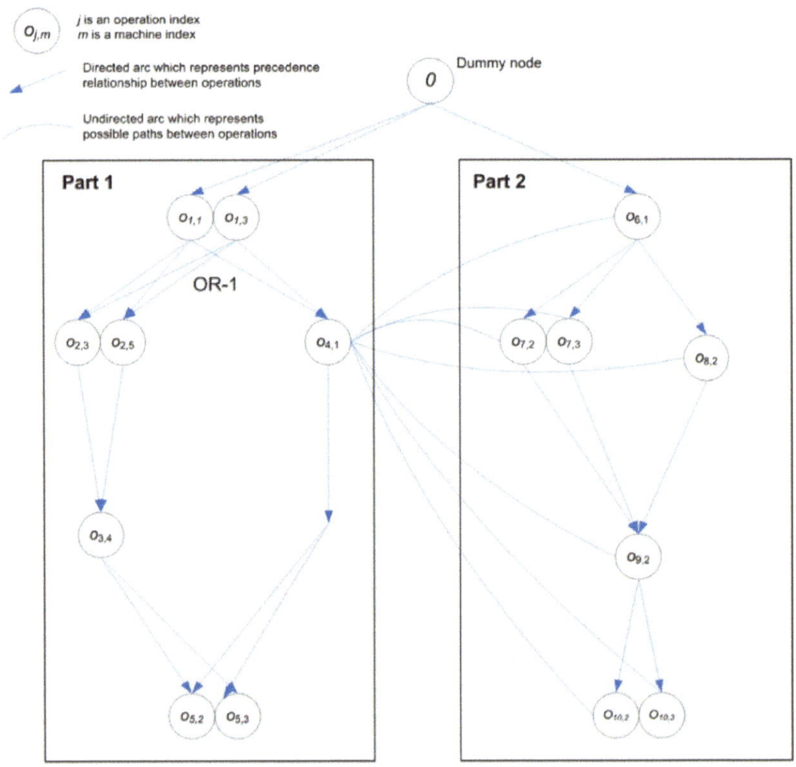

Fig. 4.13 An example of IPPS problem with ten operations and two parts (after Leung et al., 2010; with permission from Elsevier B.V.).

The ant systems algorithm for solving IPPS can then be designed as follows, following the original algorithm described in the previous section (Leung et al., 2010):

1) Pheromone trails, number of ants, and other necessary algorithmic parameters are defined.
2) All ants start their tours from the nest.
3) Each ant probabilistically selects the next operation to visit.
4) Each ant continues this procedure until a schedule is generated.
5) All schedules are reported to the supervisory agent.
6) The supervisory agent identifies the ant with the best schedule, and updates the global pheromone accordingly.
7) The procedure stops once an acceptable schedule is generated.

This procedure can be efficiently applied to any agent-based e-Manufacturing system, owing to its distributed and scalable architecture, i.e., independent software agents can be dynamically added to and removed from the system. The architecture applied here is *mediator* (see Figure 4.4), where the supervisory control mediates the initiation and termination of the algorithm, and controls the local actions and interactions of agents.

Example 2: Control through PROSA Architecture

PROSA is a class of HMS architectures (see Table 4.5) built upon three types of agents (or holons): Order, resource, and product agents. The agents are responsible for the control of internal logistics, process planning, and resource management (a similar case was discussed in Chapter 1; Figure 1.20). Sign-based stigmergy based on foraging behavior of real ants can be applied, as a communication means (information carrier) between the three types of agents, to integrate shop floor control systems (note that here, the focus is on systems engineering rather than optimization techniques discussed so far). The manufacturing environment must then be designed in accordance with the ant systems mechanism, supporting the following modules (see Hadeli et al., 2004): (1) A set of "locations" (blackboards) for the ants to reside, retrieve, or contribute information; (2) A mechanism to store and maintain pheromone (i.e., information) at different locations; (3) A propagation mechanism that enables ants to move between different locations. Artificial ants are then able to communicate through moving among different nodes in the network (Product and resource agents: Upstream; Order agents:

Downstream). Major benefits of such e-System are automated control, less complexity due to limited exposure of individual agents to the system's complexity, higher flexibility in face of abrupt changes and disturbances, and lower likelihood of congestion and conflicts.

We explain this procedure for the resource agents, which send ants upstream during the manufacturing process to make sure that theirs products visit the required processing steps in the right sequence. An example is used for sample product P1, which needs to undertake process steps 1, 3,5, and 6, in that order, and there are nine resources R1 to R9 (Hadeli et al., 2004). The procedure is as follows (Figure 4.14):

1) Propagation starts from R9 (last resource, e.g., shipment stage).
2) Product agent creates an ant agent with some initial pheromone, i.e., *subnet capacity information* (*p*, *s*, *d*), where *p*, *s*, and *d* respectively denote product ID, step ID, and destination resource ID. The initial subnet capacity information is (ProductID, ∅, ∅).
3) The product ant agent clones itself and sends one copy upstream through every possible path; to the last processing step (i.e., 6) resources, i.e., R7 and R8.
4) Upon arrival to a resource node, ants deposit the initial subnet capacity pheromone at the blackboard of the resource.
5) Ants update their pheromone information to (ProductID, lastStepID, currentResourceID).
6) This procedure continues towards the upstream node.
7) Ants retrieve the previous steps relative to the current step from the process plan, and for each previous steps, checks:

   - If the current resource is capable of processing the previous step, it updates its pheromone information to that of the previous steps, and then clones itself.
   - If the current resource is capable of processing the current step, it updates its pheromone information to that of the previous steps, and then clones itself.
   - If the current resource is not the arrival resource (i.e., R1 in this case), it clones itself and sends ants upstream from all possible paths.

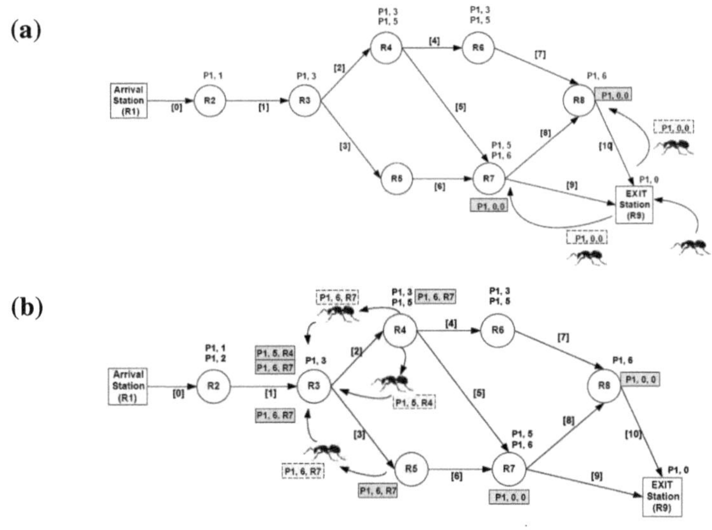

Fig. 4.14 Product ant agents: (a) initial propagation at last step resource (R7, R8); (b) higher level propagation from R4 and R5 to R3 (after Hadeli et al., 2004; with permission from Elsevier B.V.).

Example 3: Collaborative Capacity-Demand Sharing

An example of evolutionary collaboration in an e-System is explained here, where collaboration is performed within a collaborative network of enterprises (CNE) and through capacity-demand sharing. The collaborative behavior of enterprises can be formulated, optimized, and controlled based on the learning and adaptation traits of real ant systems. Consider a CNE where enterprises are allowed share their excessive capacity/demand, with the objective of improving the overall profit of the CNO. Let $N_C(\tau)$ and $N_D(\tau)$ denote the number of capacity sharing proposals and demand sharing proposals at time τ, respectively, where each proposal belongs to *one* enterprise. The design mechanism involves a best matching problem, where the elements to be matched are the demand and capacity sharing proposals. To represent the problem based on the constructs of ant systems, an artificial ant is assigned to each capacity sharing proposal. Each ant carries a capacity sharing proposal and moves through different nodes, representing demand sharing proposals (Figure 4.15). Paths are selected with respect to the ant's capacity and the nodes demands. Upon consuming all capacity by travelling between different demand sharing proposals, the ant goes back to the nest, leaving behind a *pheromone trail*. The distances between the nodes represent the fixed and variable costs of collaboration. The pheromone trails and evaporation rates enable a learning mechanism for the next generation ants to find the optimal path (i.e., minimum costs of collaboration).

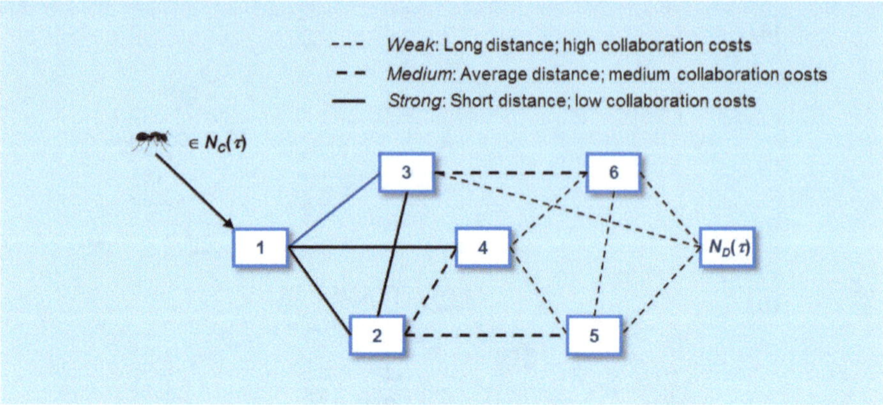

Fig. 4.15 Disjunctive graph of a DCS (Demand-Capacity Sharing) problem for one capacity sharing ant (∈ $N_C(\tau)$) with $N_D(\tau)$ demand sharing proposals (ant system); Each ant selects a certain path and leaves its capacity while passing the nodes (demand sharing proposal); Ants move simultaneously at a given time τ.

4.3.3 Evolution: Natural Selection

Evolutionary algorithms are inspired by biological evolution, the selection mechanisms in the nature where biological organisms become more/less common over time depending on their differential reproductive success in their interactions with the environment. Genetic algorithm was introduced by John Holland in 1975, as a ground-breaking evolutionary programming technique anchored in biological evolution mechanisms. Genetic algorithm is known as a promising metaheuristic for solving a wide range of optimization problems. In this section, however, we will discuss some applications of this concept in agent-based control and automation, after briefly reviewing the basics of evolutionary algorithms in general, and genetic algorithms in particular.

Theory

Evolutionary search mechanisms draw inspiration from natural selection processes, which cause the survival of the fittest individuals, based on their *adaptation* to changes in the environment. The difference in their adaptation is rooted in their unique features controlled by their basic units; their genes. A combination of genes creates a chromosome, as the key enabler of organisms in their competition for survival. Evolution then takes place as a series of changes in the unique features of *the fittest* organisms, generating a diverse pool of genes that are recombined from parents and lead to generation of offsprings. The reproduction is based on three primary mechanisms: *Selection*, *crossover*, and *mutation*. Crossover is a key mechanism in biological evolution, where genes of two parents are combined generating two offsprings. Mutation is another evolutionary mechanism that makes spontaneous random changes in a parent to generate an offspring.

Genetic algorithms imitate the aforementioned mechanism. The elementary step is *encoding*, i.e., representing a potential solution to the problem as a chromosome. Depending on the nature of the problem, various types of encoding schemes may be chosen, from simple binary strings to multidimensional matrices. Upon defining the encoding scheme, the first population of chromosomes is generated, and their *fitness values* are calculated. A portion of the fittest chromosomes are *selected* and moved directly to the next population, while the remaining ones undergo random changes through crossover and mutation operators. This procedure continues until a termination condition is reached, if no considerable improvements are achieved in a certain number of iterations. A simple pseudo-code of genetic algorithm is presented below. *SR*, *CR*, and *MR* denote the rates of selection, crossover, and mutation in each population of chromosomes, where $SR + CR + MR = 1$.

A Simple Genetic Algorithm

Require: Encoding scheme; Initial population P; Reproduction rates and parameters

repeat
  Set $P_S = \varnothing$, $P_C = \varnothing$, $P_M = \varnothing$
  Calculate fitness values
  Sort P by fitness values
  Partition set P as $P_S \mid P_C \mid P_M$, such that $|P_S| \approx SR \cdot |P|$; $|P_C| \approx CR \cdot |P|$; $|P_M| \approx MR \cdot |P|$
  if crossover condition satisfied **then**
    Select parents from P_S
    Perform crossover and generate offsprings
    Replace parents with offsprings in P_S
  end if
  if mutation condition satisfied **then**
    Select a parent from P_S
    Perform mutation and generate offspring
    Replace parent with offspring in P_M
  end if
  Generate new population $P \leftarrow P_S \cup P_C \cup P_M$
until termination condition is satisfied
output fittest individual

Example 1: Intelligent Negotiation System

One of the emerging challenges of e-Commerce is to cope with the volatility of markets and increasing competition through automation of negotiation systems in electronic transactions. Genetic algorithms can enhance the intelligence of negotiation agents by enabling negotiating agents to learn the preferences of their opponents and take actions considering the utility of all parties and intelligently collaborate with each other. Such autonomous agents are able handle negotiations on behalf of traders, which are beneficial in resolving potential conflicts, increasing the profit, reducing costs of human-based negotiations, and increasing the availability of the negotiation system (24/7). An agent-based negotiation system developed by Choi et al. (2001) is used to explain how agents can emulate natural selection mechanisms to improve their abilities to negotiate.

Negotiation processes involve several offering/counter-offering rounds, where all parties involved in the negotiation try to maximize their own and/or the overall payoff. Agent-based negotiation systems follow a similar procedure: In each round, a negotiating agent initiates the negotiation by proposing an offer, then evaluates the opponent's proposal, and finally suggests a counter-offer if the opponent's proposal is not acceptable. Genetic algorithm helps negotiating agents proactively predict their owner's and the opponent's needs. The intelligent negotiation system is then based on the following procedure (see Figure 4.16):

1) Each agent is defined as either a buyer or a seller, and receives an offer from its opponent.
2) If the offer is not acceptable, the agent uses the following procedure to generate a counter-offer:
   - Randomly generate a population of chromosomes, each representing a candidate offer for the current round of negotiation.
   - Include in the population a specific number of the best previous offers by the agent and counter-offers by the opponent (to ensure convergence of the process to a mutually acceptable result).
   - Check the feasibility of the generated offers: (1) None of them can be better than last offer; (2) all should yield positive payoff.
3) The fitness values of all chromosomes are evaluated and offsprings are generated using genetic operators.
4) The old population is replaced by the new one, and the negotiation process continues until a consensus is achieved.

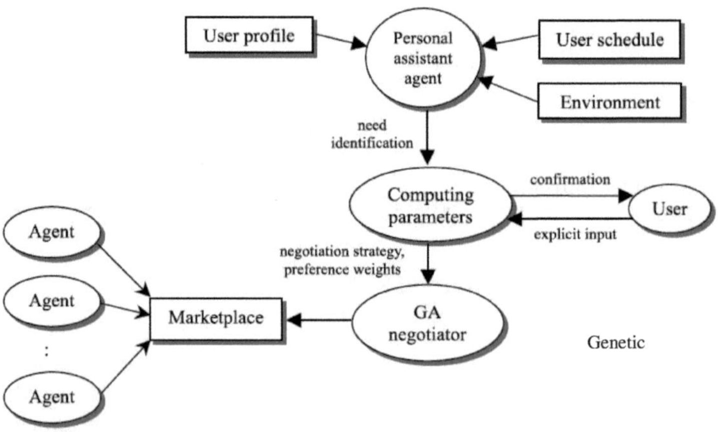

Fig. 4.16 Intelligent agent-based negotiation system based on genetic algorithm (after Choi et al., 2001; with permission from Elsevier B.V.).

Example 2: Agents Coordinate Supply Networks (MIT Beer Game)

MIT Beer Game is a business simulation game developed at MIT in 1960's, with the purpose of learning principles of supply network coordination through experiment[1]. The game involves (at least) four players, each representing a different role in supply network: Manufacturer, wholesaler, distributor, and retailer. The goal is to meet the customer demand for cases of beer, with minimum long-term inventory and backorder (or stockout) costs. The decisions of each player on how much to order depend on his/her prediction of future demand. Interestingly, the results are usually far from optimal from a global point of view, and lead to *bullwhip effect*, a phenomenon in forecast-driven distribution channels where variance of orders amplifies upstream. We briefly introduce an agent-based mechanism developed by Kimbrough et al. (2002), where human players are replaced by intelligent agents, whose intelligence is enabled by genetic algorithm. The purpose of this experiment was to show that intelligent agents can perfectly deal with deterministic and stochastic environments (in terms of demand and lead time), cooperate, learn, and mitigate the bullwhip effect even better than human players.

MIT Beer Game: Consider n players ($n \geq 4$). Let $IN_i(t)$, $C_i(t)$, $S_i(t)$, and $D_i(t)$ respectively denote the initial inventory, cost, received shipment, and demand received from downstream corresponding to player i at the beginning of period t (retailer receives demand from customers). Also let H_i and B_i denote holding and backordering costs, respectively (in the MIT Beer Game, H_i = \$1 and B_i = \$2). The total cost, i.e., total cost of all players in all periods, is then calculated as follows:

$$TC = \sum_i \sum_t C_i(t), \tag{1}$$

where

$$C_i(t) = \begin{cases} IN_i(t) \cdot H_i, & \text{if } IN_i(t) \geq 0 \\ |IN_i(t)| \cdot B_i, & \text{otherwise} \end{cases}, \quad \forall i,t,$$

and

$$IN_i(t) = IN_i(t-1) + S_i(t) - D_i(t), \quad \forall i,t.$$

The agent-based system is developed based on genetic algorithm, where the ordering rules are generated and examined using genetic operators. The interesting results shown in Figure 4.17 prove the superiority of intelligent agents over human players in minimizing the total cost, thanks to their ability to collaborate and resolve potential conflicts. The procedure is as follows:

1) The first population of rules is randomly generated.
2) Agents pick a rule from the current generation and play the Beer Game according to the selected rule until the end of the game period.

---

[1] See Chapter 10 for more experiential e-Learning games.

3) The total average cost of the entire team is calculated using Cost Function (1) and considered as the fitness value of the selected rule.
4) Steps (2) and (3) are repeated to calculate the fitness values of all of the rules in the current population.
5) Genetic algorithm is applied to generate a new population.
6) Steps (2) to (5) are repeated until the maximum number of generations is reached.

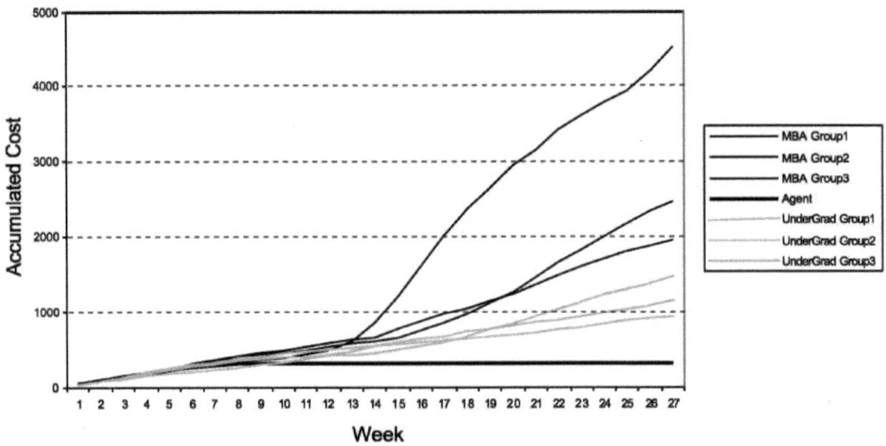

Fig. 4.17 Comparison between the performance of agents and different groups of students (MBA and undergraduate) in playing the MIT Beer Game (after Kimbrough et al., 2002; with permission from Elsevier B.V.).

Example 3: Evolutionary Robotics

Evolutionary robotics is an emerging field of research, which is based on the theory of evolution through natural selection. The goal is to enable robots to automatically develop their own control and even morphological systems. The evolutionary mechanism can provide a level of intelligence to the robots to learn from their environment and from past experiences, i.e., Phylogenetic and Ontogenetic adaptation (Floreano et al. 2008). Evolution is a form of adaptation by capturing environmental changes through several generations. Because of these capabilities, evolutionary robotics facilitates the detection of errors and conflicts through intelligent inspection. Learning, on the other hand, allows an individual to adapt to unpredictable environmental modifications in different generations. This feature enables evolutionary robots to effectively react to errors/conflicts, automatically perform the required maintenance, and recover error/conflicts. The general procedure is similar to the mechanics of evolutionary algorithms discussed earlier (see Figure 4.18):

1) An initial population of chromosomes is generated, each representing a specific control system or morphology.
2) Each chromosome is decoded into a real control system or morphology, and downloaded into the robot's processors.

3) The robot operates according to the given chromosome, and real feedback is recorded as the fitness value.
4) The fittest chromosomes are selected, some of which are directly imported to the next generation, and the rest are manipulated by genetic operators, i.e., crossover and mutation.
5) The new population is tested in a similar manner, and the process continues for a number of generations until an individual is born which satisfies the system's requirements.

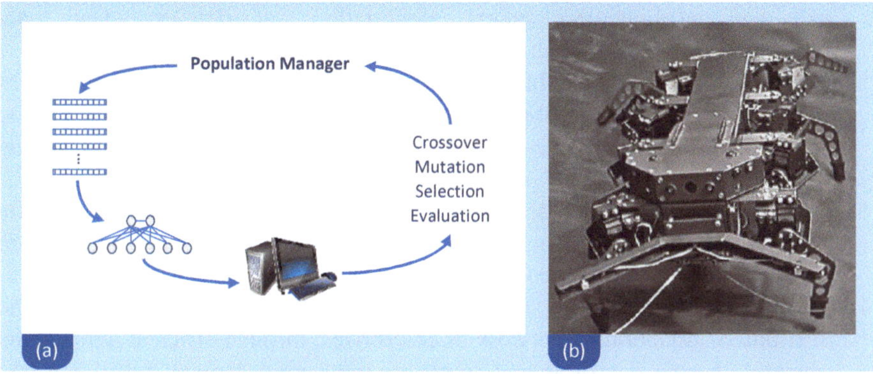

Fig. 4.18 (a) Evolutionary experiments on a robot. Chromosomes are decoded into neuro-controllers reading sensory information and sending motor commands to the robot every 300 ms. Their fitness values are automatically evaluated for reproduction. (b) Octopod robot (Courtesy of Applied AI Systems Inc.); The gaits are evolved using eight similarly organized genes in a chromosome, each coding motion characteristics of one leg such as delay and coordination with others. After a considerably few number of generations, the octopod learns how to smoothly move.

4.4 Optimal Design of Protocols and Workflows

Implementation of agent-based mechanisms in an e-System requires effective coordination platforms enabling agents to achieve their individual and common goals. These requirements are addressed by coordination protocols, a.k.a. task administration protocols, as the second e-Dimension of collaborative e-Work, through administration and synchronization of distributed processes for higher coherency and integrity. Common examples of coordination and collaboration protocols are:

(1) *Time-out protocol*: improves task allocation in client-server work-cells task allocation in agent-based systems (Peralta et al., 2003), rationalizing resource-sharing in assembly-and-test facilities with TestLAN (Williams et al., 2002).

(2) *Fault-tolerant time-out communication protocol*: designs distributed micro-flow agent arrays and networks to enhance the reliability and fault tolerance of the system (Liu and Nof, 2004; Jeong anf Nof, 2008).
(3) *Negotiation protocol*: enhances production on demand, competition power, and reaction to the market via cooperation-based agent-based paradigm (Reaidy et al., 2006).
(4) *Order management protocol*: regulates the interactions between agents through ordering processes, including outsourcing, order acceptance, coordination between supplier and customer, order confirmation, material request, canceling the orders, negotiating on delivery dates, and subcontracting (Huang et al., 2010).
(5) *Contract net protocol*: enables agents to act as managers or bidders thus form a holarchy based on establishment of contracts between a set of product holons, i.e., bidders, and a set of resource holons, i.e., managers (Hsieh and Chiang, 2011).
(6) *Best matching protocol*: designs assembly processes based on the existing information on equipment, parts and suppliers, cooperation requirement planning, part matching, and supplier selection (Velásquez and Nof, 2008).
(7) *Demand-capacity sharing protocol*: enables effective collaboration within distributed collaborative network of enterprises, through efficient demand and capacity sharing decisions and protocols improving demand fulfillment rate and the total profit (Yoon and Nof, 2010).

The purpose of this section is then not to discuss the details and step-wise mechanisms of coordination and collaboration protocols, but to present mechanisms for optimal design of protocols in terms of *responsiveness*, such that the agility and flexibility of agent-based systems discussed are preserved or even improved. These design considerations are discussed in Section 5-a below.

This section also discussed optimization mechanisms for better design of workflows, the third e-Dimension of collaborative e-Work. Workflow technology captures the systematic assignment and processing of agents activities in dynamic, distributed environments, thus enabling process scalability, reconfigurability, availability, and performance reliability. Workflow technology has received a lot of attention in recent years since it enhances management of information and operations. Section 5-b later discusses the fundamental design considerations for achieving more effective workflows, taking into account the parallel and distributed nature of emerging production, manufacturing, and service systems.

4.4.1 Protocol Optimization

A key issue concerning the design of coordination protocols is the communication *latency* and high traffic loads, due to technological constraints such as limited bandwidth, and server delays (e.g., file access delays, memory copies, and data processing). This is indeed an optimization problem aiming to minimize the communication latency of coordination protocols, taking into account technological constraints of the communication network (i.e., bandwidth and server delays). In this context, protocol optimization refers to the use of detailed protocol knowledge (i.e., how a specific protocol functions) in order to accelerate its response time. The common methods for solving protocol optimization problems are discussed below.

Compression

One of the methods for dealing with communication latency in coordination protocols is compression, i.e., reduction of data size without losing any part of it by removing redundant patterns such that space and transmission time are minimized. This method is basically applied for performance improvement of HTTP (Hyper Text Transfer Protocol) by compressing WebPages before sending them to the client, which in turn decompresses the data for use by the end user. In general, compression is measured by the following ratio (Ayyash and Khatatneh, 2012): $CR = (CS/OS)$, where CR, CS, and OS denote the compression ratio, compressed size, and original size, respectively. Compression can be performed in a variety of ways (Witten et al., 1999):

- *Lossy compression*: creates smaller files by discarding some information about the original file. Primary application is in the transfer of image files.
- *Lossless compression*: keeps all components of the original file. This method is applied in situations where no loss of any data component is allowed and the application becomes disturbed when the data files are corrupted.
- *Static compression*: is based on predefined compression encoding schemes constructed in advance and used without real-time modification.
- *Adaptive compression*: uses the preceding parts of the message text as a basis for predicting the forthcoming message text, and compression is affected in a single pass. The encoding scheme is updated in real-time as each component of the message is transferred to the recipient.
- *Streaming compression*: compresses the data as it is written. The objective is to minimize buffering delays, improve memory efficiency and scalability of applications that generate and consume streams of data (Lindstrom and Isenburg, 2006).

Object Pipelining

This optimization method is based on anticipation of user requests and provision of data/information in advance, and is efficient for reducing the delays caused by remote Web-based applications, or high-latency protocols, e.g., MAPI (Messaging Application Programming Interface), that function inefficiently through a WAN (Wide Area Network). Object pipelining method (Blue Coat Systems®, 2007) is based on the fact that communication latency is often due to the number of "round-trips" required to satisfy a request. Accordingly, anticipating the user requests, the delays associated with data retrieval are minimized; since upon receiving a request from a client, the data can be served at LAN (Local Area Network) speed rather than WAN speed.

The notion of object pipelining is based on the intelligent termination of user requests, in case the exact same request made *repeatedly* by the client (which correspond to majority of delays), and placement of requests on behalf of the user. Indeed, object pipelining is an object-retrieval method, which instead of retrieving objects serially, the algorithm opens as many simultaneous TCP (Transmission Control Protocol) connections as the server allows, and retrieves objects in parallel. Application of this method for optimization of MAPI is depicted in Figure 4.19. The method includes two major functions: (1) Batching, and (2) Keep-Alive. The first function combines multiple RPC (Remote Procedure Call) messages into one batch, in order to minimize the number of required round-trips across the WAN. Using the second function, the optimization method is enabled to retrieve the users' email even after they logged off, in order to minimize the retrieval delay and improve the response time.

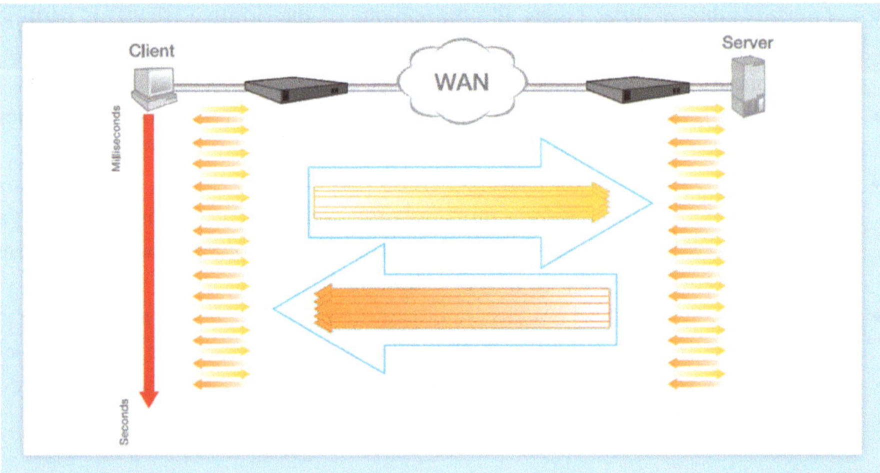

Fig. 4.19 Application of object pipelining in MAPI optimization; Batching and Keep-Alive functions (Courtesy of Blue Coat Systems®).

Transmission Sequencing

This method minimizes protocol latency by finding the optimal communication synthesis and transmission sequencing in sequential communication media, where the reading and writing sequence must follow the same order. Consider a set of processes $P = \{p_1, ..., p_n\}$, where the behavior of each process p_i is characterized by a control data flow graph G_i, containing basic blocks (representing operations) connected by control edges (representing data dependency). Each process is assigned to a processing element PE_i with certain characteristics (e.g., delay, area, power). Given a set of data $D = \{d_1, ..., d_m\}$ that must be transmitted on each communication channel, the problem is to find the optimal transmission sequence based on the control data flow graphs (G_i) of all processes such that the overall latency is minimized, subject to the design constraints and platform specifications. Figure 4.20 depicts an example of such instances, associated with a matrix multiplication algorithm (Part a). Upon system level synthesis, the algorithm is decomposed into two processes of (1) generating arrays A and B, and (2) matrix multiplication (Part b). The processes are then mapped into two processing elements, and transmit data following FIFO (First-In-First-Out) discipline (Part c). However, the sequence can be improved by sending the two matrices in an interleaved fashion, and calculate the product based on the data received (Part d). This method (Cong et al., 2004) minimizes latency, transforms the behavior description based on the computation order, and automatically generates drivers and glue logics by detecting the optimal computation and transmission sequence.

The optimization problem is basically divided into three sub-problems (Cong et al., 2004): (1) Sequencing; (2) Code transformation; (3) Interface generation. Our focus, however, is on the first sub-problem, and the other two are left as further reading. The sequencing algorithm is composed of the following three steps:

Step 1: Generate a general control data flow graph by combining all G_i ($i = 1,...,n$) into a single process network.

Step 2: Map each data element d_j ($j = 1,...,m$), transmitted by a PE_i, into a special type of operation denoted by o_j, and setup the correct data dependencies by linking the definitions of d_j to o_j.

Step 3: Solve the corresponding resource-constrained scheduling problem.

It can be proven that solving the resource-constrained scheduling problem yields the optimal solution to the transmission sequencing problem (Cong et al., 2004). Figure 4.20 shows an example of a simple process network consisting of two processes and three data elements. Following the original sequence (1, 2, 3), the latency is equal to seven cycles (Part b). However, by modifying the sequence to (1, 3, 2) considering data dependencies, the overlap of computation and communications is maximized, and the latency is then reduced to five cycles (Part c).

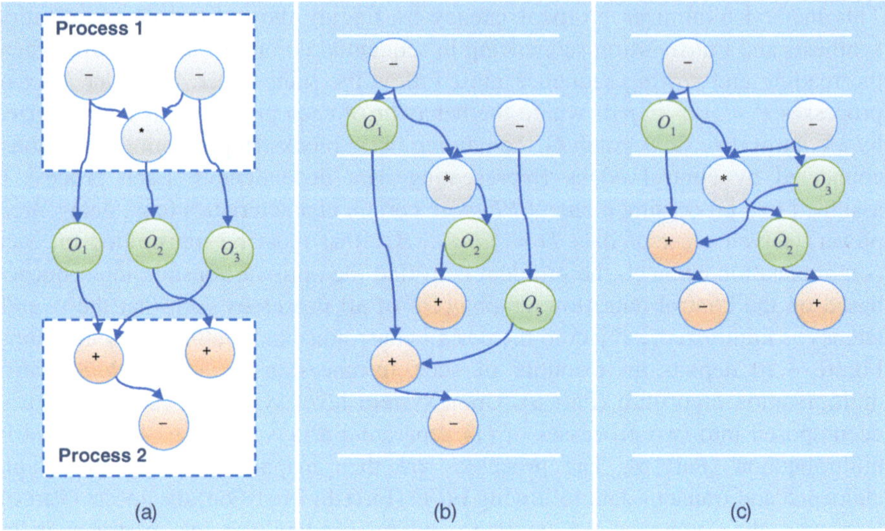

Fig. 4.20 (a) Merged control data flow graph with two processes and three data elements; (b) Results for sequence (1, 2, 3); (c) Results for sequence (1, 3, 2).

Layering and Event Tracing

Layered protocols provide many advantages such as modularity and layered protocol suites, compared to non-layered monolithic protocols. Protocol layers represent the units of development in a communication system implementing the functionalities associated with non-layered protocols. Layered protocols have several unique advantages compared to non-layered protocols. Layered protocols are modular, can be combined in different ways depending on the application requirements, and are efficient in rapidly development and testing of high level protocols. The suites of layered protocols can also be flexibly selected based on their expected workload. In addition, layered protocols are much easier to verify than large non-layered protocols. A typical layering model is composed of the following six components (Figure 4.21-a):

1. *Event*: represents the records used by protocol layers for transferring information related to a message.
2. *Event queue*: is created for passing the events between layers following the FIFO discipline.
3. *Protocol layer*: implements a small protocol as an event-driven automaton, interacting with the environment (i.e., adjunct layers) through the event queues.
4. *Protocol stack*: is created by the protocol layers and is typically visualized as linear vertical stacks of protocols.
5. *Application/Network*: communicates with the top/bottom of the protocol stack for sending messages downward/upward via *send* and *receive* events.
6. *Scheduler*: determines the execution order of events (FIFO), and ensures that each layer is executing at most one event at a time.

Layered protocols, however, may lead to critical deficiencies in terms of latency, caused by the abstraction barriers between the layers. This drawback can be resolved by event tracing, i.e., finding the common execution paths of the communication system such that the delays associated with layering are minimized, while its advantages are preserved. The idea is based on extracting the sequences of basic units of operation and execution, called *event traces*, as occur in protocol stacks, through adding the following three components to the layering model (Hayden and van Renesse, 1997):

- *Event trace*: begins with the arrival of a new event in the protocol stack, and moves through various layers where other event traces may be spawned up or down.
- *Trace condition*: refers to particular conditions under which an event trace is being executed, including predicates on the status of different layers and forthcoming events.
- *Trace handler*: refers to the sequence of operations executed in a particular event trace.

An example is shown in Figure 4.21-b, where a message is received from the network layer and passed to the routing layer, which in turn forwards a copy down to the next destination and passes a copy to the network. Upon extraction of the event traces, they can be optimized in through compression (e.g., eliminating intermediate data structures, compressing message headers) and transmission sequencing, as discussed earlier.

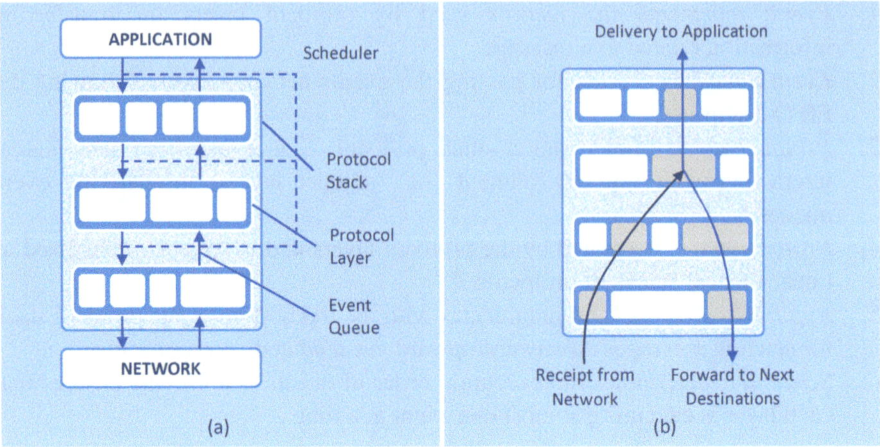

Fig. 4.21 (a) Basic elements of a protocol layering model; (b) A complex and non-linear trace in a routing protocol stack.

Parallelization

This method focuses on deriving optimal parallel schedules from sequential protocol specifications. The idea is based on the concept of parallel computing, i.e., distribution of protocol functions over multiple parallel processors, subject to the constraints imposed by control/data dependencies. The parallelization method consists of the following steps (Leue and Oechslin, 1993):

(1) *Decomposition*: The specification of the protocol stack is decomposed into operations, which can be either statements or decisions creating branches.
(2) *Common path analysis*: The branches are divided into two classes of common and uncommon. The common branches are those taken with stochastic certainty, while uncommon branches are those with probabilities below a certain statistical confidence value. The uncommon branches are then dropped from the decision nodes. The decision nodes with multiple outgoing branches are called *branching nodes*.
(3) *Composition of common data graph*: The common path graphs are then combined to form a monolithic common path graph for the entire protocol stack (Figure 4.22). The composition starts from a *root node* (receiving input from the environment) and ends with an *end node* (sending outputs to the environment).
(4) *Data dependency analysis*: According to the common path graph, the dependencies between data are analyzed using classic data flow analysis techniques, and the maximum possible parallelization of the protocol is then determined (only data independent operations can be executed in parallel).

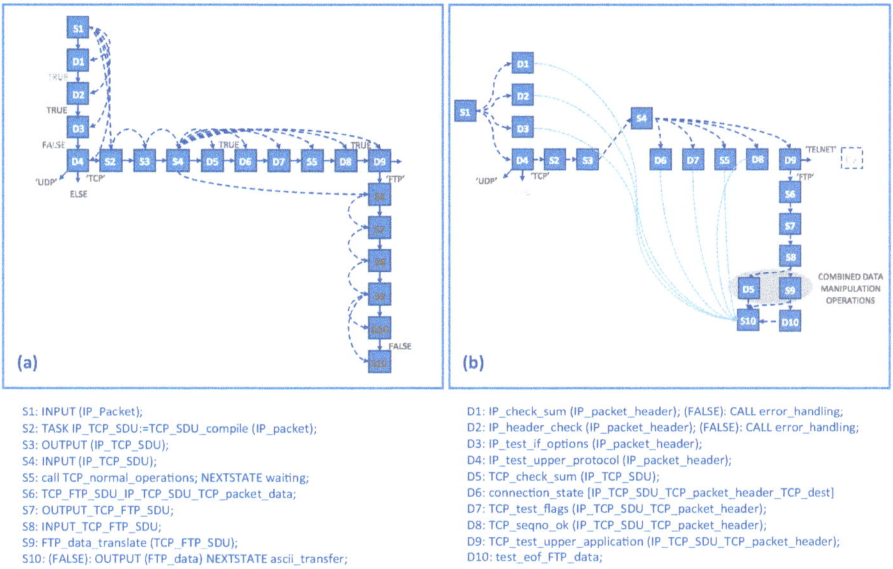

Fig. 4.22 (a) Common path graph with control (solid arrows) and data (dashed arrows) dependency relations; (b) The common path graph with relaxed control dependencies; From D4 to S2, and D9 to S6, and the edges rooting from D1, D2, D3, D5, D6, D7, D8, and S5 to the exit node S10.

(5) *Relaxation of control dependencies*: The control dependency relations that connect neighboring program statements are replaced based on control dependency relations, i.e., every non-branching node only depends on its closest predecessor. Indeed, the dependencies along the common path graph are transitive (see Figure 4.22-b), and thus, when a non-branching node transitively depends on a branching node, there is no need for a new dependency edge (Figure 4.22-b).
(6) *Lazy message processing*: The notion of lazy message processing lies in the integration of data manipulation operations, and reducing the execution time by minimizing the time attributed to fetching the operands with negligible processes. The minimization is performed by combining the data manipulation operations with identical operands.
(7) *Consistency analysis*: The operations at the exit node are modified to test all the operation results against the predicted results. In case of any mismatch, the entire process must be restarted following the classic non-optimized code version of the protocol stack.

4.4.2 Workflow Optimization

In collaboration engineering, task administration protocols serve as controllers of workflow in complex, interdependent systems. In a way, they are conductors of several orchestras playing autonomously yet attempting to play harmoniously. Workflows describe the sequence and relations of tasks in an e-System, and store the answers to two questions: (1) Which agent will benefit from the task when it is completed by one or more given agents? (2) Which task must be finished before other tasks can begin? Workflows are specific to a given system, and are managed by workflow management systems (which include the task administration protocols). Workflow systems support the following functions:

1. Classification, excavation, and removal of information carriers in an archive.
2. Physical transport of information carriers.
3. Recognition of documents and transmission to the appropriate agent.
4. Connecting of incoming documents with previous information (for example, an incoming document will be connected with previously saved data about the client).
5. Deadline monitoring and resubmission, capacity exchange (for example, forwarding systems).
6. Updating of data in the course of the process of individual processing steps (e.g., verification of the inventory when sending out orders).

Efficient organization of automated work processes, however, requires the structure of the associated workflow to be optimized, as it influences the work process of each individual agent by allocating individual procedure steps and forwarding them after processing. In general, workflow lifecycle is composed of the following phases (Holl et al., 2013):

(1) *Design and refinement*: involves the design of a new workflow or refinement of an existing workflow, through selection and composition of components, and definition of data and component dependencies.
(2) *Planning and sharing*: involves the creation of an executable workflow from the abstract design, through mapping the abstract elements into concrete applications, defining data sources, and sharing the designed workflow with each individual agent.
(3) *Execution*: refers to the application of workflow engine to map the executables to appropriate execution environment by retrieving information about software, resource, and data sources, monitor the execution process, and send the execution results to the users.
(4) *Analysis and learning*: involves examination of the results and comparison with the expected results. If the results are not satisfactory, the cycle restarts considering modification recommendations concerning the parameters, the choice or even the execution order of components.

With the rising complexity of e-Systems, it has become almost impractical to optimize workflows through such *trial and error* approaches, and thus, workflow optimization must be included as a new distinct phase in the conventional workflow life cycle (Figure 4.23). The optimization phase is placed between Phases 2 and 3 (i.e., planning and sharing, and execution) to ensure early validation of the prototype workflow before the computationally intensive execution of the full workflow. The optimization phase may involve different levels of the workflow description, from parameter optimization (i.e., identifying and incorporating application-specific constraints and dependencies between parameters) to the choice of components (i.e., finding the best algorithms for fulfilling of a given subtask) and/or workflow pattern (e.g., sequential, parallel, synchronized, exclusive choice). In e-Work, the principle of workflow parallelism has deeper implications due to the distributed nature of the e-Activities and the fact that interactions take place over human and software spaces and can include (1) human-human interactions, (2) human-machine and human-computer interactions, and (3) machine-machine and computer-computer interactions.

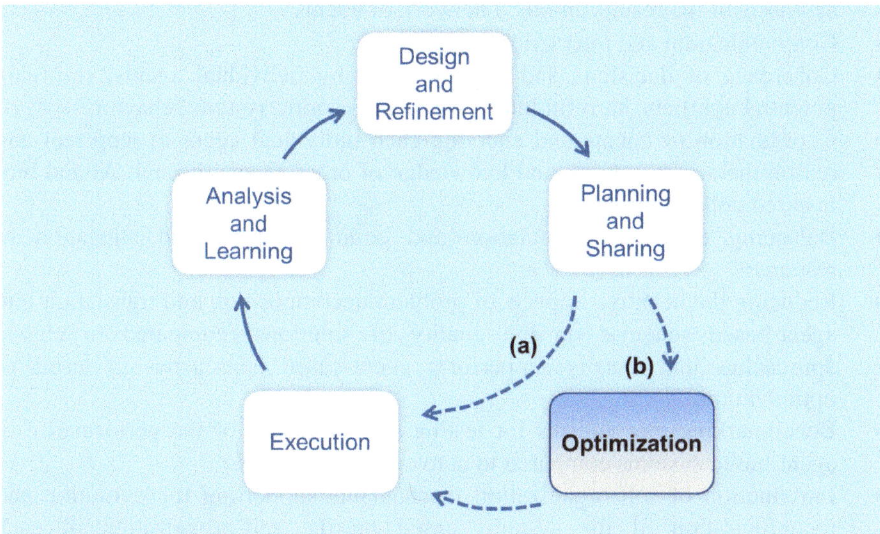

Fig. 4.23 (a) Conventional workflow lifecycle; (b) The extended workflow lifecycle including workflow optimization.

4.5 Emerging Trends and Challenges

Compared to classic monolithic systems, agent-based architectures provide several unique advantages addressing the fundamental needs of emerging e-Systems:

Autonomy and intelligence, which in turn revolutionize interaction and collaboration as the main drivers of adaptability, flexibility, reconfigurability, changeability, robustness, and scalability. Agent-based control systems enable distributed and collaborative solution mechanisms that do not rely on a central control function. This provides flexible, programmable, and dynamic control architectures based on hierarchical and heterarchical distribution of intelligence at different system levels. Such emerging control systems are highly responsive to disturbances, and adequate for low-high volume operations with medium-high variability. Agent-based systems are pluggable systems (i.e., insensitive to adding/removal of entities), which is critical for satisfying the emerging requirements for reconfigurability and scalability.

In spite of potential advantages of agent-based systems for optimization and control in collaborative e-Work, there are still several challenges concerning the design and implementation of e-Systems, mostly concerned with the development of effective protocols and workflows that enable/support:

- Formulation, description, decomposition, and allocation of problems, and synthesis of the results among a network of agents.
- Communication and interaction mechanisms.
- Coherence of decisions and actions taken by individual agents, resolving potential conflicts, harmful interactions, and chaotic system behavior.
- Coordination of agents and enabling each individual agent to represent and reason the actions, plans, and knowledge of other agents through AI and bio-inspired collaboration.
- Balancing of local computation and communication, and allocation of resources.
- Reducing the negative impacts of problem decomposition and translation into agent-based schemas on the quality of solutions (compared to classic approaches that usually outperform agent-based approaches in terms of optimization).
- Benchmarking mechanisms for testing and validation of the performance of agent-based systems compared to conventional systems.
- Introduction of self-organization mechanisms supporting the evolution and reconfiguration of the systems based on the self-organization of each individual agent through swarm intelligence and evolutionary mechanisms.
- Combination of agent-based systems with service-oriented architectures (SOA) to enhance the orchestration of services and provide high-level interface for the composed processes.
- Identification, diagnosis, and recovery of exceptions and disturbances, prediction of the occurrence of future disturbances, and mitigating their impacts.

References

Ahn, H.J., Lee, H.: An Agent-based Dynamic Information Network for Supply Chain Management. BT Technology Journal 22(2), 18–27 (2004)

Akanle, O.M., Zhang, D.Z.: Agent-based model for optimizing supply-chain configurations. International Journal of Production Economics 115, 444–460 (2008)

Ayyash, A.A., Khatatneh, K.: HTTP Protocol Optimization. Working paper, Balqa Applied University (2012)

Azevedo, A., Torscano, C., Sousa, J.P.: An order planning system to support networked supply chains. In: Proc. of PRO-VE 2002, pp. 237–244 (2002)

Babanov, A., Collins, J., Gini, M.: Asking the right question: risk and expectation in multi agent contracting. AIEDAM 17(3), 173–186 (2003)

Baker, A.D.: Manufacturing control with a market-driven contract net. PhD Thesis, Rensselaer Polytechnic Institute, NY, USA (1991)

Barbati, M., Bruno, G., Genovese, A.: Applications of agent-based models for optimization problems: A literature review. Expert Systems with Applications 39, 6020–6028 (2012)

Barber, S., White, E., Goel, A., Han, D., Kim, J., Li, H., Liu, T.H., Martin, C.E., McKay, R.: Sensible agent problem-solving simulation for manufacturing environments. In: Proc. of AI & Manufacturing Research Planning Workshop, Albuquerque, NM, pp. 1–9 (1998)

Barry, J., Aparicio, M., Durniak, T., Herman, P., Karuturi, J., Woods, C., Gilman, C., Lam, H., Ramnath, R.: NIIIP-SMART: an investigation of distributed object approaches to support MES development and deployment in a virtual enterprise. In: Proc. of EDOC 1998, La Jolla, CA, pp. 366–377 (1998)

Berry, N.M., Kumura, S.: Evaluating the design and development of Reagere. Working Notes of the ABM (Agent-Based Manufacturing) Workshop, Minneapolis, MN, pp. 5–13 (1998)

Böcker, J., Lind, J., Zirkler, B.: Using a multi-agent approach to optimise the train coupling and sharing system. European Journal of Operational Research 131(2), 242–252 (2001)

Bremer, C.F., Molina, W.M.: Global virtual business – a systematic approach for exploiting business opportunities in dynamic markets. IJAM 1(12) (1999)

Brennan, R., Balasubramanian, S., Norrie, D.H.: Dynamic control architecture for advanced manufacturing systems. In: Proc. of International Conference on Intelligent Systems for Advanced Manufacturing, Pittsburgh, PA, pp. 213–223 (1997)

Bruckner, S., Wyns, J., Peeters, P., Kollingbaum, M.: Designing agents for the manufacturing process control. In: Proc. of AI and Manufacturing Research Planning Workshop, Albuquerque, NM, pp. 40–46 (1998)

Budenske, J., Ahamad, A., Chartier, E.: Agent-based architecture for exchanging modeling data between applications. Working Notes of the ABM Workshop, Minneapolis, MN, pp. 18–27 (1998)

Burke, P., Prosser, P.: The distributed asynchronous scheduler. In: Zweben, M., Fox, M.S. (eds.) Intelligent Scheduling, pp. 309–339. Morgan Kaufman Publishers, San Francisco (1994)

Butler, J., Ohtsubo, H.: ADDYMS: Architecture for distributed dynamic manufacturing scheduling. In: Famili, A., Nau, D.S., Kim, S.H. (eds.) Artificial Intelligence Applications in Manufacturing, pp. 199–214. The AAAI Press (1992)

Chan, Y.S., Lee, J.K.: Case-based modification for optimization agents: AGENT-OPT'. DSS 36(4), 355–370 (2004)
Choi, S.P.M., Liu, J., Chan, S.P.: A genetic agent-based negotiation system. Computer Networks 37, 195–204 (2001)
Colombo, A., Schoop, R., Neubert, R.: An agent-based intelligent control platform for industrial holonic manufacturing systems. IEEE Transactions on Industrial Electronics 53(1), 322–337 (2006)
Cong, J., Fan, Y., Han, G., Jiang, W., Zhang, Z.: Behavior and communication co-optimization for systems with sequential communication media. In: 43rd ACM/IEEEDesign Automation Conference, pp. 675–678 (2006)
Cost, R.S., Finin, T., Labrou, Y., Luan, X., Peng, Y., Soboroff, I., Mayfield, J., Boughannam, A.: An agent-based infrastructure for enterprise integration. In: Proc. of ASA/MA 1999, Palm Springs, CA, pp. 219–233 (1999)
Dorigo, M., Maniezzo, V., Colorni, A.: Ant System: Optimization by a Colony of Cooperating Agents. IEEE Transactions on Systems, Man, and Cybernetics-Part B: Cybernetics 26(1), 29–41 (1996)
Dorigo, M., Gambardella, L.M.: Ant colonies for the travelling salesman problem. Bio Systems 43, 73–81 (1997)
Elliman, D.G., Youssef, S.M.: A New Intelligent Agent-based Strategy for Constrained Multiple Destination Routing Problems. Computer Journal 47(6), 708–727 (2004)
Esche, E., Müller, D., Kraus, R., Wozny, G.: Systematic approaches for model derivation for optimization purposes. Chemical Engineering Science (2013), http://dx.doi.org/10.1016/j.ces.2013.11
Fischer, T., Gehring, H.: Planning vehicle transhipment in a seaport automobile terminal using a multi-agent system. European Journal of Operational Research 166(3), 726–740 (2005)
Floreano, D., Husbands, P., Nolfi, S.: Evolutionary Robotics. In: Siciliano, B., Khatib, O. (eds.) Springer Handbook of Robotics, vol. 61, pp. 1423–1451 (2008)
Fordyce, K., Sullivan, G.G.: Logistics management system (LMS): integrating decision technologies for dispatch scheduling in semiconductor manufacturing. In: Zweben, M., Fox, M.S. (eds.) Intelligent Scheduling, pp. 473–516. Morgan Kaufman Publishers, San Francisco (1994)
Fox, M.S.: Issues in Enterprise Modelling. In: Nof, S.Y. (ed.) Information and Collaboration Models of Integration, pp. 219–234. Kluwer Academic Publishers (1993)
Fox, M.S., Chionglo, J.F., Barbuceanu, M.: The integrated supply chain management system. Internal Report, Dept. of Industrial Engineering, Univ. of Toronto (1993)
Frey, D., Stockheim, T., Woelk, P.-O., Zimmermann, R.: Integrated multi-agent-based supply chain management. In: Proc. of 2003 WET ICE, Linz, Austria, pp. 24–29 (2003)
Giret, A., Botti, V.: Holons and Agents. Journal of Intelligent Manufacturing 15, 645–659 (2004)
Goldsmith, S.Y., Interrante, L.D.: An autonomous manufacturing collective for job shop scheduling. In: Proc. of AI & Manufacturing Research Planning Workshop, Albuquerque, NM, pp. 69–74 (1998)
Gupta, A., Whitman, L., Agarwal, K.: Supply chain agent decision aid system (SCADAS). In: Proc. of the 2001 Winter Simulation Conference, Arlington, VA, pp. 553–559 (2001)
Hadeli, Valckenaers, P., Kollingbaum, M., Van Brusse, H.: Multi-agent coordination and control using stigmergy. Computers in Industry 53, 75–96 (2004)

Hayden, M., van Renesse, R.: Optimizing layered communication protocols. In: Proceedings of the Sixth IEEE International Symposium on High Performance Distributed Computing, pp. 169–177 (1997)

Hayes, C.C.: MAPP: An agent organization for process planning. Working Notes of the ABM Workshop, Minneapolis, MN, pp. 32–40 (1998)

Ho, Y.C., Moodie, C.L.: Solving cell formation problems in a manufacturing environment with flexible processing and routing capabilities. International Journal of Production Research 34(10), 2901–2923 (1996)

Holl, S., Zimmermann, O., Palmblad, M., Mohammed, Y., Hofmann-Apitius, M.: A new optimization phase for scientific workflow management systems (2013), http://dx.doi.org/10.1016/j.future.2013.09.005

Hsieh, F.S., Chiang, C.Y.: Collaborative composition of processes in holonic manufacturing systems. Computers in Industry 62(1), 51–64 (2011)

Huang, C.Y., Chen, W.L., Huang, W.J.: Communication Protocols for Collaborative Production Planning. In: The 11th Asia Pacific Industrial Engineering and Management Systems Conference and the 14th Asia Pacific Regional Meeting of International Foundation for Production Research Melaka, December 7-10 (2010)

Huang, C.Y., Nof, S.Y.: Formation of Autonomous Agent Networks for Manufacturing Systems. International Journal of Production Research 38(3), 607–624 (2000a)

Huang, C.Y., Nof, S.Y.: Autonomy and Viability – Measures for Agent-Based Manufacturing Systems. International Journal of Production Research 38(17), 4129–4148 (2000b)

Jennings, N.R., Sycara, K., Wooldridge, M.: A Roadmap of Agent Research and Development. Autonomous Agents and Multi-Agent Systems 1, 275–306 (1998)

Jeong, W., Nof, S.Y.: Performance evaluation of wireless sensor network protocols for industrial applications. Journal of Intelligent Manufacturing 19, 335–345 (2008)

Jha, K.N., Morris, A., Mytych, E., Spering, J.: MADEsmart: Agents for design, analysis, and manufacturbility. Working Notes of the ABM Workshop, Minneapolis, MN, pp. 57–63 (1998)

Kaihara, T.: Multi-agent based supply chain modelling with dynamic environment. International Journal of Production Economics 85(2), 263–269 (2003)

Karageorgos, A., Mehandjiev, N., Weichhart, G., Hammerle, A.: Agent-based optimisation of logistics and production planning. EAAI 16(4), 335–348 (2003)

Kimbrough, S.O., Wu, D.J., Zhong, F.: Computers play the beer game: can artificial agents manage supply chains? Decision Support Systems 33, 323–333 (2002)

Koestler, A.: The Ghost in the Machine. Hutchinson (Penguin Group), London (1967) ISBN 0-14-019192-5 (1990 reprint Ed.)

Leitão, P.: Agent-based distributed manufacturing control: A state-of-the-art survey. Engineering Applications of Artificial Intelligence 22(7), 979–991 (2009)

Leue, S., Oechslin, P.: Optimization Techniques for Parallel Protocol Implementation. In: Proceedings of the Fourth Workshop on Future Trends of Distributed Computing Systems, pp. 387–393 (1993)

Leung, C.W., Wong, T.N., Mak, K.L., Fung, R.Y.K.: Integrated process planning and scheduling by an agent-based ant colony optimization. Computers & Industrial Engineering 59, 166–180 (2010)

Lin, G.Y.-J., Solberg, J.J.: Integrated shop floor control using autonomous agents. IIE Transactions: Design and Manufacturing 24(3), 57–71 (1992)

Lin, F., Lin, S.: Enhancing the Supply Chain Performance by Integrating Simulated and Physical Agents into Organizational Information Systems. In: Proceedings of the Fourth Workshop on Knowledge Economy and Electronic Commerce, Taiwan (2006)

Lindstrom, P., Isenburg, M.: Fast and Efficient Compression of Floating-Point Data. IEEE Transactions on Visualization and Computer Graphics 12(5) (2006)

Liu, Y., Nof, S.Y.: Distributed micro flow sensor arrays and networks: Design architectures and communication protocols. International Journal of Production Research 42(15), 3101–3115 (2004)

Marik, V., McFarlane, D.: Industrial adoption of agent-based technologies. IEEE Intelligent Systems 20(1), 27–35 (2005)

Marler, R.T., Arora, J.S.: Survey of multi-objective optimization methods for engineering. Structural and Multidisciplinary Optimization 26, 369–395 (2004)

Maturana, F., Shen, W., Norrie, D.H.: MetaMorph: an adaptive agent-based architecture for intelligent manufacturing. International Journal of Production Research 37(10), 2159–2174 (1999)

McEleney, B., O'Hare, G.M.P., Sampson, J.: An agent-based system for reducing changeover delays in a job-shop factory environment. In: Proc. of PAAM 1998, London, UK, pp. 591–613 (1998)

Mehra, A., Nissen, M.: Intelligent supply chain agents using ADE. In: Proc. of AI & Manufacturing Research Planning Workshop, Albuquerque, NM, pp. 112–119 (1998)

Miyashita, K.: CAMPS: A constraint-based architecture for multi agent planning and scheduling. Journal of Intelligent Manufacturing 9(2), 147–154 (1998)

Mönch, L., Stehli, M., Zimmermann, J.: FABMAS: An agent-based system for production control of semiconductor manufacturing processes. In: Mařík, V., McFarlane, D.C., Valckenaers, P. (eds.) HoloMAS 2003. LNCS (LNAI), vol. 2744, pp. 258–267. Springer, Heidelberg (2003)

Narzisi, G., Mysore, V., Mishra, B.: Multi-Objective Evolutionary Optimization of Agent-Based Models: An Application to Emergency Response Planning. In: Proceedings of Computational Intelligence Conference, USA (2006)

NIST, Advanced Technology Program (1998), http://www.atp.nist.gov/

Nof, S.Y.: Intelligent, Collaborative Agents. In: Yearbook 2000, McGraw-Hill Encyclopedia of Science and Technology (2000)

Pan, J.Y.C., Tenenbaum, M.J.: An intelligent agent framework for enterprise integration. IEEE TSMC 21(6), 1391–1408 (1991)

Parunak, H.V.D.: MASCOT: A virtual factory for research and development in manufacturing scheduling and control. Technical Memo 93-02, Industrial Technology Institute (1993)

Parunak, H.V.D., Baker, A., Clark, S.: The AARIA agent architecture: From manufacturing requirements to agent-based system design. Working Notes of the ABM Workshop, Minneapolis, MN, pp. 136–145 (1998)

Pechoucek, M., Marik, V., Stepankova, O.: Coalition formation in manufacturing multi-agent systems. In: Proc. of DEXA 2000, London, UK, pp. 241–246 (2000)

Peng, Y., Finin, T., Labrou, Y., Chu, B., Long, J., Tolone, W.J., Boughannam, A.: A multi-agent system for enterprise integration. In: Proc. of PAAM 1998, London, UK, pp. 155–169 (1998)

Peralta, J., Annusornnitisarn, P., Nof, S.Y.: Analysis of a time-out protocol and its applications in a single server environment. International Journal of Computer Integrated Manufacturing 16(1), 1–13 (2003)

Putnik, G., Sluga, A., ElMaraghy, H., Teti, R., Koren, Y., Tolio, T., Hon, B.: Scalability in manufacturing systems design and operation: State-of-the-art and future developments roadmap. CIRP Annals - Manufacturing Technology 62, 751–774 (2013)

Rabelo, R.J.: Interoperating standards in multi-agent agile manufacturing scheduling systems. IJCAT 18(1-4), 146–159 (2003)

Reaidy, J., Massotte, P., Diep, D.: Comparison of negotiation protocols in dynamic agent-based manufacturing systems. International Journal of Production Economics 99(1-2), 117–130 (2006)

Rosenschein, J.S., Ephrati, E.: New approaches to multi-agent planning. In: Nof, S.Y. (ed.) Information and Collaboration Models of Integration, pp. 340–364. Kluwer Academic Publishers (1993)

Ross, A., Rhodes, D., Hastings, D.: Defining Changeability: Reconciling Flexibility, Adaptability, Scalability, Modifiability, and Robustness for Maintaining System Lifecycle Value. Systems Engineering 11(3), 246–262 (2008)

Sacile, R., Paolucci, M., Boccalatte, A.: The MAKE-IT: Manufacturing agents in a knowledge-based environment driven by internet technologies. In: Proc. of the 2000 Academia/Industry Working Conference on Research Challenges, Buffalo, NY, pp. 281–291 (2000)

Sadeh, N., Hildum, D.W., Kjenstad, D.: Agent-based e-supply chain decision support. Journal of Organizational Computing and Electronic Commerce 33(3-4), 225–241 (2003)

Shen, W., Maturana, F., Norrie, D.H.: MetaMorph II: an agent-based architecture for distributed intelligent design and manufacturing. Journal of Intelligent Manufacturing 11(3), 237–251 (2000)

Sycara, K.P., Roth, S.F., Sadeh, N., Fox, M.S.: Resource allocation in distributed factory scheduling. IEEE Expert 6(1), 29–40 (1991)

Trentesaux, D.: Distributed control of production systems. Engineering Applications of Artificial Intelligence 22, 971–978 (2009)

Van Dyke Parunak, H.: "Go to the ant": Engineering principles from natural multi-agent systems. Annals of Operations Research 75, 69–101 (1997)

Van Leeuwen, E.H., Norrie, D.H.: Intelligent manufacturing: holons and holarchies. Manufacturing Engineer 76(2), 86–88 (1997)

Velásquez, J.D., Nof, S.Y.: A best-matching protocol for collaborative e-Work and e-Manufacturing. International Journal of Computer Integrated Manufacturing 21(8), 943–956 (2008)

Vrba, P., Tichy, P., Marık, V., Hall, K.H., Staron, R.J., Maturana, F.P., Kadera, P.: Rockwell Automation's Holonic and Multi-agent Control Systems Compendium. IEEE Transactions on Systems, Man, and Cybernetics—Part C: Applications and Reviews 41(1), 14–30 (2011)

Wang, Y., Xiao, T., Duan, G., Wang, X.: Research on CAPP/scheduling integration multi-agent system model and implementation. CJME 16(4), 348–351 (2003)

Wiendahl, H.-P., ElMaraghy, H., Nyhuis, P., Zah, M.F., Wiendahl, H.-H., Duffie, N., Brieke, M.: Changeable Manufacturing – Classification, Design and Operation. CIRP Annals Manufacturing Technology 56(2), 783–809 (2007)

Williams, N.P., Liu, Y., Nof, S.Y.: The TestLAN approach and protocols for the integration of distributed assembly and test networks. International Journal of Production Research 40(17), 4505–4522 (2002)

Witten, I.H., Moffat, A., Bell, T.C.: Managing Gigabytes: Compressing and Indexing documents and images, 2nd edn., 519 p. Morgan Kaufmann (1999)

Wooldridge, M., Jennings, N.: Intelligent agents: Theory and practice. Knowledge Engineering Review 10(2), 115–152 (1995)

Yen, B.P.C., Wu, O.Q.: Internet scheduling environment with market driven agents. IEEE TSMC-A 34(2), 281–289 (2003)

Yoon, S.Y., Nof, S.Y.: Demand and capacity sharing decisions and protocols in a collaborative network of enterprises. Decision Support Systems 49(4), 442–450 (2010)

Yu, C.Y., Huang, H.P.: Development of the order fulfillment process in the foundry fab by applying distributed multi-agents on a generic message-passing platform. IEEE/ASME Transactions on Mechatronics 6(4), 387–398 (2001)

Yu, R., Iung, B., Panetto, H.: A multi-agent-based E-maintenance system with case-based reasoning decision support. EAAI 16, 321–333 (2003)

Review and Challenge Questions

1. After revisiting Figure 1.1 (in Chapter 1) explain (i) several roles autonomous agents have to play; (ii) how they would enable effective collaboration; (iii) illustrate your answers for one of the e-Work examples.
2. Consider a computer aided design system used by several distributed designers. Explain the types of agents that would be necessary to optimize the collaborative design process. (Hint: Apply resource agents and task agents, and any other types of agents; differentiate between autonomous and moderated approaches.)
3. Following Question 2, describe some of the decisions the agents of each type would need to make for: (i) Scheduling; (ii) Assigning; (iii) Preempting; (iv) Assuring timely completion; (v) Allocating; (vi) Resolving conflicts; (vii) Negotiating.
4. (i) Based on Tables 4.1, 4.2, 4.3, and 4.4, explain some of the similarities and differences in the roles of agents in each table. (ii) Justify what characteristics of autonomous agents make them relatively more appropriate for e-Work optimization. (iii) Discuss some other real-life applications of agent-based optimization in e-Manufacturing, e-Business, and e-Service.
5. Using the example of intelligent negotiation system, explain several reasons why in this case collaboration planning and control must be accomplished by multiple agents, and not by (i) an algorithm such as linear or dynamic programming; (ii) computer simulation.
6. (i) Differentiate between standard communication protocols, and collaboration specific task administration protocols. (ii) Briefly explain the unique features and objectives of each protocol optimization method discussed in Section 4.5.1.
7. Explain briefly about protocols: (i) Their common purpose; (ii) The role of layers in protocol design; (iii) The role of pipelining; (iv) The role of compression; (iv) The role of parallelization; (v) Any other features by which they contribute to collaboration engineering.
8. Explain and compare the functions of: (i) The 7-layer communication protocol design vs. the 3-later and 5-layer Internet protocol design; (ii) IP, TCP, HTTP.

9. What is the relation between agents and protocols in the context of optimizing collaborative e-Work? Illustrate for the case of (i) a supply network; (ii) multi-robot team; (iii) evolutionary robots.
10. Select two of the typical coordination and collaboration protocols listed in this chapter, and (i) explain their specific purpose and role; (ii) describe their main optimization objectives; (iii) compare and contrast between them.
11. What would a bio-inspired protocol design mean to you? How would it address some of the challenges described in this chapter?
12. (i) How does a workflow management system improve (a) the e-Work? (b) the e-Business? (c) e-Service? (ii) Discuss the shortcomings of traditional optimization methods in the design and organization of (a) e-Work, and (b) e-Business, and (c) e-Service solutions.
13. (i) With the workflow optimization process being iterative by nature, explain the optimization drivers that may guide the necessary continuous improvement. (ii) Consider the practice of optimizing dynamic systems with a static perspective, and discuss the effects it may have on the actual outcome of system performance.

Lab Questions

14. Construct a simple task administration protocol describing all the decisions needed to (i) issue a purchase order to a supplier; (ii) communicate a change in this purchase order; (iii) negotiate the price for this purchase; (iv) authorize payment for this purchase; (v) cancel this purchase after payment.
15. What would be changed in the protocol of Question 14 when it is designed to be (i) adaptive; (ii) learning?
16. For the e-Work lab system you selected (in Question 1.13) explain: (i) How can bio-inspired optimization and control improve its design and operation? (ii) What is the value, or potential value of GPS to the functions of your e-Work lab system?
17. Visit the website of an e-Service. Describe the role of agents and protocols in this e-Service.
18. Choose one of the three examples discussed in Section 4.4 a. (Swarm Intelligence: Ant Systems), and then: (i) Generate a simple numerical example; (ii) Develop a simple ant algorithm; (iii) Choose a proper e-Criterion (or a set of e-Criteria); (iv) Solve the algorithm and compare the performances obtained from the swarm with an equivalent random behavior. (v) Briefly explain the impact of swarm intelligence on the selected e-Criterion (or a set of e-Criteria).
19. Choose one of the three examples discussed in Section 4.4 b. (Evolution: Natural Selection), and then: (i) Suggest a proper encoding scheme; (ii) Develop proper crossover and mutation operators; (iii) Select a proper e-Criterion (or a set of e-Criteria); (iv) Develop and solve a genetic algorithm, and indicate its convergence. (v) Briefly explain how the evolutionary behavior of the selected system is handled by the developed genetic algorithm.

Chapter 5
Tools for e-Work

Abstract. The transformation from classic monolithic systems to distributed, more intelligent and autonomous e-Work networks has resulted in the development of many tools for supporting e-Systems over the last decade. In Chapter 4, we discussed agents and agent-based systems, as a basic cornerstone of collaborative e-Work along with the principal techniques for the optimal design and control of e-Systems based on them. This chapter introduces decision support systems (DSS) and communication protocols, as the other essential supporting tools for collaborative e-Work, along with more details on various types of software agents (Fig. 5.1). The human-/computer-integrated tools, i.e., computer integrated manufacturing (CIM) and human-computer interaction (HCI), will be discussed in the next chapters. Decision support systems refer to an interactive computerized system that helps people make decisions based on information that is culled from a wide range of sources. Protocols provide an efficient and effective platform for autonomous entities to interact and be productive. Software agents enable the automation and integration of flexible, adaptive and heterogeneous activities, components, and tasks in a collaborative network.

Fig. 5.1 e-Work support tools.

S.Y. Nof et al., *Revolutionizing Collaboration through e-Work, e-Business, and e-Service*, Automation, Collaboration, & E-Services 2, DOI: 10.1007/978-3-662-45777-1_5

5.1 Decision Support Systems

DSS is generally designed as an "ancillary" or "auxiliary" interactive computer-based system that enables the use of computer communications, data, documents, knowledge, and models to identify and solve problems, complete decision process tasks, and make decisions. The terms ancillary or auxiliary have an important implication; A DSS is not intended to replace skilled decision-makers but to support them and augment their skills. And it does not automate decision-making. A DSS serves the management, operations, and planning levels of an organization, helping make decisions, which may be rapidly changing and not easily specified in advance. Some DSSs provide structured information directly to the managers. Other DSSs help managers and staff specialists analyze situations using various types of quantitative and informatics models. Some DSSs store knowledge and make it available to the managers, and some DSSs support decisions being made by small and large groups. Companies even develop DSS to support the decision-making of their customers and suppliers (Power and Sharda, 2009).

A primary element of any DSS is the knowledge-based system. Indeed, a properly designed DSS is an interactive software-based system that helps decision-makers to compile useful information from a combination and mining of raw data, documents, personal knowledge, or business models so as to recognize, identify and solve problems, and make decisions. The common types of information that an e-Business DSS application may involve are:

- Inventories of information assets (e.g., legacy and relational data sources, data cubes, data warehouses, and data marts).
- Comparative sales figures in different periods.
- Projected revenue figures based on product sales assumptions.

Naturally, the type and definition of information may vary in specific e-Business applications and also in e-Service (e.g., DSS in e-Healthcare, e-Banking, e-Exploration, e-Cleaning). Recently, a number of academic disciplines provide the substantive foundations for decision support systems development and research:

- *Database research* has contributed tools and research on managing and mining data and documents.
- *Management science* and *operations research* have developed mathematical models for use in building model-driven DSS and provided evidence on the advantages of modeling in problem solving.
- *Cognitive science*, especially *behavioral decision-making research*, has provided descriptive and empirical information that has assisted in DSS design and has generated hypotheses for decision support research.

Some other important fields related to DSS include *artificial intelligence*, *HCI*, *software engineering*, and *telecommunications*. In fact, evolution of the Internet and information and communication technologies (ICT) has expedited the advancements in DSS and has made it difficult to keep up with the rapid changes in DSS capabilities. Yet, clearly DSSs are fundamental to collaboration engineering, augmentation and optimization.

- Before looking at the details and various structures of DSSs, a brief history of the DSS, its advent and evolution seems to be worthwhile to discuss. The first DSS was born through a pioneering study in 1960's on the development of *interactive systems*; theory development by Simon and colleagues (Simon, 1960), and Morton's (Morton, 1971) research created a foundation for building innovative DSS. In the early 1970s, BRANAID (Little, 1975) and reports of Morton's (Morton, 1971) management decision system research demonstrated the feasibility of DSS. One of the co-authors of this book pioneered in his doctoral dissertation (Nof, 1976) the integration of database and simulation in an original decision-support system for concurrent facility planning and design (Nof, 1980) later enhanced with a knowledge-based approach for integration (Nof, 1984).

By the late 1970s, a number of companies developed interactive information systems that used data and models to help managers analyze *semi-structured* problems. These diverse systems were all called DSSs. The key conceptual books on the evolving DSS include, for instance, those by Alter (1980), Sprague and Carlson (1982), and Bonczek et al. (1981). A recent example (Seok et al., 2012) describes the design of a DSS based on the Collaborative Control Theory which is explained in this book.

DSS, from the beginning, was recognized as a tool that could be designed to support decision-makers at any level in an organization. DSS could support operations, financial management, and strategic decision-making. Over the years, many of the more interesting DSS have been targeted for middle and senior managers. DSS are also often designed for specific types of systems and e-Systems, such as hospitals, banks, and insurance companies. Those specialized DSSs are sometimes referred to as vertical-market or industry-specific DSS. Today, DSSs can be both off-the-shelf, packaged applications and custom-designed systems. DSS may support a decision-maker using a single personal computer or a large group of decision-makers in a networked client-server environment. The latter systems are often called *enterprise-wide DSS*.

5.1.1 Characteristics of DSS

The term DSS has various connotations. However, all types and applications of DSS have three common characteristics (Alter, 1980): (1) specifically facilitating decision processes; (2) supporting rather than automating decision-making; (3) responding quickly to the dynamic needs of decision-makers. From a more in-depth outlook, the characteristics of DSS can be viewed as *knowledge repository*, *requests*, *operations*, *presentation*, and *coordination* (Table 5.1) to support decision-making communications, knowledge, and processes. Many terms have been used interchangeably for specific types of DSS, e.g., *business intelligence* (BI), *collaborative systems*, *data warehousing*, *model-based systems*, *knowledge management*, *online analytical processing*, and so on. Software vendors use these further specified terms for both descriptive and marketing purposes. However, the terminology used for a system or software package is a secondary concern. Our primary concern is indeed to find software and support systems that meet the decision-maker's requirements and provide appropriate information accordingly.

Table 5.1 Characteristic of DSS and their role in augmenting collaboration (adapted from Holsapple and Whinston, 1996).

Type	Functions
Knowledge repository	★ Identify and solve problems ❖ Facilitate interactions among decision-makers ➢ Define, document, and regulate the actions of decision-makers ★ ❖ ➢ Private repositories under the access control of individuals ★ ❖ ➢ Public repositories; shared access and share knowledge
Requests	❖ Customize requests based on specific needs ★ Enhance flexibility in timing of requests
Operations	★ Provide knowledge to meet unanticipated demands ★ Generate knowledge via automated calculation/analysis/reasoning
Presentation	❖ Customized presentation of results based on specific needs
Coordination	❖ Facilitate internal/external communication among decision-makers ➢ Structure and regulate individual/group decision-making tasks ➢ Structure and regulate interrelated decisions

❖ Characteristics to support communication among decision-makers
★ Characteristics to support knowledge for decision-making
➢ Characteristics to support decision-making processes

5.2 Decision Support Information System

Decision-makers need to consider and define the information and analyses needed to support their decisions. Computerized information systems provide transaction summary information and can help decision-makers understand different operations and performance issues. For example, a computerized system can enhance e-Business decisions through helping managers understand the status of operations, monitor business results, review customer preference data, and investigate competitor actions. An inevitable requirement in all these situations is information, which indeed should comprise several key characteristics. Information must be *timely* and *current*, i.e., up to date and always available to the decision-maker. It also needs to be *accurate*, *relevant*, and *complete*, and in a format *compatible* to the requirements of the decision-making process. (It should be noted that techniques to support decisions even when information is incomplete have also been developed.)

Moreover, information must be *summarized* and *concise*, and any DSS must provide an option for the decision-maker to access more *detailed information*. Figure 5.2 shows a schematic view of a decision support information system used for assessment of urban thermal landscape characteristics based on remote sensing data.

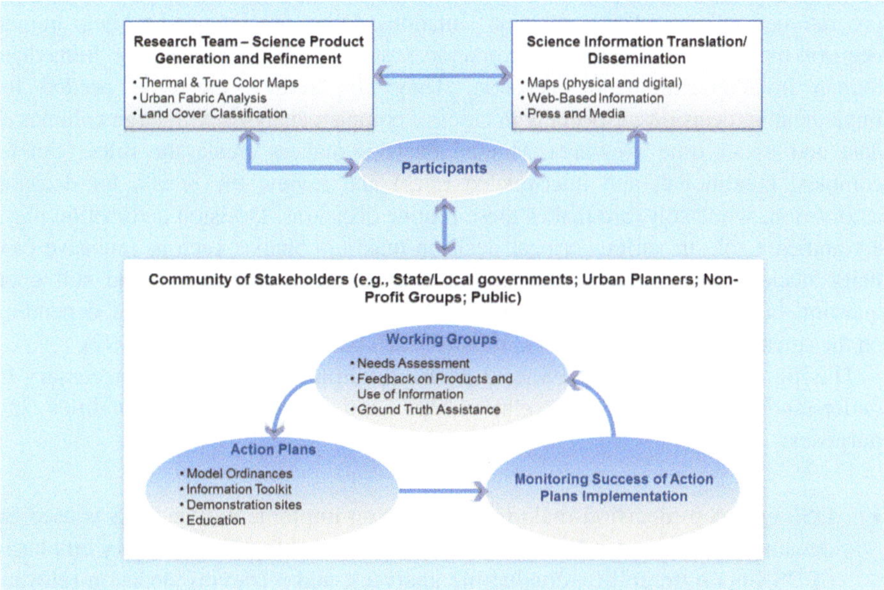

Fig. 5.2 A decision support information system used to convey the required information (based on thermal infrared data) to geographically distributed NASA research teams, coordinators, stakeholders, and policy-makers for implementation of action plans for an urban landscape management system (adapted from Quattrochi et al., 2000).

A DSS must provide information in a proper format, easy to understand and manipulate by human user. The source of the information presented by a DSS may be the analysis of transaction data, a decision model, or external sources. DSS can present internal and external facts, informed opinions, and forecasts generated for decision-makers. A serious challenge, however, is to meet all the mentioned requirements of a decision support information system. Advances in technology have enabled new decision support capabilities; however, much learning and discussion need to occur to successfully exploit the technological possibilities. Decision support systems differ in many ways from operating systems that process business transactions. A common example is enterprise resource planning (ERP). ERP is not a DSS although the objective of both is to support, provide suggestions, and improve decision-making processes. ERP is indeed an integrated *transaction processing system* (TPS) that facilitates the flow of information through all functional areas of e-Business. DSS then helps decision-makers analyze the data gathered throughout ERP systems. On the other hand, ERP systems facilitate the development of a wide variety of DSS functions.

As mentioned earlier, DSS is not intended to automate decision-making. However, it does not designed to overlook the related automation technologies. *Decision automation* refers to the use of information technologies to improve decisions and implement programmed decision processes. Decision automation is generally considered most appropriate for well-structured, clearly defined, routine or programmed (Simon, 1960) decision situations. DSSs are intended to help human decision-makers while decision automation systems reduce the need for immediate human involvement in the decision. They also reduce the time needed for implementing decisions, especially in cases of complex analyses with large volumes of data and under time pressures. Human decision-makers create the rules (can be complex, conditional, and inter-related rules) and review the results for decision automation, while software makes these routine decisions. Decision automation plays a significant role in various critical decision-making contexts such as intensive care units, nuclear power plants, and aircraft cockpits. The challenging and still open question, however, is "does automation bias decision-making?"; and if yes, depending on the situation, "how to handle and mitigate such biases" (Skitka et al., 1999).

Having discussed DSS, TPS, and decision automation, it is next necessary to understand their differences, relationships, and respective functionalities and purposes:

- DSS enhances decision-making and decision implementation. DSS is used on demand when it is needed to support decision-making. The primary emphasis of DSS is on flexibility, conducting analyses, and retrieving decision-relevant information and knowledge.
- TPS is designed to expedite and automate transaction processing, record keeping, and simple business reporting of transactions. TPS is related to the design of DSS because transaction databases often provide data for decision-oriented reporting systems and data warehouses. TPS typically provides standard reports on a periodic basis and supports the operations. TPS records current information and maintains a database of transaction information. TPS emphasizes data integrity and consistency.
- Decision automation uses software to make decisions. Decision automation often interacts with TPS. On the other hand, the results of automated decisions lead to some changes in the TPS. Decision automation can routinize specific decisions 24/7.

Decision-makers typically initiate each instance of DSS use, either by using the DSS themselves, or by requesting an agent intermediary to use the DSS. DSS generally utilizes internal and external historical data for analysis, which, in turn, may be applied for quantitative and qualitative analyses, and modeling current/future scenarios.

For example, many manufacturing systems use manufacturing resource planning (MRP-II) software. MRP-II is a specific operational-level DSS that supports master production scheduling, purchasing, and materials requirements planning (MRP-I). More recent MRP-II systems support what-if analysis and simulation capabilities. DSS supports quality improvement and controls decisions. Monsanto, FedEx, and most transportation companies use DSS for scheduling trucks, airplanes, and ships. The Coast Guard uses a DSS for procurement decisions. Companies such as Wal-Mart have large data warehouses and use data-mining software. BI and knowledge management systems (KMS) are increasingly common. On the World Wide Web, one can find DSS as part of e-Services that helps track and manage stock portfolios, choose stocks, plan trips, and suggest books, movies, and gifts. DSS supports distributed decision activities using *groupware* and *corporate Intranet.*

5.2.1 Categories of DSS

The most common categories of DSSs (Power, 2001) are listed below. Some DSSs are integrated or hybrid systems with more than one major DSS subsystem. The focus is on one major capability dimension with five categories and three secondary dimensions. The term "driver" is used as a common or shared descriptive adjective, which refers to the capability, tool or component that is providing the dominant functionality to the DSS. Note that in all categories, the DSS entails collaboration through communication, sharing, interaction, and integration among diverse data resources, analysis tools, and the human parties. Note also that while most DSSs support collaboration and better quality and timeliness of decisions, collaboration engineering techniques are necessary to augment and optimize collaboration processes and their outcomes.

1. *Communications-Driven and Group DSS*: Group decision support systems (GDSS) and groupware emerged in the mid-1980's; however, nowadays a broader category of communications-driven DSS is emerging. This type of DSS includes communication, collaboration, and decision support. A group DSS is an interactive computer-based system intended to facilitate the solution of problems by decision-makers working together as a group; such systems often derive functionality from a model more than from supporting collaboration and hence are model-driven DSS. Groupware supports electronic communication, scheduling, document sharing, and other group productivity and decision support enhancing activities. The technologies in use are collaborative group DSS, two-way interactive video, white boards, bulletin boards, chat, and email systems.
2. *Data-Driven DSS*: emphasizes analysis of data. These systems include file drawer and management reporting systems, data warehousing and analysis systems, executive information systems (EIS), and data-driven spatial decision support systems (SDSS). EIS are targeted at senior managers and SDSS display spatial data for decision support. BI systems are also examples of data-driven DSS. A data-driven DSS provides access to and manipulation of large databases of structured data and especially a time series of internal company data and external data. Simple file systems accessed by query and retrieval tools provide

the most elementary level of functionality. Data warehouse systems that allow the manipulation of data by computerized tools tailored to a specific task and setting or by more general tools and operators provide additional functionality. Data-driven DSS with online analytical processing (OLAP) provides the highest level of functionality and decision support that is linked to analysis of large collections of historical data (Dhar and Stein, 1997).

3. *Document-Driven DSS*: as an evolving paradigm, helps decision-makers retrieve and control unstructured documents and web pages. A document-driven DSS integrates a variety of storage and processing technologies to provide complete document retrieval and analysis. The web provides access to large document databases including databases of hypertext documents, images, sounds, and video. Examples of documents that would be accessed by a document-driven DSS are policies and procedures, product specifications, catalogs, and corporate historical documents, including minutes of meetings, corporate records, and important correspondence. A search engine is a powerful decision-aiding tool associated with a document-driven DSS. Some authors call this type of system a knowledge management system.
4. *Knowledge-Driven DSS*: recommends actions to decision-makers. This type of DSS has specialized problem-solving expertise. The expertise consists of knowledge about a particular domain, understanding of problems within that domain, and skill at solving some of these problems.
5. *Model-Driven DSS*: includes systems that use accounting, financial, representational, and optimization models. Model-driven DSS emphasize access to and manipulation of a quantitative model. Simple statistical and analytical tools provide the most elementary level of functionality. Some OLAP systems that allow complex analysis of data may be classified as hybrid DSS systems providing modeling, data retrieval, and data summarization functionality. Model-driven DSS use data and parameters provided by decision-makers to aid them in analyzing a situation, but they are not usually data intensive. Very large databases are usually not needed for model-driven DSS, but data for a specific analysis may need to be extracted from a large database.

All five categories of DSS are (or can be) web-based, i.e., build upon the Internet. A web-based DSS is a computerized system that delivers decision support information/tools to the decision-maker using a thin-client web browser (e.g., Firefox or Internet Explorer). The computer server hosting the DSS application is linked to the user's computer by a network with the transmission control protocol/Internet protocol (TCP/IP). In many applications, a web-based DSS is synonymous with an intranet or enterprise-wide DSS. Web technologies are powerful tools for creating DSS, and especially inter-organizational DSS that support the decision-making of customers and suppliers. Web 2.0 technologies are the leading edge for building DSS, but some DSS will continue to be built using mainframe and client-server enabling technologies.

5.2.2 *Design of DSS*

Classic DSS architectures are typically composed of five key components: (1) model management server; (2) data server; (3) user interface; (4) client computer; (5) DSS architecture and network (Sprague and Carlson, 1982). These components collectively create the overall architecture of a DSS. This classic view is still valid due to identifying commonalties between different types of DSS and providing an initial perspective for understanding the complexity of DSS architectures. The extended DSS framework (Power, 2001) is based on the different emphases placed on DSS components when a specific system is actually constructed (Fig. 5.3).

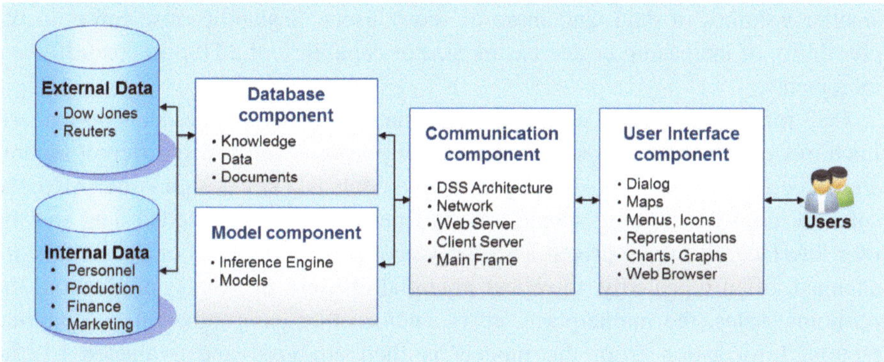

Fig. 5.3 Basic DSS components.

A major, and in many ways the most important, component in the architecture of a modern and collaborative DSS is the user interface. The tools commonly used for building DSS user interfaces include DSS generators, query and reporting tools, and front-end development packages. DSS user interfaces can be distributed to clients or delivered over a network using web pages or Java applets through *thick-client architectures*. Thin-client architectures provide many advantages where a user interacts through a web browser, but until recently the sophistication of the user interface was limited compared with a thick-client architecture where a program resides on a DSS user's computer.

A DSS database is a collection of data organized for easy access and analysis. Large databases in enterprise-wide DSS are often called *data warehouses* or *data marts*. Document or unstructured data is stored differently than structured data. Web servers provide a powerful platform for unstructured data and documents. The architecture for a data-driven DSS often involves databases on multiple servers, specialized hardware, and in some cases both multidimensional and relational database software. However, the extraction, transformation, loading, and indexing of structured DSS data is sometimes very difficult.

Quantitative models are an important part of DSS, especially model-driven DSS. Model management software can be centralized on a server with a database, or specific models can be distributed to client computers. Java applets and JavaScript programs provide a powerful new means to deliver models to users in *thin-client architecture*. The DSS network component refers to how hardware is organized, how software and data are distributed in the system, and how components of the DSS are integrated and physically connected. An ongoing issue is whether a specific DSS should only be available using thin-client technology on a company intranet or available on the global Internet. This should depend on the needs analysis and a feasibility study. Scalability is also an important DSS issue. Scalability refers to the ability to scale hardware and software to support larger or smaller volumes of data, and more or fewer users. Scalability also refers to the possibility of increasing or decreasing size or capability of a DSS in cost-effective increments.

DSS models should be used and manipulated directly by the decision-makers. Each model-driven DSS has a specific set of purposes and hence different models are needed and used. Choosing appropriate models is a key design issue. Also, the software used for creating specific models needs to manage needed data and the user interface. In model-driven DSS the values of key variables or parameters are changed, often repeatedly, to reflect potential changes in supply, production, the economy, sales, the marketplace, costs, and/or other environmental and internal factors. Information from the models is then analyzed and evaluated by the decision-maker. Knowledge-driven DSS use special models, an inference engine, for processing rules or identifying relationships in data.

Data-driven, document-driven, and knowledge-driven DSS require specialized database components. A model-driven DSS may use a simple flat-file database with fewer than 1000 records, but the model component is very important. Experiences and empirical evidences indicate that design and implementation issues vary for data-driven, document-driven, model-driven, and knowledge-driven DSS. Collaborative systems such as group and inter-organizational DSS create complex implementation issues. For instance, when implementing a data-driven DSS a designer should be especially concerned about the user's interest in applying the DSS in unanticipated or novel situations. Competition DSS and DSS for competitiveness have also been developed, e.g., in strategic planning, e.g., (Sauter, 2005); competition for resources, e.g., (van de Giesen et al., 2001); competitive intelligence, e.g., (Dai et al., 2011); and service enterprise competitions, e.g., (Xinping et al., 2010).

A data-driven DSS database is often a collection of current and historical structured data from a number of sources that have been organized for easy access and analysis. The data component is expanded to include unstructured documents in document-driven DSS and knowledge in the form of rules in knowledge-driven

DSS. Large databases of structured data in enterprise-wide DSS are often called data warehouses or data marts. DSS usually use data that has been extracted from all relevant internal and external databases. Managing information often means managing a database. Supporting management decision-making means that computerized tools are used to make sense of the structured data or documents in a database.

In creating an accounting or financial simulation model, for instance, a developer should attempt to verify that the initial input estimates for the model are thoughtful and reasonable. In developing a representational or optimization model, an analyst should be concerned about possible misunderstandings of what the model means and how it can or cannot be used (Alter, 1980). Networking issues create challenges for many types of DSS, but especially for communications-driven systems with many participants. Today, architecture and networking issues are increasingly important in building DSS.

When it comes to collaborative (group/multi-participant) decision-making, DSS architectures become even more complicated, since it has to handle some additional dimensions besides defining the domain of the problem: Definition of roles, interactions, and communication among multiple decision-makers. The communications component refers to how hardware is organized, how software and data are distributed in the system, and how components of the system are integrated and connected. A major issue today is whether DSSs should be available using a web browser on a company intranet and also available on the global Internet. Managers and information systems staff both need to develop an understanding of the technical issues and the security issues related to DSS architectures, networks, and the Internet. In this context, networking technology is the key driver.

A collaborative DSS involves three types of knowledge (Holsapple and Sena, 2003) (see also Table 5.1):

(1) *System knowledge*: corresponds to the distributed decision-making roles, relations, and regulations.
(2) *Domain knowledge*: is related to the subject of decision-making, and includes descriptive, procedural, and reasoning knowledge.
(3) *Relational knowledge*: characterizes the distributed decision-makers and relations between roles, and regulate decision-making activities.

Fig. 5.4 presents a generic DSS architecture for collaborative decision-making (for both private and public repositories). The procedure is composed of making requests to recall knowledge, acquiring knowledge, deriving knowledge, clarifying prior responses, accepting knowledge, routing messages, and providing help if required. The presented architecture is a basic architecture for ERP systems (Holsapple and Sena, 2003).

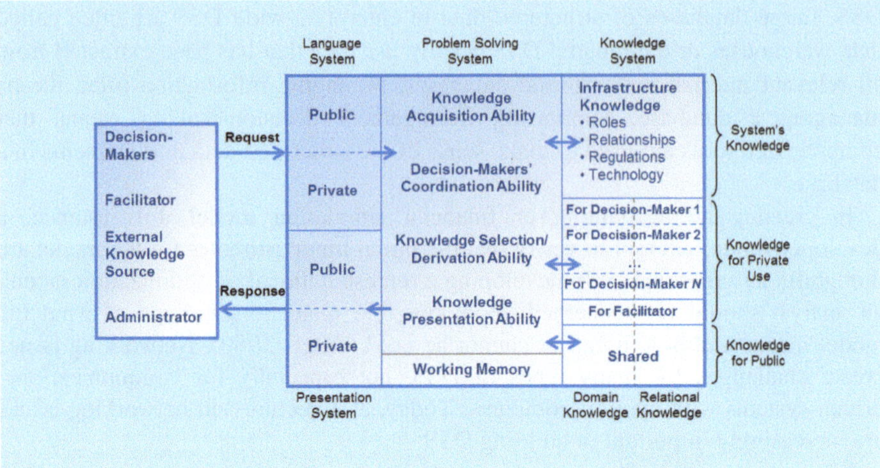

Fig. 5.4 A collaborative DSS architecture (adapted from Holsapple and Sena, 2003).

DSS should be built and implemented based on appropriate process. Many small, specialized model-driven DSS are built quickly. Large, enterprise-wide DSS are built through sophisticated tools and systematic and structured system analysis and development approaches. Communications-driven and group DSS are usually purchased as off-the-shelf software and then implemented. Creating enterprise-wide DSS environments remains an iterative and evolutionary task. An enterprise-wide DSS grows and inevitably becomes a major part of the overall information systems infrastructure of an organization. Despite the significant differences created by the specific task and scope of a DSS, all DSS have similar technical components and share a common purpose: supporting decision-making.

Computerized DSS assume that decision-makers need to be kept in the decision-making loop; but they need and want help to improve their decision-making. When it is possible and appropriate to automate decision-making using software, then we have this option to remove a human decision-maker from the decision-making process. Decision-makers choose to automate, create computerized support or make unaided decisions.

5.3 Communication Protocols

In order for autonomous agents to efficiently and effectively interact and collaborate within an e-Work network, a proper platform needs to be developed. In the previous chapters, several coordination and collaboration protocols with various functionalities were introduced. The role of those protocols (aka, Task Administration Protocols) is to allocate and administrate distributed tasks and resources according to the fundamental design principles of CCT, the Collaborative Control Theory. A practical concern, however, is how to enable distributed agents (e.g., sensors) synchronically interact and communicate to accomplish their tasks and achieve their goals. Such *telecommunicating* issues are targeted by *communication protocols*. In telecommunication domain, a communication protocol is a mechanism with certain digital rules that enables exchange of information between distributed agents within an e-Work network. In Chapter 4, several techniques were introduced for the optimal design of communication protocols. In this section, we review the main wired and wireless communication protocols, in more detail, which have a variety of applications mostly in industrial automation, e-Manufacturing, and shop floor control.

5.3.1 Industrial Communication Protocols

In e-Manufacturing and autonomous shop-floor control, digital communication is a well established concept. Proprietary communication systems within supervisory control and data acquisition (SCADA) systems have been supplemented, and partially displaced by *Fieldbus* and *sensor bus systems*. Fieldbus systems emerged in line with the evolution of e-Work; paradigms that deploy distributed, autonomous, and intelligent agents, emphasizing decentralized decision-making and control loops (e.g., COBS, the cooperative objects for autonomous shop-floor control). Wired Fieldbus systems are already standardized as the most important communication systems used in commercial control installations. Analogous to Fieldbus systems is the *Ethernet*, a *non-time-critical communication system* that is the most commonly used communication technology within the office domain with low component prices caused by mass production. This interesting property has led to an increasing interest in adapting Ethernet for industrial applications. Ethernet-based solutions are dominating as a merging technology.

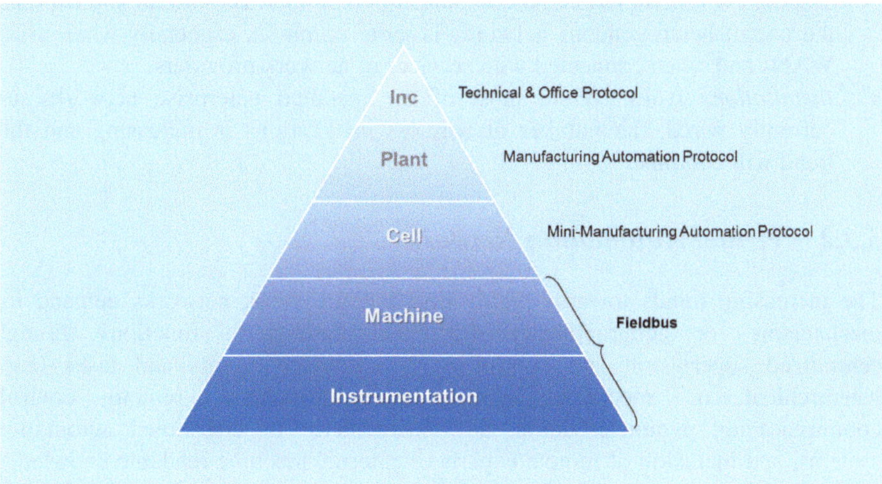

Fig. 5.5 e-Manufacturing communication architecture (adapted from Thomesse, 2005).

To further highlight the role of these industrial communication technologies, Fig 5.5 depicts the different layers of an e-Manufacturing system, from *technical and office protocol* (for communication between business and technical offices; e.g., Boeing) to *manufacturing automation protocol* (for communication between design offices and plants, or between machine tools and robots; e.g., General Motors) and at the lower hierarchy, the Fieldbus (Thomesse, 2005).

In line with the advances in Ethernet-based industrial protocols, the use of wireless technologies in the industrial domain has also been increasingly investigated. Following the notion of *merging automation and office networks*, heterogeneous networks (virtual automation networks (VAN)), consisting of local and wide area networks (WAN), as well as wired and wireless communication systems, are becoming important (Neumann, 2004). Industrial communication systems can be classified based on their capabilities as follows:

- *Real-time behavior*: Within the automation domain, real-time requirements are of uttermost importance. The focus is on the response time behavior of data packets. There is also a class non real-time, which means systems without real-time requirements.
- *Distribution*: The most important achievement of industrial communication systems are local area communication systems, consisting of sensor/actuator networks (see Chapter 8 for more details), Fieldbus systems, and Ethernet-based local area networks (LAN). Of increasing importance is the use of WAN (e.g., telecommunication networks, Internet). Thus, it should be advantageous to consider WANs as part of an industrial communication system, mostly within the upper layers of an enterprise hierarchy.
- *Homogeneity*: There are homogeneous parts (e.g. standardized Fieldbus systems) within an industrial communication system. But in real applications, the use of heterogeneous networks is more common, especially when using WANs and when connected with services of network providers.
- *Installations types*: While most of the installed enterprise networks are currently wired, the number of wireless installations is increasing and this trend will continue.

5.3.2 Virtual Automation Networks

The increasing trends towards highly distributed e-Work networks demand for mechanisms for geographically distributed automation functions through centralized supervision and control of decentralized agents and tasks (e.g., heterarchical or mediator agent-based architecture), remote control, commissioning, parameterization, and maintenance of distributed automation systems, and inclusion of remote experts or external machine-readable knowledge (e.g., for plant operation and maintenance, asset management, and condition monitoring).

Heterogeneous networks, consisting of (partially homogeneous) local and wide areas, as well as wired and wireless communication systems, will play a pivotal role in the near future. Following a unique design concept, regarding the objects to be transmitted between geographically distributed communication end points, a heterogeneous network becomes a VAN (Neumann et al., 2007; VAN, 2006). VAN characteristics are defined for domains where the expression domain is widely used to address areas and devices with common properties/behavior, common network technology, or common application purposes.

Within an automation and communication environment, a VAN domain covers all devices that are grouped together on a logical or virtual basis to represent a complex application (e.g., a plant). Therefore, the encompassed networks may be heterogeneous and devices can be geographically distributed over a physical environment, which shall be covered by the overall application. But all devices that have to exchange information within the scope of the application (equal to a VAN domain) must be VAN aware or VAN enabled devices. Otherwise, they are VAN independent and are not a member of a VAN domain.

Devices related to a VAN domain may reside in a homogeneous network. Depending on the application, however, additional VAN relevant devices may only be reached by crossing other network types (e.g., WAN type communication) or they need to use proxy technology to be represented in the VAN domain view of a complex application. A VAN network consists of various communication paths and network transitions. Figure 5.6 depicts the required transitions in heterogeneous networks.

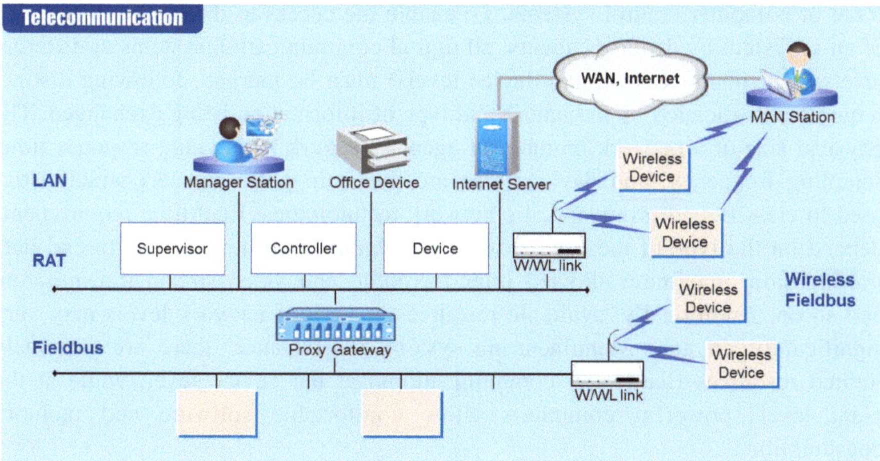

Fig. 5.6 Network transition (local area networks (LAN), wireless LAN (WL), wired/wireless (W/WL), real-time Ethernet (RTE), metropolitan area network (MAN), WAN).

Depending on the network and communication technology of the single path, there will be differences in the addressing concept of the connected network segments. Besides, communication paths have different communication line properties and capabilities. Therefore, for the path of two connected devices within a VAN domain the following views are possible:

- *Logical view*: Properties/capabilities of the whole communication path.
- *Physical view*: Detailed properties/capabilities of technology-dependent communication paths.
- *Behavioral view*: Various cyclic/acyclic temporal behavior of segments.

A communication path between network segments/devices (or their combinations) can be built in a variety of ways: Ethernet line (with/without transparent communication devices), wireless path, telecommunication network (1:1), public networks (n:m, provider-oriented), VPN tunnel, gateway (without application data mapping), proxy (with application data mapping), VAN access point, and IP mapping. All networks, which cannot be connected via an IP-based communication stack, must be connected using a proxy. For connecting non-nested/cascaded VAN sub-domains via public networks, VAN access point is preferred.

5.3.3 Wired Industrial Communication Networks

Over the last three decades, wired digital communication has been a key driving force of computer control systems. To enable the access to data in various layers of an e-System by different agents, all digital communication systems at different levels (e.g., plant, control, and device levels) must be merged, following distinct requirements dictated by the nature and type of information being exchanged. The physical size of a network, number of agents, network bandwidth, response time, sampling frequency, and payload size are the main performance characteristics used to classify and group specific network technologies. Real-time requirements depend on the type of messages to be exchanged: Deadlines for end-to-end data transmission, maximum allowed jitter for audio and video stream transmission, and so on. Additionally, available resources at various network levels may vary significantly. In an e-Manufacturing system, for instance, there are extremely limited resources (hardware, communications) at the device level, while at the plant level, powerful computers allow comfortable software and memory consumption.

Considering the aforementioned requirements, different types of industrial communication systems may be developed, as part of a hierarchical automation system, e.g.,

- *Sensor/actuator networks*: at the field (sensor/actuator) level (HART, 2009).
- *Fieldbus systems*: at the field level, collecting/distributing process data from/to sensors/actuators, communication medium between field devices and controllers/PLCs/management consoles (IEC 61158/61784-1, 2003). Fieldbus systems are standardized and widely used in industrial automation. The IEC 61158 and 61784 standards (IEC 61158/61784-1, 2003) contain ten different Fieldbus concepts. Seven of these concepts have their own complete protocol suite: *PROFIBUS* (Siemens, PROFIBUS International); *Interbus* (Phoenix Contact, Interbus Club); *Foundation Fieldbus H1* (Emerson, Fieldbus Foundation); *SwiftNet* (B. Crowder); *P-Net* (Process Data); *WorldFIP* (Schneider, WorldFIP). Three of these concepts are based on the Ethernet functionality; high speed Ethernet (HSE) (Emerson, Fieldbus Foundation).
- *Controller networks*: at the controller level, transmitting data between powerful field devices and controllers as well as between controllers. Such networks require powerful communication technology such as Ethernet technology.
- *WAN*: at the enterprise level, connecting networked segments of an enterprise automation system.

Vendors of industrial communication systems offer a set of fitting solutions for each level of the automation/communication hierarchy.

5.3.4 Wide Area Communication Networks

With the emergence of remote automation mechanisms (e.g., remote supervisory, operation, service) through WANs, the stock of existing communication technology has become broader including features such as:

- All appearances of the Internet (mostly supporting best effort quality of services).
- Public digital wired telecommunication systems: either line-switched [e.g. integrated services digital network (ISDN)] or packet-switched [such as asymmetric/symmetrical digital subscriber line (ADSL, SDSL)].
- Public digital wireless telecommunication systems (GSM-based, GPRS-based, UMTSbased).
- Private wireless telecommunication systems, e.g. trunk radio systems.

The transition between different network technologies is facilitated by multiprotocol label switching (MPLS) and synchronous digital hierarchy (SDH). There are several private protocols (over leased lines, tunneling mechanisms, etc.) that have been used in the automation domain using these technologies. Most of the wireless radio networks can be used in non-real-time applications, some of them in soft real-time applications; however industrial environments and industrial, scientific, and medical (ISM) band limit the applications.

The end-to-end connection behavior via these telecommunication systems depends on the quality of service (QoS). It strongly limits the use of these systems within the application domains. To apply a WAN paradigm to geographically distributed e-Systems, however, some basic design decisions must be made:

- A VAN is an infrastructure for standard LAN-based distributed industrial automation concepts (e.g. PROFINET or other) in an extended environment. The productive functions and applications are described by their object models used in existing industrial communications. The application service elements (ASEs), as they are specified in the IEC 61158 standard, can additionally be used.
- The establishment of the end-to-end connections between distributed objects within a heterogeneous e-Work network is based on web services. Once this connection has been established, the runtime channel between these objects is equivalent to the runtime channel within the local area by PROFINET (or similar) runtime mechanisms.
- The VAN addressing scheme is based on names to avoid the use of IP and MAC addresses during establishment of the end-to-end path between logically connected applications within a VAN domain. Therefore, the IP and MAC addresses remain transparent to the connected application objects.
- Since there is no new Fieldbus or real-time Ethernet protocols, no new specified application layer is necessary. Thus, the well-tried models of industrial communications (as they are specified in the IEC 61158 standard) can be used. Only the additional requirements caused by the influence of WANs have to be considered and they lead to additional functionality following the above mentioned design guidelines.

To guarantee the QoS for data transmission between two application access points via WAN, written agreements between customer and service provider [service level agreements (SLA)] must be contracted. In cases where the provider cannot deliver the promised QoS, an alternative line must be established to hold the connection for the distributed automation function (this operation is fully transparent to the application function). This line should be available from another provider, independent from the currently used provider. The automation devices (so called VAN access points (VAN-APs)] should support functions to switch (either manually or automatically) to an alternative line.

There are different mechanisms to realize a connection between remote agents:

- *VAN switching connection*: the logical connection between two VAN-APs over a WAN or a public network. One VAN switching connection owns one or more communication paths. VAN switching line is defined as one physical communication path between two VAN-APs over a WAN or a public network. The endpoints of a switching connection are VAN-APs.

- *VAN switching line*: the physical communication path between two VAN-APs over a WAN or a public network. A VAN switching line has its provider and own QoS parameter. If a provider offers connections with different warranted QoS each of these shall be a new VAN switching line.
- *VAN switching endpoint access*: the physical communication path between one VAN-AP and a WAN or a public network. This is a newly introduced class for using the switching application service elements of virtual automation networks for communication via WAN or public networks.

These mechanisms are pivotal for VANs in heterogeneous e-Work networks. Depending on the priority and importance of the data transmitted between distributed communications agents, the kind of transportation service and communication technology is selected based on economical aspects.

5.3.5 Wireless Communication Networks

Wireless communication networks are increasingly penetrating the application area of wired communication systems, and have been faced with the requirements of industrial automation. Wireless technology has been introduced in automation as wireless local area networks (WLAN) and wireless personal area networks (WPAN). Currently, the wireless sensor networks (WSN) is an emerging technology, particularly for process automation (For specific application aspects of wireless communications, see Chapter 8 on Factory Sensor and RFID Networks). The basic standards are the following (ABI, 2009; Stallings, 2004; Willig and Matheus, 2005):

- *Mobile communications standards*: GSM, GPRS, and UMTS wireless telephones (DECT).
- Lower layer standards (IEEE 802.11: Wireless LAN (2009), and 802.15 (2009): personal area networks) as a basis of radio-based local networks (WLANs, Pico networks and sensor/actuator networks).
- Higher layer standards (application layers on top of IEEE 802.11 and 802.15.4, e.g. WiFi, Bluetooth (2009), wireless HART, and ZigBee (2009)).
- Proprietary protocols for radio technologies (e.g. wireless interface for sensors and actuators (WISA, 2009).
- Upcoming radio technologies such as ultra wide band (UWB) and WiMedia.

The increasing advances of ICT, collaboration engineering, and networked devices with increasing computing capabilities are rapidly transforming plants to decentralized industrial automation systems. In this context, wireless communication networks play a key role (see Fig. 5.7). This situation will lead to an increased interest in autonomic systems with capabilities such as self-configuration, self-organization, self-optimization, self-healing, and so on.

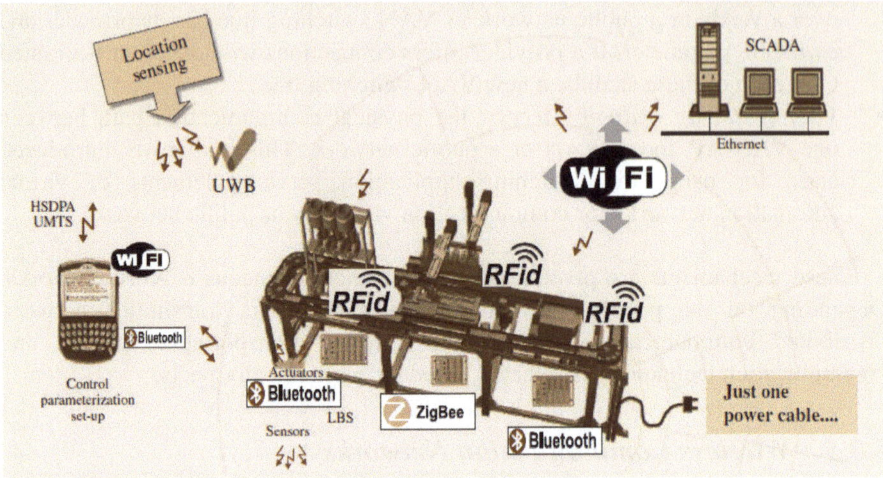

Fig. 5.7 The wireless factory (adapted from Zuehlke, 2008).

The idea is to develop autonomous and intelligent agents that are able to manage themselves given high level objectives (see also the discussion on Emerging Bio-Inspired Solutions in Chapter 4). In the following, the most popular wireless communication networks are discussed.

5.3.5.1 WLAN

WLAN is a wireless version of the Ethernet used to build computer networks for office and home applications. The original standard (IEEE802.11) specified an infrared, a direct sequence spread spectrum (DSSS) and a frequency hopping spread spectrum (FHSS) physical layer. There is an approval for WLAN to use special frequency bands, however it has to share the medium with other users. The Wi-Fi Alliance was founded to assure interoperability between WLAN clients and access points of different vendors. Therefore, a certification procedure and a Wi-Fi logo are provided. WLANs use a license-free frequency band and no service provider is necessary.

WLAN is a mature technology that can be easily implemented on PCs, laptops, and PDAs. Modules for embedded systems development are also available. WLAN can be used almost worldwide. Embedded WLAN devices need a powerful microcontroller. WLAN enables wireless access to Ethernet based LANs and is helpful for the vertical integration in an automated manufacturing environment. It offers high speed data transmission that can be used to transmit productive data and management data in parallel. The WLAN propagation characteristics fit into a number of possible automation applications. WLAN enables more flexibility and a cost effective installation in automation associated with mobility and localization. The transition to Ethernet is simple and other gateways are possible. The largest part of the implementation is achieved in hardware; however improvements can be made above the MAC layer.

5.3.6 Wireless Sensor/Actuator Networks

Features such as time synchronized operation, frequency hopping, self-organization (with respect to star, tree, and mesh network topologies), redundant routing, and secure data transmission are of concern in the design of WSN concepts (Akyildiz, 2005; Industrial wireless technology, 2002). Process automation requirements can be generally fulfilled by two mesh network technologies, which use the standard IEEE 802.15.4 low-rate wireless personal area network (WPAN, 2009), specifying the physical layer and parts of the data link layer (medium access control):

- *ZigBee* (2009): distinguishes between three device types:
  - *Coordinator ZC*: root of the network tree, storing the network information and security keys. It is responsible for connection the ZigBee network to other networks.
  - *Router ZR*: transmits data of other devices.
  - *End device ZED*: automation device (e.g. sensor), which can communicate with ZR and ZC, but is unable to transmit data of other devices.

An enhanced version allows one to group devices and to store data for neighboring devices. To save energy, there are full-function devices and reduced-function devices. The ZigBee application layer (APL) consists of three sublayers: application support layer (APS) (containing the connection lists of the connected devices), an application framework (AF), and Zigbee device objects (ZDO) (definition of devices roles, handling of connection requests, and establishment of communication relations between devices). For process automation, the ZigBee application model and the ZigBee profiles are very interesting. The application functions are represented by application objects (AO), and the generic device functions by device objects (DO). Each object of a ZigBee profile can contain one or more clusters and attributes, transferred to the target AO (in the target device) directly or to a coordinator, which transfers them to one or more target objects. Fig. 5.8 shows an overview of the ZigBee protocol stack.

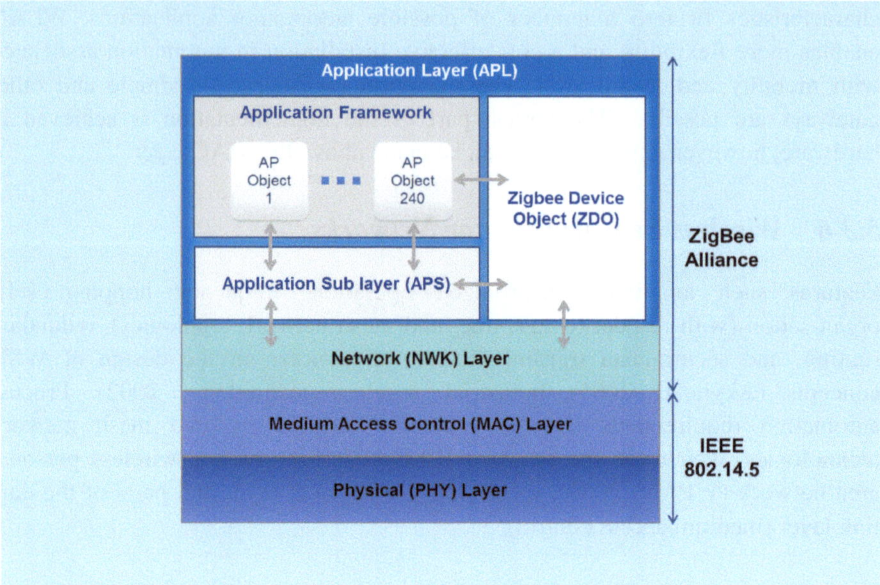

Fig. 5.8 The functional layers and protocol stack of ZigBee (after Baronti et al., 2007; with permission from Elsevier B.V.).

- *Wireless HART* (Griessmann, 2007;HART 7, 2007): the revision 7 of HART protocol includes the specification of Wireless HART (2007). The mesh type network allows the use of redundant communication paths between the radio-based nodes. The temporal behavior is determined by the time synchronized mesh protocol (TSMP) (Dust Networks, 2006). TSMP enables a synchronous operation of the network nodes (called motes) based on a time slot mechanism. It uses various radio channels (supported by the MAC layer) for an end-to-end communication between distributed devices. It works comparably with a frequency hopping mechanism missed in the basic standard IEEE 802.15.4. TSMP supports star, tree, as well as mesh topologies. All nodes have the complete routing function (contrary to ZigBee). A self-organization mechanism enables devices to acquire information of neighboring nodes and to establish connections between them. The messages have their own network identifier. Thus, different networks can work together in the same radio area. Each node has its own list of neighbors, which can be actualized when failures have been recognized. To support security, TSMP uses mechanisms for encryption (128 bit symmetric key), authentication (32 bit MIC for source address), and integrity (32 bit MIC for message content). Additionally, the frequency hopping mechanism improves the security features. For detailed information see (Sycara, 1998). Fig. 5.9 shows a generic Wireless HART network.

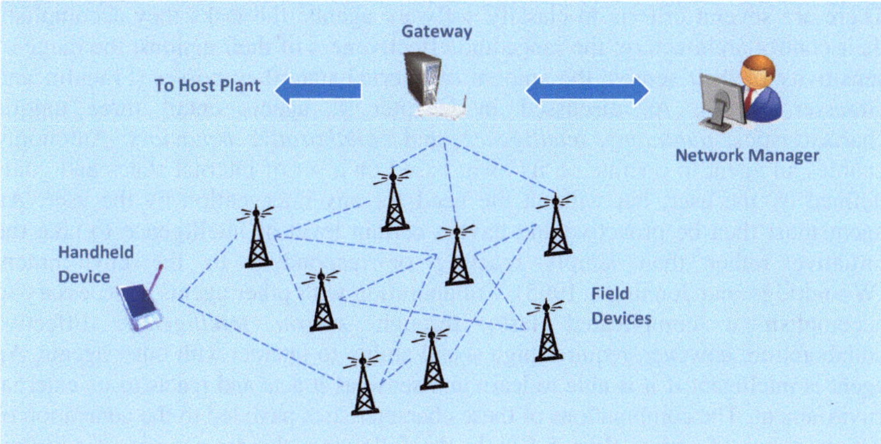

Fig. 5.9 A Wireless HART network and gateway implementation with multiple network access points.

5.4 Software Agents

Agents provide a convenient and powerful way to describe a complex software entity that is capable of acting with a certain degree of autonomy and at the same time to effectively accomplish tasks on behalf of their users (Nwana, 1996). In Chapter 4, we thoroughly discussed agents and agent-based systems, as a promising technology to enhance automation, optimization, and control of emerging e-Systems based on various collaboration engineering concepts. Agent-based systems are composed of multiple interacting intelligent agents, which can be used to solve problems that are difficult or impossible for an individual agent or monolithic system to solve (e.g., online trading, disaster response, and modeling social structures). Agents provide several important and interesting characteristics such as autonomy, local view, and decentralization (see Chapter 4 for more details). That is, each individual agent has incomplete information or capabilities for solving the problem and, thus, has a limited viewpoint. There is no system global control, data are decentralized, and computation is asynchronous.

Research on agent-based systems is mainly concerned with software agents. However, the agents in a multi-agent system could equally well be robots, humans, human teams, or human-agent teams. Multi-agent systems can manifest self-organization and complex behaviors even when the individual strategies of all their agents are simple. Agents can share knowledge using any agreed language, within the constraints of the system's communication protocol (Wooldridge, 2002; Shoham and Leyton-Brown, 2008; Sycara, 1998). Nevertheless, unlike objects, which are defined in terms of methods and attributes, an agent is defined in terms of its behavior. Agents are not all the same, and have a variety of classes with specific characteristics that are important to accomplish their own types of tasks.

There are several criteria to classify software agents: the tasks they accomplish; their control architecture; the range and effectiveness of their actions; the range of sensitivity of their senses; the amount of internal state they posses (Franklin and Graesser, 1996). As discussed in Chapter 4, agents entail three unique characteristics: *autonomy*, *intelligence*, and *collaborative behaviors*. Autonomy enables an agent to operate on its own, based on a set of internal states and goals defined by the user, but without the need for any intervention by the user. An agent must then be proactive and have a certain level of intelligence to take the initiative rather than simply reacting or responding to the environment (Wooldridge and Jennings, 1995). Collaboration with other agents is necessary to accomplish a complicated task; through *swarm intelligence*. Effective collaboration, however, requires high social ability to interact with other agents. An agent is intelligent if it is able to learn and sense as it acts and reacts to its external environment. The combinations of these characteristics have led to the generation of various types of agents (Fig. 5.10). In the following, the main types of software agents are introduced along with their unique features and applications.

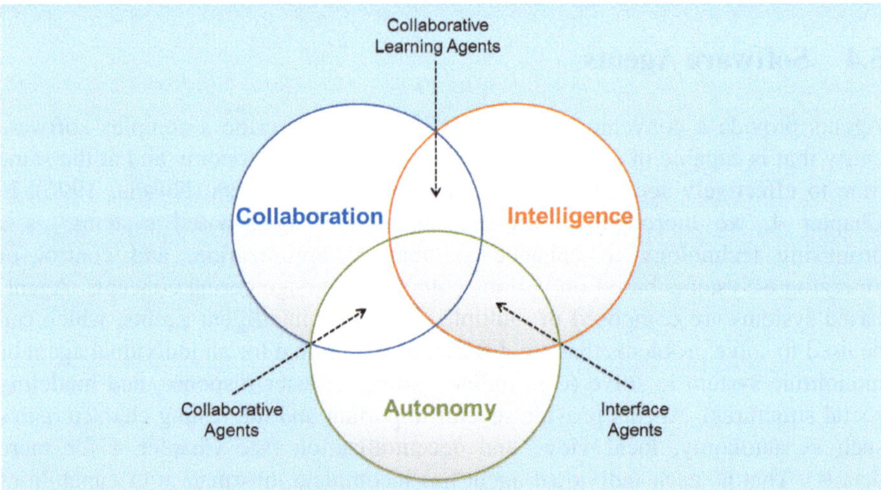

Fig. 5.10 Classification of software agent (adapted from Franklin and Graesser, 1996).

5.5 Reactive Agents

Reactive agents act and respond in a stimulus-response manner (Ferber, 1996) to the present state of the environment in which they are embedded. Reactive agents are typically characterized by the following key features (Maes, 1991):

- *Emergent functionality*: the dynamics of the interaction leads to emergent complexity.
- *Task decomposition*: a reactive agent is viewed as a collection of modules which operate autonomously and are responsible for specific tasks (e.g., sensing, computation).
- They tend to operate on representations that are close to raw sensory data.

Applications of reactive software agents are not very diverse. A good example is entertainment industry, e.g., Philips Corp., where researchers are already working on digital video and 3D graphics-based reactive agent animation (Nwana, 1996).

5.6 Interface Agents

The focus of interface agents is on autonomy and learning (Koda and Maes, 1996). They support and provide assistance to the user learning to use a particular application (such as a spreadsheet). The interface agent observes the action being carried out by the user, learns new shortcuts, and then provides suggestions for more efficient processing of the task. The key metaphor underlying interface agents is that of a personal assistant who is collaborating with the user in the same work environment. The collaboration between the interface agent and the user is inspired from the human-human collaboration; e.g., two mechanics working on the same car engine together (Fig. 5.11). The learning procedure for an interface agent is based on one or a combination of the followings (Maes, 1997):

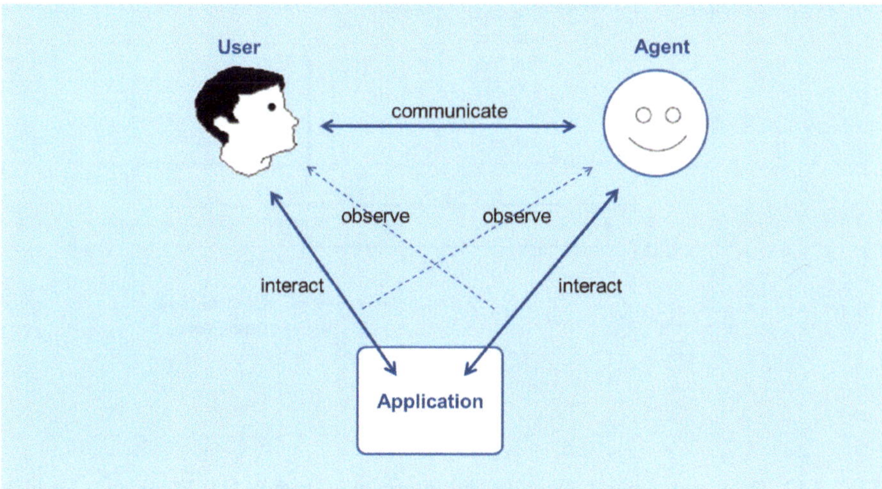

Fig. 5.11 A simple interface agent paradigm (adapted from Rich and Sidner, 1998).

- Observing and imitating the user.
- Receiving feedback (positive or negative) from the user.
- Receiving explicit instructions from the user.
- Asking other agents for advice (i.e., learning from peers).

The manipulation metaphor for interface agents (Maes and Schneiderman, 1997) may need to be modified in a way that facilitates delegation of some tasks to the software interface agents so to accommodate novice users. Current computer interfaces interact with the user through receiving commands from the keyboard, mouse, microphone, touch-screen, etc. Nonetheless, the computer will no longer be known as a merely passive entity waiting for commands to execute them. It must be able to provide the user with any help whenever she wishes to perform some complex tasks. In short, the main role of interface agents (coupled with information agents) is to *sense* users' interests and act autonomously on their behalf.

5.7 Information Agents

Advances in ICT and the Internet, and emergence of the *Internet of Things* have given a rise to information agents (aka, Internet agents). The main role of an

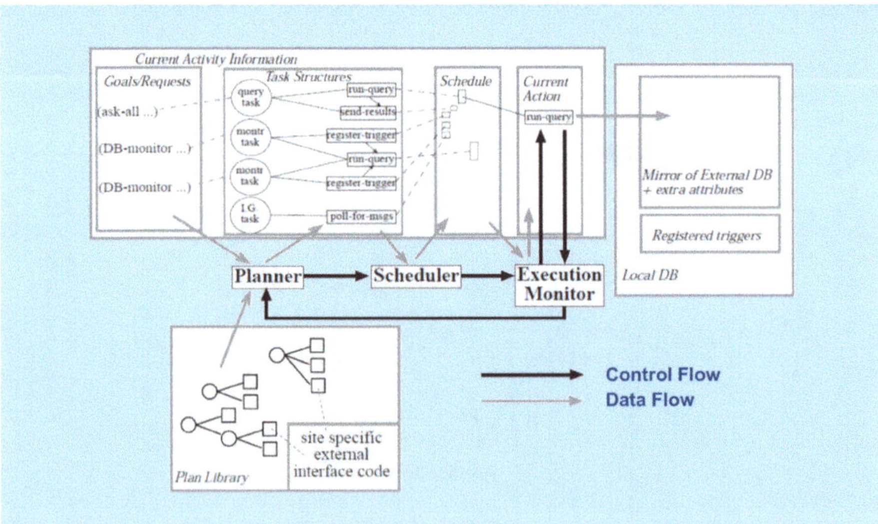

Fig. 5.12 Data and control flow in an information agent; (1) *Planning* – achieve global/local objectives; (2) *Scheduling* – sequencing and allocation of tasks defined by the plans; (3) Execution monitoring – real-time control of plans' execution to detect (potential) changes in the system and modify the plans/schedules accordingly (adapted from Decker and Sycara, 1997).

information agent is to help the user manage, manipulate, and/or collate information from highly distributed resources (Maes, 1995). With the explosive growth of information, *the big data*, and the need for analytical tools to manage and evaluate the massively growing data and transform them to meaningful information and knowledge, however, the goals of agents may overlap. Thus, to distinct between interface and information agents, it is important to know that information agents are defined by what they do, in contrast to the interface agents that are defined by what they are. In an e-Finance system, for instance, information agents are expected to handle a variety of tasks ranging from prediction of future workloads, future changes in the current information queries, future failures and changes in the structure of their network (e.g., addition/removal of agents or data sources), and so on (Decker and Sycara, 1997). Figure 5.12 shows a generic structure for information agents.

Information agents are most useful on the Web where they can help users with mundane tasks. For example, the user carries out actions that may take long time (e.g., searching the Web for information). Information agents effectively enhance processing of such tasks through carrying out the task for the user and later on presenting her the results. An information agent may be static or mobile.

5.8 Collaborative Agents

Collaboration between the individuals (agents) of a virtual society (e-Work network) requires the communication links to be established and used effectively. Distributed Artificial Intelligence (AI) is concerned with a virtual society of agents, as problem-solvers, interacting to solve a common problem (Green et al., 1996). Implementing efficient ways of cooperation among agents is actually one of the focal concerns of agent-based systems (Moulin and Chaib-Draa, 1996). One of the motivations for having collaborative agents is to provide solutions to inherently distributed problems. Emerging collaborative agent paradigms are mainly based on bio-inspired solutions, specifically swarm intelligence and evolutionary techniques, as discussed in Chapter 4 in detail. Collaborative agents have a wide variety of applications. Examples include distributed sensor networks (Durfee and Montogomery, 1987), air-traffic control, process planning and scheduling, collaborative control of e-Manufacturing processes, enterprise collaboration, online negotiation systems, and many other applications (see Chapter 4 for details).

5.9 Hybrid and Mobile Agents

A hybrid agent is an agent who is designed as a combination of two or more types of agent philosophies (e.g., mobile, interface, information, collaborative) within a singular agent (Nwana, 1996). Hybrid agents are mainly deployed in cases where the benefits of having a combination of design philosophies within a single agent are greater than the benefits of having several single-type agents, each with a different design philosophy. A good example are collaborative interface agents (Lashkari et al., 1994).

A software agent is considered to be mobile if it is able to migrate from one host to another in a heterogeneous environment. Thus, mobile agents are only applicable in environments that are compatible to their *mobility*, i.e., mobile agent environments where software systems are distributed over a network of heterogeneous computers. That is, the environment must support the mobile agents to run (execute), not only by the agent's transports, but also its state. When the mobile agent reaches a new host, it should be able to operate appropriately in its new environment (Fig. 5.13).

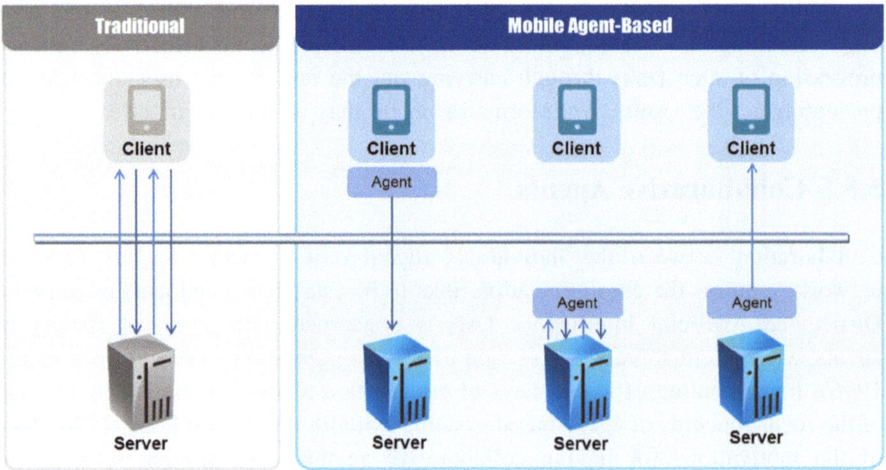

Fig. 5.13 Traditional client-server system (left) and mobile agent-based system.

With the rising advancements of ICT and the Internet, mobile agents are now becoming more and more popular (Fig. 5.14). The reason is that they provide a generic and powerful framework that supports all distributed and information-oriented applications, and increase the time and flexibility required by the developers to provide more useful applications and features (Kotz and Gray, 1999). Mobile agents have various applications such as

- *Data processing in unreliable networks*: In an unreliable network, the low-reliability network is enabled to transfer agents, rather than a chunk of data, from one place to another. In such paradigms, agents dynamically move between the nodes on the network, process the information on the respective nodes (without the risk of network disconnection), and then return to home.

- *e-Commerce*: Instead of spending a huge amount of time going through online bookstores to find the best deal on a book, firing up an agent to perform this task would save the user a considerable amount of time. The agent would be programmed to visit a number of bookstores and find the best deals.
- *Network management*: In heterogeneous network environments, network operators are required to have extensive knowledge regarding the various networks that need to be managed. Such management requires the collection of large amount of data from nodes on the network. Mobile agents are an ideal tool for collection and analysis of such data on behalf of the user.

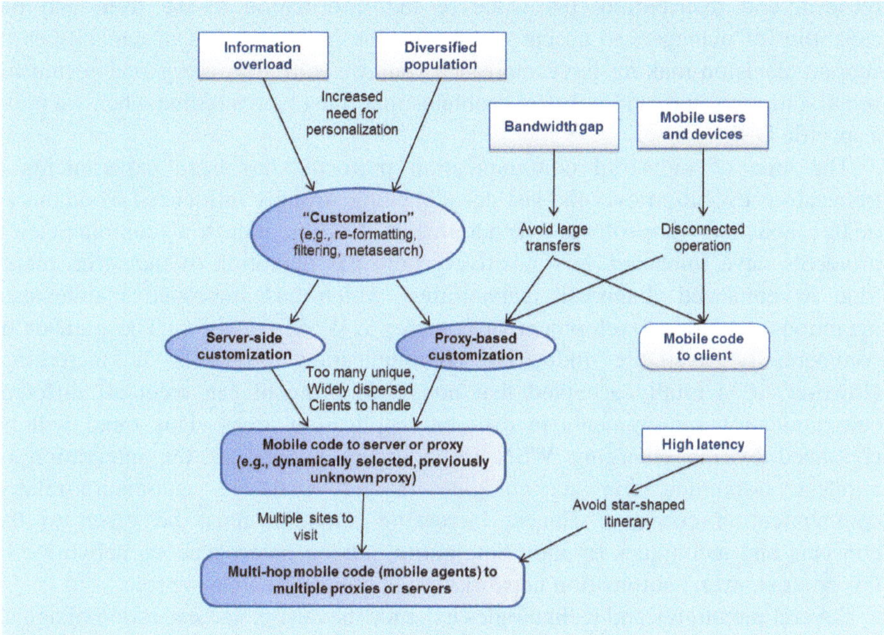

Fig. 5.14 The emerging trends towards mobile agents (adapted from Kotz and Gray, 1999).

5.10 Emerging Trends and Challenges

Various paradigms associated with the three fundamental tools for the development of collaborative e-Systems were discussed in this chapter, including: DSS, as a supportive concept for the Active Middleware technology, and an augmentation tool of collaboration; and communication protocols and software agents, as two active technologies corresponding with the first wheel of collaborative e-Work (see Chapter 1).

The technologies associated with DSS evolve rapidly and continuously, and decision-makers must be aware of this evolution in the technological and social context of DSS. Such innovations are strongly supported by the emergence of new hardware/software tools, Web and mobile technologies. DSS, however, is not a panacea for improving business decisions. Most people acknowledge that managers need good information to manage effectively, but a DSS is not always the solution for providing good information. A DSS can provide a competitive advantage and a company may need computerized decision support to remain competitive, but decision support capabilities are limited by the data that can be obtained, the cost of obtaining, processing, and storing the information, the cost of retrieval and distribution, the value of the information to the user, and the capability of managers to accept and act on the information. Our capabilities to support decision-making have increased, but we still have very real technical, social, interpersonal, and political problems that must be overcome when we build a specific DSS.

The area of industrial communication protocols has been experiencing a tremendous evolution over the last decade, being strongly influenced by advances in ICT and hardware/software developments. Current industrial communication protocols have impacted very positively both the operation of industrial plants (due to enhanced diagnostic capabilities, which has improved maintenance operations) and the development of complex e-Work networks. The number of commercially available industrial communication protocols is increasing. However, it is totally accepted that no single protocol can meet all different communication requirements in different application areas. This trend will be continued by the emerging WSN technologies as well as the integration of wireless communication technologies in all available automation-related communication concepts. Hence, increasing attention must be given to the concepts and techniques to allow integration among heterogeneous networks. In this context virtual automation networks are playing an increasing role.

Several paradigms and technologies exist for the design and execution of agents and agent-based systems in various e-Systems, as thoroughly discussed in Chapter 4 and also in this chapter. The great challenge associated with agent-based systems, however, is the design complexity, from the interaction and communication mechanisms to integration, intelligence, flexibility, fault tolerance, and most importantly, collaboration engineering. Another concern is the lack of systematic and experimentally-proven toolkit; a flexible set of tools to specify the problem-solving behaviors of agents, and interaction and error and conflict resolution mechanisms. The other challenging issue is how to build trust in the users of an agent-based system to delegate tasks with different levels of criticality to an agent, and how to define the optimal level of agents' autonomy.

References

ABI Research Forecast and Information on Emerging Wireless Technologies, http://www.abiresearch.com (last accessed April 6, 2009)

Akyildiz, I.F.: Key wireless networking technologies in the next decade. In: IFAC Conf. Fieldbus Technol., FeT 2005, Puebla (2005)

Alter, S.L.: Decision Support Systems: Current Practice and Continuing Challenge. Addison-Wesley, Reading (1980)

Baronti, P., Pillai, P., Chook, V.W.C., Chessa, S., Gotta, A., Hu, Y.F.: Wireless sensor networks: A survey on the state of the art and the 802.15.4 and ZigBee standards. Computer Communications 30, 1655–1695 (2007)

Bonczek, R.H., Holsapple, C.W., Whinston, A.B.: Foundations of Decision Support Systems. Academic, New York (1981)

Bluetooth: The Official Bluetooth Membership Site, https://www.bluetooth.org (last accessed April 6, 2009)

Dai, Y., Tuomo, K., Erkki, S.: MinEDec: a decision-support model that combines text-mining technologies with two competitive intelligence analysis methods. International Journal of Computer Information Systems and Industrial Management Applications 3, 165–173 (2011)

Decker, K.S., Sycara, K.: Intelligent Adaptive Information Agents. Journal of Intelligent Information Systems 9(3), 239–260 (1997)

Dhar, V., Stein, R.: Intelligent Decision Support Methods: The Science of Knowledge. Prentice Hall, Upper Saddle River (1997)

Durfee, E.H., Montogomery, T.A.: Coherent Cooperation among Communicating Problem Solvers. IEEE Transactions on Computers, 1275–1291 (November 1987)

Dust Networks: TSMP Seminar. Online-Präsentation (2006)

Dust Networks: Technical Overview of Time Synchronized Mesh Protocol (TSMP), White Paper. Dust Networks (2006)

Ferber, J.: Reactive Distributed Artificial Intelligence: Principles and Applications. In: O'Hare, G.M.P., Jennings, N.R. (eds.) Foundations of Distributed Artificial Intelligence, pp. 287–317. John Wiley & Sons, Inc. (1996)

Franklin, S., Graesser, A.: Is it an agent, or just a Program? A Taxonomy for Autonomous Agents. In: Jennings, N.R., Wooldridge, M.J., Müller, J.P. (eds.) ECAI-WS 1996 and ATAL 1996. LNCS, vol. 1193, pp. 21–35. Springer, Heidelberg (1997)

Green, S., Hurst, L., Nangle, B., Cunningham, P., Somers, F., Evans, R.: Software Agents: A review. Technical Report, Trinty College Dublin, Ireland (May 1996)

Griessmann, J.-L.: HART protocol rev. 7 including WirelessHART. ATP Int.-Autom. Technol. Pract. 2, 21–22 (2007)

HART, http://www.hartcomm2.org/ (last accessed April 6, 2009)

HART 7: HART Protocol Specification, Revision 7. HART Communication Foundation (2007), see also http://www.hartcomm.org/ (last accessed April 6, 2009)

Holsapple, C.W., Whinston, A.B.: Decision support systems: A knowledge-based approach. West, St. Paul (1996)

Holsapple, C.W., Sena, M.P.: The Decision-Support Characteristics of ERP Systems. International Journal of Human-Computer Interaction 16(1), 101–123 (2003)

IEC 61158 Ser., Edition 3: Digital data communication for measurement and control – Fieldbus for use in industrial control systems (2003)

IEC 61784-1: Digital data communications for measurement and control – Part 1: Profile sets for continuous and discrete manufacturing relative to fieldbus use in industrial control systems (2003)
Industrial wireless technology for the 21st century, Industrial Wireless Workshop, San Francisco (2002)
Kim, A.N., Hekland, F., Petersen, S., Doyle, P.: When HART Goes Wireless: Understanding and Implementing the Wireless HART Standard. In: IEEE International Conference on Emerging Technologies and Factory Automation, ETFA 4638503, pp. 899–907 (2008)
Koda, T., Maes, P.: Agents with Faces: The Effects of Personification of Agents. In: Proceedings of HCI 1996, London, UK (August 1996)
Kotz, D., Gray, R.S.: Mobile agents and the future of the internet. ACM SIGOPS Operating Systems Review 33(3), 7–13 (1999)
Lashkari, Y., Metral, M., Maes, P.: Collaborative Interface Agents. In: Proceedings of the Twelfth National Conference on Artificial Intelligence, vol. 1. AAAI Press, Seattle (August 1994)
Little, J.D.C.: Brandaid, an online marketing mix model, Part 2: Implementation, calibration and case study. Oper. Res. 23(4), 656–673 (1975)
Maes, P. (ed.): Designing Autonomous Agents: Theory and Practice from Biology to Engineering and back. MIT Press (1991)
Maes, P.: Intelligent Software. Scientific American 273(3), 84–86 (1995)
Maes, P.: Agents that Reduce Work and Information Overload. In: Bradshaw, J.M. (ed.) Software Agents, pp. 145–164. AAAI Press/MIT Press (1997)
Maes, P.: Schneiderman, B.: Direct Manipulation vs. Interface Agents: a Debate. In: Interactions, vol. IV(6). ACM Press (1997)
Morton, M.S.S.: Management Decision Systems; Computer-Based Support for Decision Making. Harvard Univ. Press, Boston (1971)
Moulin, B., Chaib-Draa, B.: An Overview of Distributed Artificial Intelligence. In: O'Hare, G.M.P., Jennings, N.R. (eds.) Foundations of Distributed Artificial Intelligence, pp. 3–55. John Wiley & Sons, Inc. (1996)
Neumann, P.: Communication in industrial automation – what is going on? In: 11th IFAC Symp. Inf. Control Probl. Manuf., INCOM 2004, Salvador da Bahia (2004)
Neumann, P., Pöschmann, A., Flaschka, E.: Virtual automation networks, heterogeneous networks for industrial automation. Atp Int. Autom. Technol. Pract. 2, 36–46 (2007)
Nof, S.Y.: Integrated Description and Performance Evaluation of Conveyorized Systems. PhD Dissertation, University of Michigan, Ann Arbor (1976)
Nof, S.Y.: A Methodology for Computer-Aided Facility Planning. International Journal of Production Research 18(6), 699–722 (1980)
Nof, S.Y.: An Expert System for Planning/Replanning Programmable Facilities. International Journal of Production Research 22(5), 895–903 (1984)
Nwana, H.S.: Software Agents: An Overview. Knowledge Engineering Review 11(3), 205–244 (1996)
Nwana, H.S.: Software Agents: An Overview. Knowledge Engineering Review 11(3), 205–244 (1996)
Personal Area Networks: IEEE 802.15: IEEE 802.15 Working Group for Wireless Personal Area Networks, Task Group 1 (TG1), http://www.ieee802.org/15/pub/TG1.html (last accessed April 6, 2009), and: Task Group 4 (TG4), http://www.ieee802.org/15/pub/TG4.html (last accessed April 6, 2009)

Power, D.J.: Supporting decision-makers: An expanded framework. In: Harriger, A. (ed.) e-Proc. Inf. Sci. Conf., Krakow, pp. 431–436 (2001)
Power, D.J., Sharda, R.: Springer Handbook of Automation: Decision Support Systems. ch. 87. Springer (2009)
Quattrochi, D.A., Luvall, J.C., Rlckman, D.L., Estes, M.G.J., Laymon, C.A., Howell, B.F.: A Decision Support Information System for Urban Landscape Management Using Thermal Infrared Data. Photogrammetric Engineering & Remote Sensing 66(10), 1195–1207 (2000)
Rich, C., Sidner, C.L.: COLLAGEN: A Collaboration Manager for Software Interface Agents. User Modeling and User-Adapted Interaction 8(3-4), 315–350 (1998)
Seok, H.S., Nof, S.Y., Filip, F.G.: Sustainability Decision Support System based on Collaborative Control Theory. Annual Reviews in Control 36(1), 85–100 (2012)
Shoham, Y., Leyton-Brown, K.: Multiagent Systems: Algorithmic, Game-Theoretic, and Logical Foundations. Cambridge University Press (2008)
Simon, H.A.: The New Science of Management Decision. Harper Row, New York (1960)
Skitka, L., Mosier, K.L., Burdick, M.: Does automation bias decision-making? International Journal of Human-Computer Studies 51(5), 991-1006 (1999)
Sprague Jr., R.H., Carlson, E.D.: Building Effective Decision Support Systems. Prentice Hall, Englewood Cliffs (1982)
Stallings, W.: IEEE 8O2.11. Wireless LANs from a to n. IT Professional 6(5), 32–37 (2004)
Sycara, K.P.: Multiagent Systems. AI Magazine 19(2), 79–92 (1998)
Thomesse, J.-P.: Document Fieldbus technology in industrial automation. Proceedings of the IEEE 93(6), 1073–1101 (2005)
van de Giesen, N.I., Andreini, M., van Edig, A., Vlek, P.: Competition for water resources of the Volta basin. IAHS Publication, 199-206 (2001)
Sauter, V.L.: Sauter: Competitive intelligence systems: qualitative DSS for strategic decision making. ACM SIGMIS Database 36(2), 43–57 (2005)
Virtual Automation Networks: European Integrated Project FP6/2004/IST/NMP/2 016696 VAN. Deliverable D02.2-1: Topology Architecture for the VAN Virtual Automation Domain (2006)
Willig, A., Matheus, K., Wolisz, A.: Wireless technology in industrial networks. Proc. IEEE 93(6), 1130–1151 (2005)
Wireless LAN: IEEE 802.11: IEEE 802.11 Wireless Local Area Networks Working Group for WLAN Standards, `http://ieee802.org/11/` (last accessed April 6, 2009)
WISA, `http://library.abb.com/GLOBAL/SCOT/SCOT209.nsf/VerityDisplay/4E478BD7490A3F8BC12571F100427DCB/File/2CDC171017K0201.PDF` (last accessed April 6, 2009)
Wooldridge, M., Jennings, N.R.: Intelligent Agents: Theory and Practice. Knowledge Engineering Review 10(2), 115–152 (1995)
Wooldridge, M.: An Introduction to MultiAgent Systems. John Wiley & Sons Ltd. (2002)
Xinping, S., Hongyun, T., Xiaowei, W.: Study on SMEs-Oriented Strategic Decision Support System Based on Competitive Intelligence. In: 2010 International Conference on Management and Service Science (MASS), pp. 1–4. IEEE (August 2010)
ZigBee, `http://www.zigbee.org/` (last accessed April 6, 2009)
Zuehlke, D.: SmartFactory from vision to reality in factory technologies. In: IFAC Congress 2008, Seoul (2008)

Review and Challenge Questions

1. (i) Explain how the e-Work support tools interact to accomplish and augment the engineered collaboration in e-Work, e-Business, and e-Service. (ii) Illustrate with specific examples from Chapters 1 through 5.
2. Describe: (i) The key components and functions of a DSS. (ii) Their role in enabling and augmenting collaboration.
3. Illustrate your answer to Question 2 in the case shown in Fig. 5.2.
4. (i) Describe the differences between and functionalities of DSS, TPS, and decision automation. (ii) Illustrate these differences for (a) emergency services; (b) piloting an airplane.
5. Select two different categories of DSS and illustrate how CCT principles and techniques can improve their use. (Examples: Reduce errors and conflicts, operate in parallel, plan collaboration requirements, etc.)
6. (i) Explain the components of a collaborative DSS and illustrate them in the case of a supply network. (ii) Describe several types of DSSs associated with competition, and explain whether or not they also include collaborative components.
7. In specific relation to DSS functionalities, what are some of the roles of (i) telecommunication protocols; (ii) task administration protocols?
8. Answer Question 7 in relation to multi-agent e-Work facilities.
9. Describe in a table the different roles of the six types of networks included in Fig. 5.6 and explain/justify whether all of them are needed for (i) an e-Work facility example; (ii) an e-Business example; (iii) an e-Service example.
10. Compare and contrast the traditional client-server design architecture, and the mobile agent-based architecture, and describe the relative advantages and weaknesses of each.
11. Explain how the three fundamental tools for collaborative e-Work correspond with the Collaborative e-Work wheels discussed in Chapter 1.
12. Outline and illustrate how several emerging trends and challenges shape the future progress of e-Work, e-Business and e-Service.

Lab Questions

13. Access and explore a particular DSS which applies to e-Work, e-Business, or e-Service. (i) Describe its main components and decision support functionalities; (ii) Identify the category of DSS for which it is designed. (iii) Identify the collaboration functions in it. (iv) Explain which task administration protocols are active, or could be engineered in this DSS to improve its collaboration abilities. (v) Briefly list some of the benefits of the protocols you discussed in (iv).

14. Access and explore an e-Work, e-Business, or e-Service enterprise. (i) Describe and identify several different agents that you understand are designed in it, or could be designed to improve its performance. (ii) For each of the agents you discuss in (i), identify its type, e.g., mobile, information, reactive, etc. (iii) Illustrate how those agents interact with the DSS and TAP included in this enterprise.

Chapter 6
e-Work in Product and Service Development

Abstract. Applications of collaborative e-Work to the product and service development processes are emerging. The development processes range from identification of customer requirements to formalization of customer benefit package (CBP) for summarizing the company's offer for resolving customers' needs and expectations, based on the product lifecycle management (PLM) framework. In this context, knowledge management is an essential for the evolution of products and services, focusing on innovation and excellence along with management of disappearing products and services, considering environmental issues. This chapter reviews state-of-the-art collaborative design techniques for the development of products and services in networked e-Systems, with special reference to CI, the Collaborative Intelligence.

6.1 PLM (Product Lifecycle Management) Systems for Product and Service Development

The global nature of emerging products and services is increasingly forcing companies to reach the global markets in a timely manner. However, the enforced urgency for shortening the reactions to the customers' needs has carried along the necessity for shortening the complete product and service lifecycles. Companies with effective and efficient lifecycle planning of their products and services have higher chances to prevail in the industry. The key performance indicators of companies excelling in global markets indicate reduced time-to-market, customer-centric offers, reduced time-to-volume by means of mass customized production, and reduced time-to-profit for gaining competitive advantages. These performance indicators are usually the result of technology-based development advancements such as agile product and service development, leading capacity and capabilities in supply and manufacturing processes, and lifecycle-based revenue analysis. This present scenario contrasts deeply with the traditional scenarios where competitive advantage usually was based on the pressure for permanently increasing the operational efficiency.

Nowadays, as the leading providers of products and services take control of over 80% of the market, the attention is set on enabling capabilities for developing innovative and customer-centric products and services. According to the notion of "economy of ideas" due to Paul Romer (1993), as customers become more knowledgeable about products and services they demand, even indirectly, the

S.Y. Nof et al., *Revolutionizing Collaboration through e-Work, e-Business, and e-Service*,
Automation, Collaboration, & E-Services 2, DOI: 10.1007/978-3-662-45777-1_6

Table 6.1 Data and information flows in PLM (adapted from Kiritsis, 2009).

Information flow	Category	Data
BOL* to MOL*	Bill of material information	Product ID and structure, part and component ID, design specification
	Maintenance/service information	Spare part ID list and prices, maintenance/service instructions
	Production information	Assemble/disassemble instruction, production specifications, production history and routing data, production plan, inventory status
BOL to EOL*	Product information	Material information, bill of material, part/component cost, disassemble instruction, assembly and remanufacturing information
	Production information	Production date, lot ID, production location
MOL to EOL	Maintenance history information	Number of breakdowns, parts IDs, installation date, maintenance engineers' IDs, list of replaced parts, aging statistics after substitution, maintenance cost
	Product status information	Quality degree of each component, performance definition
	Usage environment information	Usage condition, user mission profile, usage time
	Updated bill of material	Updated bill of material after repairing or changing parts
MOL to BOL	Maintenance and failure information (for design improvement)	Ease of maintenance/service, reliability problems, maintenance date, frequency of maintenance, mean time between failure/to repair, failure rate, critical component list, root causes
	Technical customer support information	Customer complaints, customer profiles, response
	Usage environment information	Usage condition, user mission profile, usage time
EOL to MOL	Recycling/reusing part information	Reuse part, remanufacturing information, quality of remanufacturing part
EOL to BOL	EOL product status information	Product/part lifetime, recycling/reuse rate of each part
	Dismantling information	Ease to disassemble, reuse or recycling value, disassembly cost, disposal cost, remanufacturing cost
	Environmental effects information	Material recycle rate, environmental hazard information

* BOL: Beginning of life; MOL: Middle of life; EOL: End of life.

innovations by the companies are becoming more concerned with providing value to customers as well as generating economic growth at both company and national levels. Thus, customer satisfaction and loyalty are other key elements of the need for PLM in emerging businesses and services, which can be realized by broadening the scope of collaboration through incorporating customers in the collaborative design and development processes. Customers are no longer restricted to the offers by local companies, due to the globalized markets where they can demand customized products and services that resolve their specific requirements. Accordingly, participation of customers in product and service customization has turned them into a collaborative party of production and service systems, especially in the product/service design stage. Such customer-company collaboration, however, relies on timely and complete information sharing (Table 6.1) in order to be useful along the value chain. Customers are the most significant source of information that must be effectively transformed to knowledge regarding the relationships between products, services, and customer requirements.

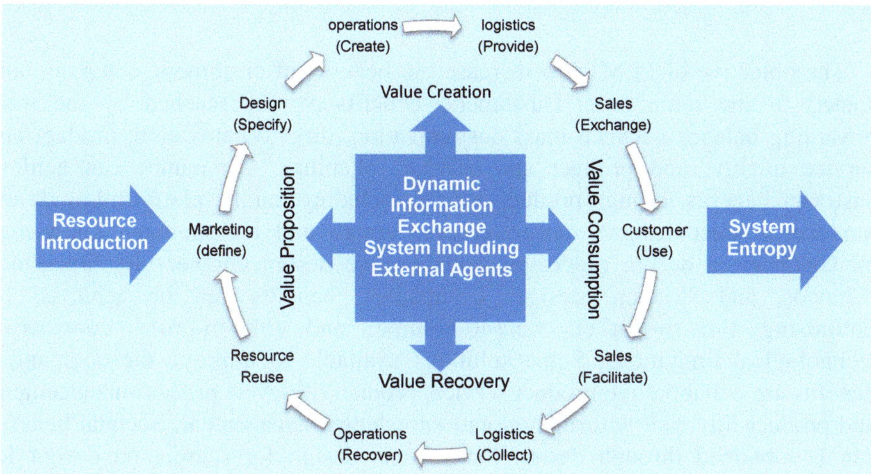

Fig. 6.1 Sustainable value cycle with exchange agents collecting for managing EOL systems (adapted from Barber et al., 2012).

Operational excellence is another concern in enhancing the competitive advantage of the company, which calls for effective PLM systems, and can be achieved by performing all tasks in the value chain at the maximum efficiency along with exploiting the tasks relationships through process integration. An alternative approach for increasing the operational efficiency is reduction of wastes in the values chain, e.g., wrong data, information/knowledge duplication, delays in searching and translating data. Therefore, value chains require systematic capturing, management, and dissemination of knowledge and elimination of inefficiencies caused by time, distance, and interacting disciplines, in order to enhance operational excellence. Figure 6.1 illustrates a conceptual value cycle framework, where the traditional value chains, starting from the companies and ending with the customers, is transformed to a value cycle, in which continuous provision of functionality to the end customers, e.g., sustainable design, reverse logistics, reuse of items, and information flow, is the main objective.

In line with the increasing trends towards mass customization (Proud and Wetzer, 2003), PLM has become a significant element of (Ming et al., 2005):

- Management and design of products and services alternatives.
- Design of processes.
- Supply of inputs to the processes.
- Introduction of new product and services to the markets.
- Service of products.
- Control of the products/services environmental impacts.

The objective of PLM is to increase the benefits of customers, company, and society at the same time. Customers' benefits can be reached by the most rewarding balance between mass customization, time-to-innovation, product and service quality, and product and service reliability. Companies can achieve customer benefits through product design (including family, platform-based, and modular product design), design process design and management, knowledge management in design processes, collaborative design engineering, functional, behavior, and structure design. Companies' benefits can be achieved by optimizing time-to-market, time-to-volume, and time-to-profit considering technological limitations. Some solutions available to improve the companies' benefits are collaborative product service, product life-cycle process management, and product life-cycle information and knowledge management. Societal benefits can be obtained through design for service, design for reuse, and design for recycle implementations. Enabling technologies for achieving these implementations are product/service co-design, collaborative early design for life-cycle efficiency, and environmentally conscious design among others.

PLM is a collaborative effort with economic perspective to manage the portfolio of products, processes, and services, from the initial concept development to the design, production, and environmental control of the disposal or recycling. Knowledge is a key element of all tools, theories, and systems supporting various stages in the lifecycle of products and services. Hence, PLM systems require knowledge management systems to support continuous improvement in the decision-making processes throughout the lifecycle of products and services. Collaborative e-Activities supporting the development of products and services, e.g., specifications of customer requirements and needs, design, production, supply chain management, etc., depend highly on information and collective experiences structured in the form of knowledge. According to Ming et al. (2005), the key requirements for the developments of products and services are:

1. Geographically distributed design teams and collaborative supply network partners.
2. Mass customized, easily configurable, and module-based products instead of static designs.
3. Extended enterprise business models.
4. Accelerated and predictive products and service development.
5. Exchange and control of product and service information for real-time management of projects and program activities.
6. Inter-enterprise data, information, and knowledge to provide design teams with dynamic virtual design spaces.

The PLM solutions to the above requirements and challenges are:

1. Provision of information continuums to deliver pervasive, real-time analytics, querying, and reporting throughout the lifecycle.
2. Provision of collaborative environments bridging multiple roles, constituents, and stakeholders in threaded discussions.
3. Enabling interactions viewing and commentary upon product development through multiple devices, channels, and systems involved with the product lifecycle.
4. Provision of open but integrated solutions supporting key enterprise value disciplines for product leadership, customer satisfaction, and operational excellence.

PLM solutions are the enablers of coordination and collaboration in activities regarding products, projects, and information processing throughout the value chain and among both internal and external parties. They support product-centric business models that unify the product and service lifecycle through enabling online sharing of product knowledge and business applications. The emergence of product-embedded information devices (PEID) (Figure 6.2) has enhanced real-time data gathering and analysis of product-related information enabling closed-loop PLM; a strategic business approach for the effective management of product lifecycle activities by using product data/information/knowledge which can compensate PLM to realize product lifecycle optimization dynamically in closed loops with the support of PEIDs and product data and knowledge management systems.

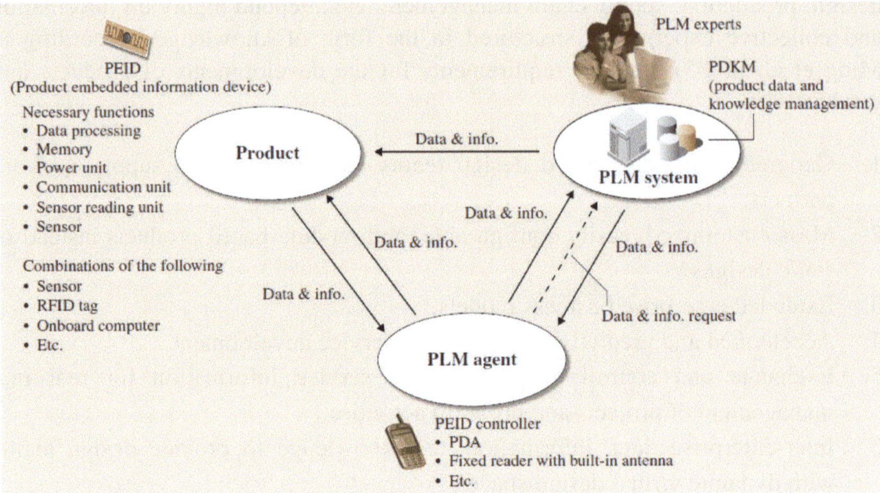

Fig. 6.2 Applications of PEIDs in PLM (after Kiritsis, 2009).

6.2 Technology Support in PLM Systems

Technological components of PLM solutions enable the creation, transformation, and exchange of knowledge throughout the lifecycle. A component of a PLM solution may be a computer system for knowledge transformation, from tacit to explicit forms, or standard communication technologies. Technological solutions within the PLM system are evaluated by their capability to manage knowledge, regardless of its complexity or originality. The elementary steps in the knowledge creation procedure are data collection and transformation into information, i.e., meaningful and useful data. The last step is then to transform the information into knowledge, as the key element for continuous improvement, shorter development cycles, and increasingly profitable lifecycles of products and services. Since the emergence of early CAD (Computer-Aided Design) and CAM (Computer-Aided Manufacturing) systems, the goal has been to obtain better, cheaper, faster, and robust design along with low variability, effective, and efficient manufacturing based on product data. Integration of CAD/CAM systems to enhance product development, however, relies on real-time information rather than raw data.

In line with the extension of the design scope, from the product to the manufacturing system and the entire value chain, PLM systems have become more concerned with knowledge than information. This requirement can be addressed only through effective extended enterprise knowledge management, where the complete product/service developments take place. Considering data mining and data warehouse, two emerging technology-based solutions, knowledge creation and management can be conceptualized as the operations over a large repository of data (either in a centralized or geographically scattered database) for generating meaningful elements of knowledge based on the trends/patterns obtained from the available data. This type of knowledge is called "explicit", since it is generated and understood by machine agents. The other type of knowledge is referred to as "tacit", generated through mental processes, and thus is not related to the technological implementations.

From the collaborative e-Work perspective, explicit knowledge is of interest, since it enables representation and usage of knowledge by means of hardware- and software-based solutions. In view of that, e-Work enablers of PLM must be capable of supporting the creation, exchange, and transformation of knowledge in the product/service lifecycle (Figure 6.3). Knowledge transformation, as the most complicated task, can be performed in various ways (Newman, 2002):

1. *Socialization*: sharing of tacit knowledge among individuals and benefiting from implementations of web meeting applications, video conferencing systems, and collaboration and visualization tools.
2. *Externalization*: transformation of tacit knowledge into explicit knowledge and benefiting from information authoring tools, ranging from complex CAD, CAM, CAE (Computer-Aided Engineering), to simple text editors.
3. *Internalization*: transformation of explicit knowledge into tacit knowledge through document management systems, change management, workflow management systems or search engines.
4. *Combination*: integration of two or more pieces of knowledge to generate new explicit knowledge, implemented through expert systems.

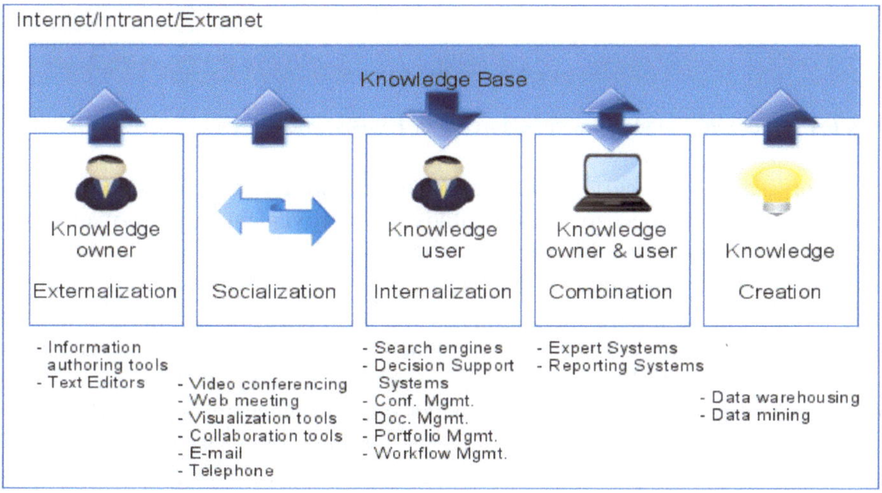

Fig. 6.3 PLM components.

The other two tasks, i.e., knowledge creation and exchange, for monitoring and managing the product information throughout the entire lifecycle, along with provision of feedback are supported by the emerging intelligent products and identification technologies such as RFID. These emerging technologies are increasing enabling companies to enhance the performance of their business models and augment their efficiency and competitiveness (Kiritsis, 2009). Figure 6.4 shows the architecture of a closed-loop PLM system, enabled by PEIDs responsible for data gathering, processing, and data storage. The architecture is composed of five layers, mainly classified into business process, software, and hardware. The middleware is the intermediate software between different application layers, PEIDs and product data and knowledge management systems. It supports complex and distributed applications such as interactions among RFID tags and business information systems, communication, coordination, and management of information, etc.

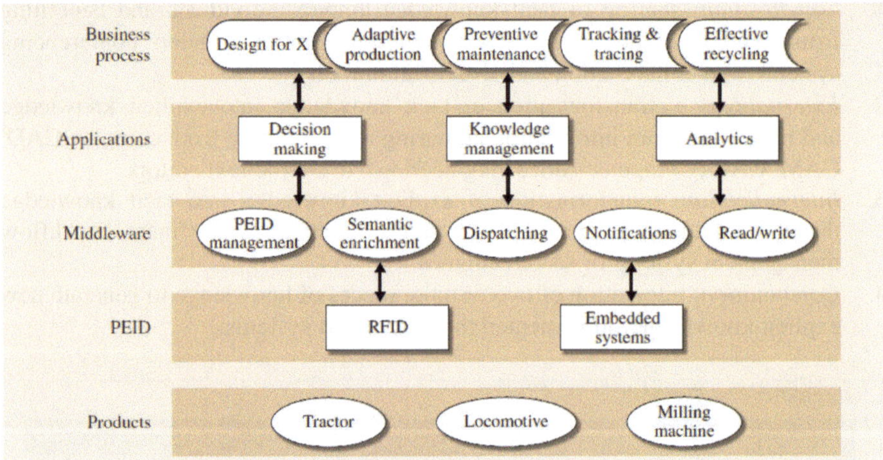

Fig. 6.4 The closed-loop PLM architecture (after Kiritsis, 2009).

According to the *S-curve theory*, the state of maturity of PLM systems can be determined, with respect to the magnitude of improvement in performance of a technological solution to the maturity of the solution and its components. At the early stages from introduction, the rate of progress in performance is relatively slow. When the technology solution is better understood, controlled, and known, the rate of technology-based improvement is improved. Finally, at its mature stage, technology reaches its performance enhancing limits (Christensen, 1992). Figure 6.5 shows the S-curves for product data management and PLM systems implementation. In Figure 6.5, while the product data management (PDM) technology is reaching its maturity stage, the PLM technology is at its early development stage from introduction. This transformation in performance is the result of moving from data to knowledge management.

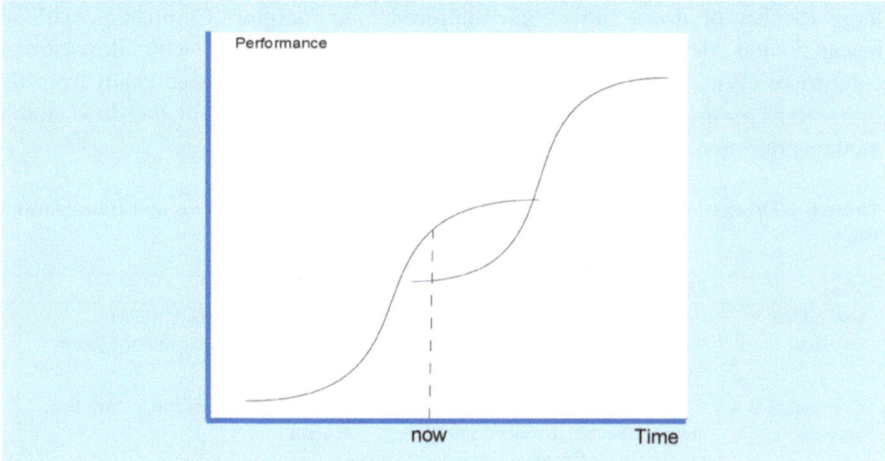

Fig. 6.5 S-curve for PDM and PLM solutions.

6.3 Collaborative e-Work Support for Product and Service Development

In line with the "*knowledge revolution*" (Sahlman, 1999), competition, distribution, and technological advancements drive out inefficiencies, force better business process reengineering, and provide knowledgeable customers with higher quality products and services. A key motivation to collaborate is then the need to exchange scientific and technological knowledge, and management expertise (Stephan, 1996). The spillover effect transfers knowledge from research centers to underdeveloped regions (e.g., new plants, service centers, shops). Collaborative manufacturing, especially in multinational companies, may have positive or negative impacts on the development of local economies (Rodriguez-Clare, 1996). There are increasing returns for knowledge-based industries (Arthur, 1996), which are contrary to the classical economic equilibrium theory, in which the larger the return, the more attracted companies will be to enter the business or start producing the product or service (up to the point of diminishing the return).

Increasing return on products and services, on the other hand, presents a positive feedback behavior, creating instability in the industry. Increasing returns enable companies on the leading edge to advance further ahead of those companies that are slower in research and development of new products and technologies. A classical example is the DOS operating system developed by Microsoft, which had a lock-in with the spreading of the IBM PC, as the most popular computer platform, which enabled Microsoft to distribute its costs over a large number of users and obtain unprecedented margins. Companies with no research and development facilities need to collaborate with downstream enterprises (Wise and Baumgartner, 1999), and look at the value chain from the customers' perspective to detect opportunities, following one of the downstream models presented in Table 6.2.

Table 6.2 Downstream models and e-Collaboration (adapted from Wise and Baumgartner, 1999).

Model	Characteristics	Example
Embedded services	Embedding downstream services into the product and freeing customers from such services	Honeywell and its Airplane Information Management System
Comprehensive services	Covering downstream services that are not possible to embed into the product, e.g., financial services	General Electric in the locomotive market
Integrated solutions	Combining products and services to address customers' needs	Nokia, array of products for mobile telephony
Distribution control	Moving forward over the value chain to take control over distribution activities	Coca-Cola's worldwide distribution network

The ability to respond quickly and effectively to the customers' requirements is essential, which depend highly on the available e-Work technologies and infrastructures. Agility, quality, traceability, and service-ability enable networked SMEs (Small-to-Medium-sized Enterprises) to enhance their performance through cooperative processes. Virtual enterprises provide SMEs with the capability of collaborative design and supply of high quality and customized products through (Yusuf and Sarhadi, 1999):

- Identifying and creating value.
- Enabling the flow of value stream.
- Allowing customers to pull value uninterrupted.
- Responding to unpredictable change.
- Forming tactical and virtual partnerships.

which necessitate collaborative processes for effective integration of technologies, people, and business processes.

6.3.1 Collaborative Processes

As explained earlier in this book, e-Work (Nof, 1999, 2000a, 2000b, 2003), refers to collaborative, computer-supported, and communication-enabled productive activities in highly distributed networks of humans and/or robots or autonomous systems. The purpose is to effectively design and control other emerging e-Systems and their respective e-Activities, namely, activities based on and executed through information technologies, including v-(virtual) Design, e-Business, e-Commerce, intelligent robotics, e-Manufacturing, v-Factories, e-Logistics, i-(intelligent) Transportation, v-Enterprises, telemedicine, e-Learning, e-Training, navigation and security monitoring by geographic information systems (GIS), and more. All these e-Activities rely on computer support and communication technologies, and all of them require collaboration in their inherent interactions between machines, humans, and computers. Under the overload of tasks and information, over wide distribution and time constraints, collaboration must be engineered and optimized among people, machines, and computers. Without effective collaboration, the potential of emerging and promising electronic work activities, e-Work, cannot be fully materialized (Nof, 2003). The view of e-Work as the foundation for e-Systems and e-Activities implies that work abilities, including design and development of products and services, can be significantly improved through collaborative processes. Examples include

1. Better connectivity and ability for farther, global reach.
2. Enhanced communication and coordination.
3. Acceleration of knowledge sharing and distribution.
4. Better interactivity, flexibility, customization-ability.
5. Lower agency cost by replacing certain human tasks with software agents' tasks.

Hence, collaborative e-Work increases the opportunities for more productive and competitive supply of products and services, improving company's competitiveness. Some fundamental motivations are:

1. Greater work complexity over distributed supply and service networks.
2. Increasing inter-dependence, e.g., among functions and suppliers.
3. Issues of integrity and trust, e.g., between companies and customers.
4. Greater need for coordination, cooperation, and synchronization.
5. Problems of mismatch, e.g., inconsistent versions.

Each of the fifteen e-Dimensions associated with the four "wheels" of e-Work, discussed in Chapter 1, can be directly linked to the collaborative processes associated with the development of products and services through collaborative e-Work (Table 6.3).

Table 6.3 Fifteen e-Dimensions of e-Work (Nof, 2003; 2007) and e-Collaboration.

	e-Dimension	Role	e-Collaboration
A: e-Work	A1. Theory and models	Augment human abilities to work	Development and realization of the theory and models for specific applications
	A2. Agents	Useful for creating systems that are flexible, adaptive to changes, and able to integrate heterogeneous components and tasks	Development of agent system and human-agent integration to assists interaction processes
	A3. Coordination Protocols	Efficient coordination processes will lead to effective collaboration, especially in large distributed systems	Coordination of multi-participant negotiations, group decision mediation, conflict management
	A4. Workflow	Enabling process scalability, availability and performance reliability	Better collaboration effectiveness, Including detecting and solving errors when they occur
B: Integration, Coordination, and Collaboration	B1. Theory and models	Helping system developers identify the link between information resources and information processing entities, to improve the performance of collaborative activities	Decision making by distributed and autonomous team members, and formulation of resolutions among distributed collaborators
	B2. Human-Computer Interaction (HCI)	Develop systems that support people in their role as learners, explorers, and workers	Beyond exploring installed bodies of information, discover new knowledge via intensive information exchange, similar to human brainstorming
	B3. Computer Integrated Manufacturing (CIM)	Integration and management of the entire product life -cycle	1) Distributed tools govern flow under time, budget, manufacturing, environmental constraints; 2) Embedding MEMS/nano sensors for intelligent real-time control, interactive decisions
	B4. Extended enterprises	Integration of business processes beyond the single enterprise, including end-to-end functions, internal operations, and entire supply chains	Real-time decision making, execution, and conflict management among collaborators over the supply network
C: Distributed Decision Support	C1. Decision Models	Reduce model construction time while increasing model quality; contribute to a wider applicability of online decision support systems	Interaction-based decision models with distributed knowledge sources enabling adaptive and more responsive decision structures
	C2. Distributed control systems	Allow parties to negotiate for their tasks and their benefits with some degree of autonomy	Systems integration with higher level of autonomy, such as supply networks

Table 6.3 (*continued*).

	e-Dimension	Role	e-Collaboration
	C3. Collaborative problem - solving	Information exchange to influence other parties' local decisions to act coherently and achieve better results for an entire organization	Consider collaboration among loosely- coupled entities where less mutual benefits are acceptable or expected by collaborators
D: Active Middleware	D1. Technology	Facilitating the development of distributed applications among heterogeneous, legacy and advanced environments	A very large scale computing platform serving highly heterogeneous, autonomous, and distributed components, beyond typical enterprise networks
	D2. GRID Computing	Computational economy and the development of grid computing architecture and its components, such as scheduler and resource allocator	Explore the potential of grid computing paradigm in both scientific and business applications (e.g., supply grid, grid optimization)
	D3. Distributed knowledge Systems	Maintain high reliability, integrity, and quality of service for large scale distributed information and knowledge networks	Better understanding of e-Trust among distributed information producers and consumers
	D4. Knowledge-Based Systems (KBS)	Construction and extraction of knowledge for variety of information needs	Assist human users in making better decisions, especially for complex, unstructured problems involving high levels of uncertainty

6.3.2 Requirements of Collaborative Processes

The competitiveness of companies in the dynamic, globalized, and geographically distributed markets depends highly on reduction of time-to-market and development cycles of products and services. Hence, enterprises frequently engage in collaborative partnerships in order to focus and coordinate their efforts, and establish themselves in the market. Collaboration results in integrated and aligned networks of enterprises, comprised of several independent companies, or subsidiaries. Along with intra-enterprise activities, companies often need to collaborate with external designers, technicians, departments, and divisions to attain the desired results in an efficient way (Figure 6.6). The increasing complexity then necessitates formation of concurrent design and management teams in order to integrate a wide range of expertise in different areas. This diversity of knowledge and insights provides the team with the view of the best processes and resources needed to handle the complexities more efficiently. Computer Supported Collaborative Work (CSCW) have been implemented largely for collaborative design and engineering (Phillips, 1998), along with more sophisticated techniques such as conflict resolution in distributed design (Nof and Huang, 1998, Huang et al., 2000).

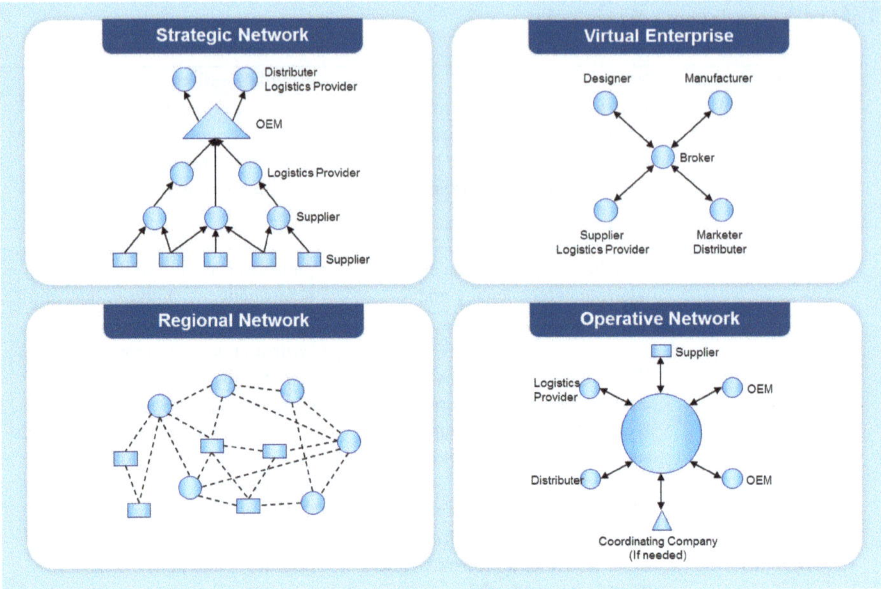

Fig. 6.6 The four general paradigms of manufacturing networks for collaborative product design and development (after Warnecke et al., 2001; with permission from John Wiley & Sons, Inc.).

Distribution of technologies among different SMEs, however, must be performed in such a way that in spite of exploiting value-creating capabilities, their core competencies are protected, especially in face of big suppliers, e.g., Bosch that is becoming bigger and stronger than its OEM (Karlsson and Sköld, 2007). Successful collaboration then requires effective classification of technologies as follows (Karlsson, 2003):

1. *Horizontal*: related to product functions, distinctive brand attributes, further related to missions of business and core competency of the company, mostly sourced from reliable and skillful suppliers at different levels.
2. *Vertical*: pure or state-of-the-art technologies, developed by company's research and development programs, mostly sourced from external research institutes.

The three building blocks of networked design and development of products (and services) are "actors", "activities", and "resources", where actors control resources and perform activities to change/exchange the resources, based on the available knowledge about the activities and resources (Figure 6.7). The OEM (Original Equipment Manufacturer) company and different tiers of suppliers are integrated and specialized actors (Brunsoni et al., 2001), performing different levels of activities on distributed resources. In this context, an "externalized enterprise", refers to the companies acting as "integrators" of information and knowledge sourced internally or externally, adapting, integrating, and reconfiguring distributed organization resources and functions in order to meet dynamic market requirements (Teece et al., 1997).

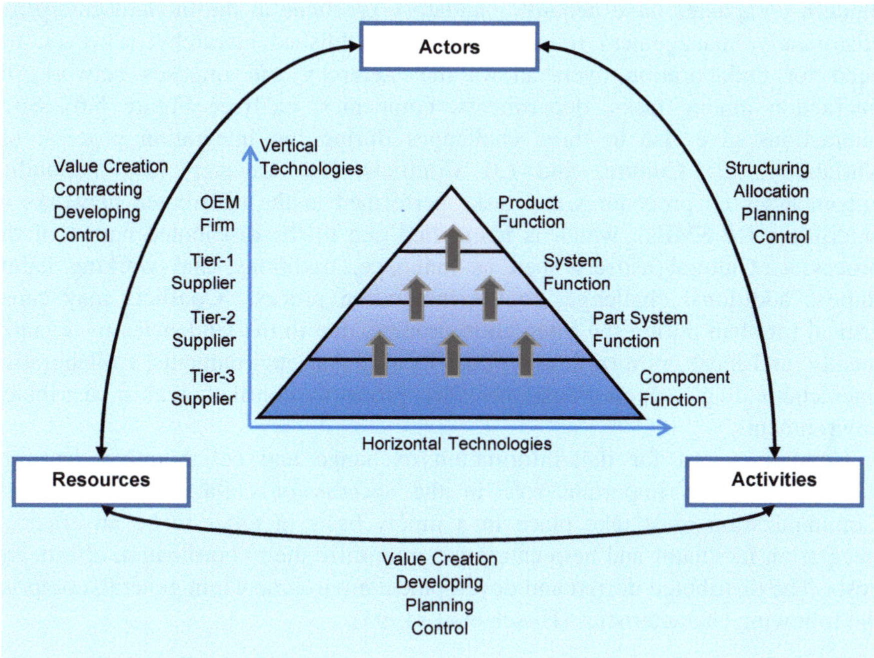

Fig. 6.7 Conceptual model for manufacturing network structuring and development with respect to horizontal and vertical technologies (adapted from Karlsson and Sköld, 2007).

Information technologies coupled with collaborative e-Work enable effective acquisition of information and objective analysis by distributed teams through computer-supported design and development of products and services. The emerging challenges, however, are information overload, lack of knowledge integration, cooperation, and coordination among team members, and utilization of collaboration-support tools. Adaptive reconfiguration is the other key aspect that must be taken into account, considering constantly changing conditions in industry. Numerous methods for quick reconfiguration of collaborative engineering initiatives have been developed, ranging from integration requirements of manufacturing activities (Khanna and Nof, 1994; Witzerman and Nof, 1995; Kim and Nof, 1997) to the concepts taken from disciplines other than manufacturing such as parallel/distributed computing (Khanna et al., 1998; Ceroni, 1999; Ceroni and Nof, 1999).

6.3.3 Framework for Collaborative Processes

The distributed environments present challenges for design, management, and operational functions. Integrated approaches for design and management of modern companies have become mandatory practices in the modern enterprise. Historically, management relied on a well-established hierarchy; however, the need for collaboration overshadows the hierarchy and imposes networks of interaction among tasks, departments, companies, etc (see Figure 6.6). Such interactions give rise to three challenges during the integration process: (1) Variability, (2) Culture, and (3) Conflicts (see Chapter 11). Variability encompasses all procedures and tasks performed in the distributed networks of enterprises (or SMEs), which is intensified due to the distributed nature of the processes. Cultural aspects such as language, traditions, and working habits impose additional challenges to the integration process. Conflicts may cause critical problem during the integration process, due to the tendencies to organize locally, and local optimizations in dual local/global environments. Collaborative interactions (e.g., supplier-customer) are prone to conflicts within distributed environments.

Communication for data/information exchange and collaborative decision-making plays an important role in the success of collaborative processes. Communication must take place in a timely basis in order to be an effective integration facilitator and help enterprises minimize their coordination efforts and costs. The distributed design and development environments, in general, comprise the following characteristics (Hirsch et al., 1995):

1. Cooperation of different/independent enterprises.
2. Shifting of project responsibilities during the product life cycle.
3. Different conditions, heterogeneity, autonomy, and independence of the participants' hardware and software environments.

In line with the above characteristics, the following requirements must be addressed in order to establish guidelines for effective integration of distributed enterprises:

1. Support of geographically distributed systems and applications in multi-site production environments, and in special cases, support of site-oriented temporal manufacturing.
2. Consideration of heterogeneity of systems ontology, software, and hardware platforms and networks.
3. Integration of autonomous systems within different enterprises (or enterprise domains) with unique responsibilities at different sites.
4. Provision of mechanisms for business process management to coordinate the information flow within the entire integrated environment.

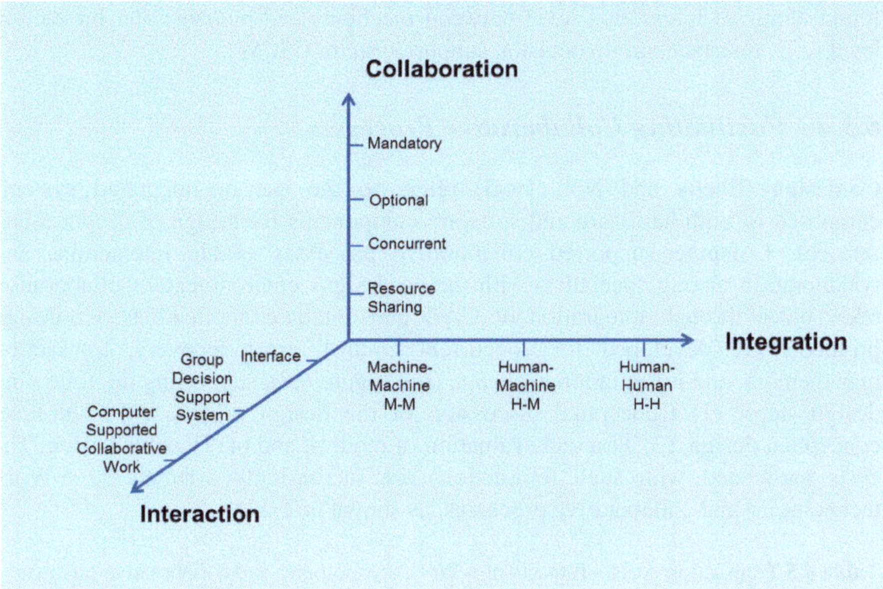

Fig. 6.8 Integration Taxonomy.

Table 6.4. Example of collaboration types for integration problems (Nof, 1994).

	Integration Problem	Example	Collaboration Type	
1	Processing of task without splitting and with sequential processing	Concept design followed by physical design	H-H*	Mandatory, sequential
2	Processing of task with splitting and parallel processing	Multi-robot assembly	M-M*	Mandatory, parallel
3	Processing of task without splitting; Very specific operation	single human planner (particular)	H-M	Optional, similar processors
4	Processing of task without splitting; General task	single human planner (out of a team)	H-M	Optional, different processors types
5	Processing of task can have splitting	engineering team design	H-H	Concurrent
6	Resource allocation	Job-machine assignment	M-M	Competitive
7	Machine substitution	Database backups	M-M	Cooperative

* H: Human; M: Machine.

Further efforts are also required to construct a framework for collaborative manufacturing, following Nof´s taxonomy of integration (Figure 6.8 and Table 6.4). The integration taxonomy classifies collaboration into four types: mandatory, optional, concurrent, and resource sharing. Each of these collaboration types is found along an integration level between machines and humans, and interaction level (e.g., interface, group decision support system, CSCW).

6.3.4 Facilitating Collaborative Processes

Co-design (Eberts and Nof, 1995) refers to the use of integrated systems composed of both hardware and software components for design of products and services. Computer-supported collaborative processes enable interactions and collaboration among specialists with the co-designs environments. Collaboration takes place through integration of CAD and database applications. Co-design protocols are developed for concurrent control, error recovery, transaction management, and information exchange (see Figure 6.8), supporting the following design steps: (1) Conceptual discussion of the design project, (2) High-level conceptual design, (3) Test and evaluation of models, and (4) Documentation. The costs associated with such transactions are increasingly affected by e-Work technologies and collaborative processes, as shown in Table 6.5.

Table 6.5 Transaction costs – Impacts of e-Work technologies and collaborative processes.

No.	Conventional (Busalacchi, 1999)	Emerging
1	Searching for suppliers/customers	Coordination between potential suppliers/customers
2	Identifying the nature of products/services	Rapid access to information about products and services in progress
3	Negotiating the terms for products/services	Means for rapid negotiation of terms between suppliers and customers
4	Selection of suppliers and vendors	Access to evaluation criteria for suppliers and vendors
5	Monitoring the design and supply to ensure quality, quantity, and timeliness	Efficient mechanisms for ensuring the quality and quantity of products and services
6	Enforcing compliance with the agreement	Efficient mechanisms for enforcing compliance with contracts and agreements

6.3.5 Enabling Collaborative Processes

Two major prerequisites for effective e-Work are autonomous agents and active protocols. Their main role is to enable efficient information exchanges at the application level, and administer tasks to ensure smooth, efficient interactions, collaboration, and communication to augment the natural human abilities (see Chapter 2). In an analogy with the massively parallel/distributed/network computing infrastructures, the Teamwork Integration Evaluator (TIE) (Nof and

Huang, 1998, Huang and Nof, 1999, 2002, Anussornnitisarn and Nof, 2000, Williams and Nof, 1995, Williams et al., 2003) enables e-Collaboration through parallel simulation of distributed, networked teams of operators (human, robots, and agents). Three versions of TIE have been implemented with the Message Passing Interface:

1. *TIE/Design*: models integration of distributed designers or engineering systems (Figure 6.9) – implemented on Intel's Paragon.
2. *TIE/Agent*: analyzes the viability of distributed, agent-based manufacturing enterprises (Figure 6.10) – implemented on networks-of-workstations.
3. *TIE/Protocol*: models and evaluates the performance of different task administration active protocols, e.g., in integrated assembly-and-test networks (Figure 6.11) – implemented on Silicon Graphics' Origin-2000.

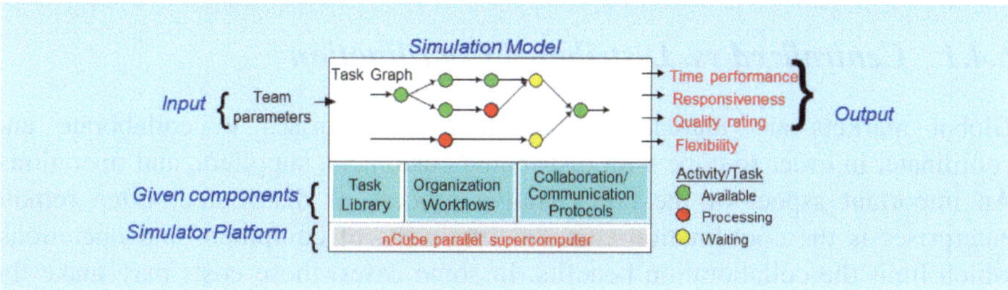

Fig. 6.9 TIE/Design.

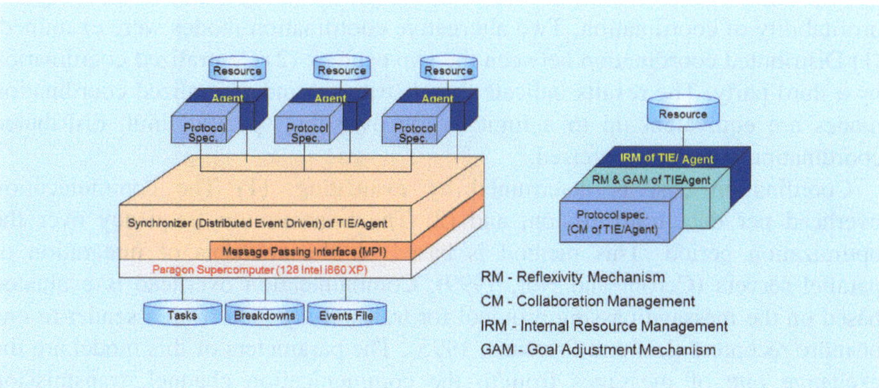

Fig. 6.10 TIE/Agent.

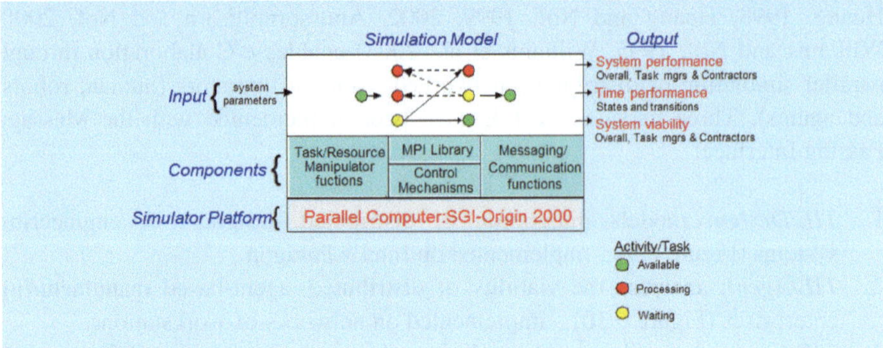

Fig. 6.11 TIE/Protocol.

6.4 Case Illustrations with Collaborative Processes

6.4.1 Centralized vs. Distributed Coordination

Global markets are increasingly demanding enterprises to collaborate and coordinate, in order to cope with distributed customers, suppliers, and operations. An important aspect of the collaborative process of distributed, often remote enterprises is the coordination cost, i.e., the costs of equipment and operations, which limit the collaboration benefits. In some cases, these costs may make the interactions among distributed enterprises non-profitable. Previous research investigated a distributed enterprise case, operating under a job-shop model with two distributed collaborating centers, one for sales and one for production. A new model incorporating the communication cost was developed (Ceroni et al., 1999; 2004), which calculates the net rewards of the entire system, determining the profitability of coordination. Two alternative coordination modes were examined: (1) Distributed coordination between the two centers; (2) Centralized coordination by a third party. The results indicate that distributed and centralized coordination modes are equivalent up to a limit; while over this specific limit, distributed coordination is always preferred.

Coordination cost is determined by evaluating: (1) The communication overhead per data transmission, and (2) The transmission frequency over the optimization period. This method is based on the concepts of integration of parallel servers (Ceroni and Nof, 1999). Communication overhead is evaluated based on the message passing protocol for transmitting data from a sender to one or more receptors (Lin and Prassana, 1995). The parameters of this model are the exchange rate of messages from/to the communication channel, transmission startup time, data packing/ unpacking time from/to the channel, and the number of senders/receptors. The coordination of distributed sales and production centers is modeled for investigating the benefits of different coordination modes (i.e., distributed and centralized).

Results show that the coordination cost and the number of negotiation iterations should be considered in the decision on how to operate the system. The numerical results indicate: a) same break-even point for the distributed and centralized modes at 400 iterations and 2 jobs per time unit; b) the lowest break-even point for the centralized mode at 90 iterations and 5 jobs per time unit; c) consistent better profitability for the centralized mode. This effect is explained by lower communication requirements and competitive hardware investment in the centralized mode; d) the distributed mode with consistent better profitability than the centralized mode at a higher hardware cost. This implies that distributed coordination should be preferred at a hardware cost less than half of that required by the centralized coordination mode.

From this analysis, the limiting factor for selecting the coordination mode is given by the hardware cost, with the distributed and centralized modes becoming comparable for a lower hardware investment in the centralized case. Coordination of distributed parties interacting for attaining a common goal is also demonstrated to be significant (Ceroni and Nof, 1999) with the inclusion of parallel servers in the system.

6.4.2 Collaboration in Distributed Design and Manufacturing

MANACO S.A. is a Bolivian subsidiary of BATA International, a Czech shoemaker company with subsidiaries in most of South American countries as well as Canada and Spain. The company comprises several enterprises inside the country along with international enterprises located in La Paz and Cochabamba (about 250 miles apart). The design process at MANACO is performed in collaboration with designers in Italy, by developing prototype products for testing in the local market. The prototypes are implemented at La Paz enterprise and then released to the production at the Cochabamba enterprise. This case study analyzes the integration of prototype production and production planning operations being performed at distributed locations by applying the Distributed Parallel Integration Evaluation Model (Ceroni, 1996; 1999; Ceroni and Nof, 2005).

The production planning operation (Enterprise A) consists of six tasks, where some of them are executed in parallel. The prototype development operation (Enterprise B) consists of four tasks, all performed sequentially. In order to compare the scenarios with and without collaborative integration, two alternative solutions are developed. The first solution considers sequential processing by the enterprises: The prototype is developed collaboratively with Italian designers and implemented at La Paz enterprise. The results are then sent to the Cochabamba enterprise for production planning. Parallelism with collaboration is included at each enterprise for reducing the individual execution cycles. The second solution considers the collaborative integration and the inclusion of parallelism into both enterprises, simultaneously. An integration model is developed for the parallelism optimization.

6.5 Integrated Optimization

The integration process of both enterprises is simplified by assuming collaborative processes between tasks pertaining to different enterprises. A collaborative process denotes an association of the tasks based on the similarities observed in the utilization of information, resources, personnel, or the pursuing of similar objectives. Task collaborations allow construction of an integrated model for the enterprises. This integrated model preserves the execution order of the tasks considering the local models.

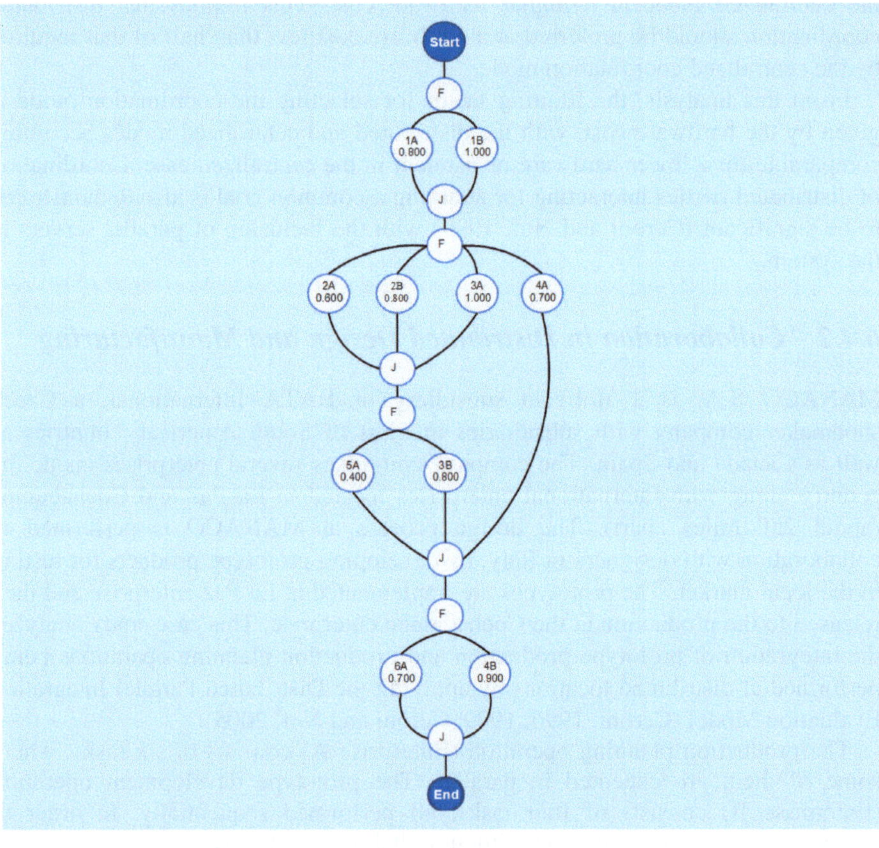

Fig. 6.12 Integrated enterprises model with collaborative processes.

Upon generation of the integrated model (Figure 6.12), parallelism analysis is performed. To evaluate the parallelism degree of the system, a measure of the communication and congestion delays needs to be estimated. The estimation of these delays is performed using the TIE 1.4 software (Khanna and Nof 1994; Huang and Nof 1998). TIE 1.4 allows the simulation of a network of distributed tasks with an Intel Paragon Supercomputer. TIE 1.4 uses a message passing mechanism for communicating data among the computer nodes simulating the tasks. The data exchange for collaborative processes can take place either synchronously or asynchronously. In the synchronic data exchange, the activity of the sending processor is stalled while waiting for confirmation from the receiver processor. In the asynchronous data exchange, the sending processor does not wait for confirmation from the receiving nodes and continues its activities.

6.6 Analysis Results

Three models are simulated: Enterprise A, Enterprise B, and Integrated Enterprise Operation, to observe in each case the production (**Π**), interaction (**T**), and total (**Φ**) times, and the degree of parallelism (**Ψ**), which is the collaborative concurrency measure for the system. The simulation of individual enterprises enables estimation of the delay times caused by communication and flow congestion, which necessitate optimization of the operations locally. The numerical results are presented in Table 6.6.

Table 6.6 Summary of results for the local and integrated scenarios.

Scenarios	**Φ**	**Π**	**T**	Number of sub-tasks
Local scenario	1.9846	1.4239	0.5607	52
Enterprise A	1.3908	1.0350	0.3558	16
Enterprise B	0.5938	0.3889	0.2049	36
Integrated scenario	1.8071	1.1666	0.6404	26

The results show a slight difference with respect to the total production time, which seems to contradict the hypothesis that the integrated scenario will benefit from a reduction of the total cycle time. However, it must be noted that the number of sub-tasks required by the local scenario for achieving an equivalent total production time is two times more than the number of sub-tasks required by the integrated scenario. This situation is the result of imposing no constraints on the maximum number of sub-tasks regarding each task in each scenario (i.e., infinite divisibility assumption for tasks). In particular, for Enterprise B, the number of sub-tasks in which each task is divided is nine, which can be considered as excessive given the total number of four tasks in this enterprise.

For evaluating the comparative performance of the integrated and local scenarios, a local scenario for each enterprise has been designed according to the number of sub-tasks in the integrated scenario. Therefore, the number of sub-tasks is set to 15 for enterprise A and 11 for enterprise B, resulting in 26 sub-tasks in both local enterprises, which is equal to the number of sub-tasks for the integrated scenario. The values for the performance parameters show a total best production time (**Π+Φ**) of 2.7769 shifts, direct-production time (**Π**) of 2.3750, and interaction time (**Φ**) of 0.4019 shifts. These values represent 54% increase in the total production time, 104% increase in the production time, and 37% decrease in the interaction time, compared to the integrated solution.

The result obtained from this illustration reinforces the hypothesis that potential benefits can be exploited during the optimization of the parallelism of integrated distributed operations. The key issues are the communication and congestion modeling and the corresponding assumptions. The modeling of time required by tasks for data exchange relates to the problem of coordination of cooperating servers. On the other hand, the congestion modeling relates to the delays resulted from task granularity (i.e., the number of activities being executed, concurrently).

6.6.1 Variable Production Networks

To achieve and protect their core competencies, and improve their business performance, enterprises are forced to collaborate closely with their suppliers and customers (Lutz et al., 1999). The supply collaboration concept requires formation and optimization of production or supply networks (Figure 10), characterized by intensive collaborative communication between the distributed entities. The goal is to allocate among the collaborating partners the production demand. This capability provides the entire network with the required flexibility to respond quickly to demand disruptions for the products and services. The FAS/net, a tool developed at the Institute of Production Systems of the Hanover University, employs basic methods of production logistics that provide procedures for the efficient use of capacity redundancies in a production network (Figure 6.13). The tool satisfies the following requirements derived from the capacity subcontracting (outsourcing) process:

1. Monitoring resource availability and order status throughout the network.
2. Internal (within the enterprise) and external (between partners) monitoring.
3. Support of different network partner perspectives (supplier and producer) for data capsulation.
4. Detection of the logistics bottlenecks and capacity problems.

A key aspect of the system is the identification of orders that the partner is unable to produce (due to capacity shortage, for example). This is accomplished by detecting the bottlenecks through the concept of "degree of demand" (a comparison of the capacity needed and available in the future, expressed as the ratio between the planned input and the capacity). All the systems with potential to create bottlenecks are identified by the FAS/net system and ranked by their degree of demand. The subcontracting of the orders is then performed by alternative criteria such as history, production costs, or throughput time. The system relies on the confidence between partners and the availability of communication channels among them. Carefully planned agreements among the partners, concerning the legal aspects, duties, responsibilities, and liability of the exchanged information are the main obstacles to implement production networks.

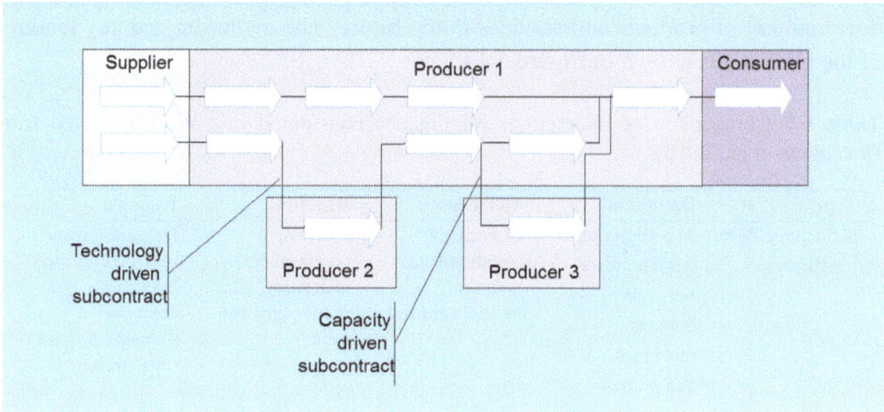

Fig. 6.13 Production networks (adapted from Lutz et al., 1999).

6.7 Collaborative Intelligence (CI)

One of the emerging issues in collaborative e-Work, in general, and collaborative design and development of products and services, in particular, is finding the best collaborators during the formation and execution stages of collaborative networks. Collaboration, by itself, does not necessarily guarantee success of the design and development of a product or service. It is critical to identify the target enterprises and optimize the timing for collaboration, according to a valid and all-inclusive set of measures. CI, the Collaborative Intelligence, is a measure proven to gauge the optimality of collaboration through assisting the decision-makers in evaluation of collaboration effects ahead of time, considering future requirements. Table 6.7 presents the main types of interaction intelligence considered in e-Work, all sharing the premise of cyber-supported interaction and capabilities of collaborators to yield better outcomes (Nof, 2003). CI aggregates all four types of interaction intelligence, i.e., (1) Understanding of the exact services/tasks requested (communication); (2) Resolving errors/conflicts and performing online and real-time maintenance (cooperative); (3) Optimizing solutions through learning from similar services/tasks (collective); and (4) Avoiding past mistakes (cumulative).

The PRISM (Production, Robotics, and Integration Software for Manufacturing & Management) center at Purdue University is developing a HUB platform with CI capabilities (HUB-CI) that supports collaborative activities such as team forming, data/information/knowledge sharing, and collaborative decision-making, and enables synchronous/asynchronous collaboration for information and tool exchange (Seok and Nof, 2011). The purpose is to enable autonomous users to identify and decide what knowledge to contribute, which tools to use for analysis, and with whom to collaborate. HUB-CI is a next generation collaborative e-Work platform anticipated to be the foundation for cyber-supported collaborative development of products and services in the future. The evolution and key features of the HUB-CI is shown in Figure 6.14.

Table 6.7 Comparison of interactions intelligence considered in e-Work (adapted from Devadasan et al., 2013).

Type	Definition	Methods	Benefits	Example
Communication intelligence	The ability to communicate between humans, machines, and human - machine systems	Effectively communicate between multiple entities	Increase communication efficiency, and reduce cost and time	Negotiation and making contracts between business partners; process outsourcing
Cumulative intelligence	The ability to accumulate knowledge and form a base for decision-making	Make decisions based on present state and also previous states	Identify positive and negative strategies from historical data	Research activities (standing on the shoulders of giants)
Cooperative intelligence	The ability to deal with cooperation among multiple partners	Prevent conflicts during real time parallel execution	Reduce down time and interruptions due to conflicting actions	Concurrent assembly operations along a conveyor
Collective intelligence	The ability to integrate intelligence from a group/organization and to act, even approximately, as a single, rational agent	Collect and combine knowledge from disparate sources	Increase the probability of better decisions	Trade promotions suggestions for client and customer
CI, Collaborative Intelligence	The ability of an entity or a group to collaborate well with others	Collaborate towards a set of common goals while keeping autonomy	Streamline the workflow of information and operations for overall welfare	Between knowledge intensive clients and e-Service providers; between retailers, manufactures & suppliers (Nestle, Wal-Mart)

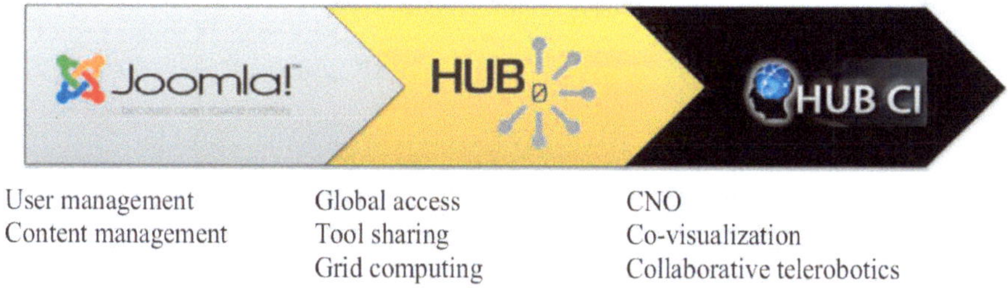

Fig. 6.14 HUB-CI evolution; Joomla! (joomla.org) enables the management of users and contents for a collaboration system; HUBzero® (hubzero.org) elevates the collaboration to a higher level with computational tool and knowledge sharing, along with accesses from throughout the globe to high performance computing clusters.

6.8 CI-Based Optimization Model

A multiobjective optimization model (Devadasan et al., 2013), is used for calculating the CI measure of knowledge-based services required for cyber-supported collaborative development of products and services. The decision-maker must assign a set of initial weights to the input parameters used for determining the set of optimal operating points. The model then assists the decision-maker to define the most suitable operating points out of a set of feasible alternatives, based on the calculated CI. The presented methodology requires the decision-makers to know about the partnership and access to future forecasts of the company relating to its infrastructure. The primary assumptions are as follows

1. All input parameters are quantifiable.
2. The decision-maker has sufficient knowledge about the industry and all potential collaborating partners.
3. Only two collaborators are considered at any point, and each company can handle only one service project.
4. All required data are accurate and available to the decision-maker.
5. The collaborators are involved in relatively long-term contracts.

The CI-based model calculates the CI level for each collaborator. One of the main indicators of collaboration performance is the "defects", as the main cause of disrupting the streamlined workflow of two collaborating entities. Let n and $D=\{d_i \mid i=1,\ldots,n\}$ denote the total number and the set of defects, respectively. Then, defining the costs associated with each resource r_j, $j=1,\ldots,m$, as

$$C = \sum_{j=1}^{m} u_j r_j, \tag{1.6}$$

where

C	Total resource cost	u_j	Unit cost of resource r_j

the following three objectives of the CI-based model must be optimized:

1. *Minimize total defects*: reduces the total number of defects, but may increase the total resource cost and implementation time

$$\min Z_1 = \sum_{i=1}^{n} E(d_i) - k_1 C - k_2 T, \tag{2.6}$$

2. *Minimize total cost*: reduces the total cost, but may have negative or positive impact on the total number of defects

$$\min Z_2 = C + k_3 \sum_{i=1}^{n} E(d_i) + k_4 T, \tag{3.6}$$

3. *Performance improvement over time*: requires increasing improvement from the collaborators and thereby increases the CI level

$$\min Z_3 = \sum_{i=1}^{n} E(d_i) e^{-k_5 T}, \tag{4.6}$$

subject to satisfaction of the following set of constraints:

$$0 \le C \le C^{\max} \tag{5.6}$$

$$0 \le T \le T^{\max} \tag{6.6}$$

where

Z_1	Total defects objective	Z_2	Total costs objective
Z_3	Performance improvement objective	$E(d_i)$	Existence function for defect d_i
T	Implementation and integration time	k	Normalization weight parameters
C^{max}	Maximum allowable value for C	T^{max}	Maximum allowable value for T

Combining objectives (2.6), (3.6), and (4.6) into one objective function as

$$f(Z_1, Z_2, Z_3) = w_1 Z_1 + w_2 Z_2 + w_3 Z_3, \tag{7.6}$$

the CI of a given collaborator can be obtained through the following formula:

$$CI = \frac{\alpha}{f(Z_1, Z_2, Z_3)}, \tag{8.6}$$

where

w	Relative weights of objectives	α	Scalar factor for standardization of CI

According to the CIs obtained from the optimization model presented above, the decision-maker can select the best set of collaborators. The presented methodology for calculating the CI, however, can be modified with respect to the defined parameters and objectives for solving different case problems of e-Collaboration. The decision-maker is enabled to tailor the model input parameters interactively, and reflect the implementation based on his/her knowledge of the market and company's policies. The decision-maker may also incorporate his/her expertise in selection of the collaborators, following the recommendations provided by the CI-based optimization model. The optimization model can be revised and updated in order to maintain the validity of the results and recommendations.

6.9 Emerging Trends and Challenges

With PLM playing a major role in companies associated with service-centric and customer-centric strategies, the role and significant benefit of collaboration-based PLM have become evident. Several companies have begun using the term CLM, Collaborative Life-Cycle Management. CLM highlights the value of PLM in the design and management of products and services, emphasizing the collaboration engineering required, and opening up the traditional PLM to embrace also e-Business and e-Service activities throughout extended, networked enterprises.

A prominent emerging trend that can be observed from the collaborative processes applied in and enabled by e-Work is the emergence of collaborative partnerships. In the past, the concept of giant, self-sufficient corporations that were active in several continents was prevailing. What is emerging and expected frequently in the years to come is agile enterprises willing to take advantage of and participate in collaborative partnerships. Hence, the times of "winning all" or "losing all" ("zero-sum game") are behind us. What now matters is to stay in business as competitively as possible. Collaborative efforts move beyond the manufacturing function downstream, where significant opportunities exist. Manufacturing enterprises are integrating and aligning with the rest of value chains, leaving behind concepts such as "product marginal profit", which controlled operations for so long. The future will be one of flexible and adaptive channels for acquiring supplies, producing, delivering, and after-sale services. Companies will continue to drive inefficiencies out and become agile, optimized systems, while forcing companies around them to move in the same direction. At the same time, they should be attentive to the efficiency and effectiveness of their e-Collaborative partnerships through continuous calculations and update of CIs, considering additional factors such as capacity and number of resources involved in collaboration, communication costs, cultural effects, etc. (see also Chapter 11 for details).

References

Anussornnitisarn, P., Nof, S.Y.: e-Work: The Challenge of the Next Generation ERP Systems. Production Planning & Control 14(8), 753–765 (2003)

Arthur, W.B.: Increasing Returns and the New World of Business. Harvard Business Review, 100–109 (July-August 1996)

Barber, K.D., Beach, R., Zolkiewski, J.: Environmental sustainability: A value chain research agenda. Production Planning and Control 1-3, 105–119 (2012)

Brunsoni, S., Prencipe, A., Pavitt, K.: Knowledge specialization, organizational coupling, and the boundaries of the firm: why do firms know more than they make? Administrative Science Quarterly 46, 597–621 (2001)

Busalacchi, F.A.: The Collaborative, High Speed, Adaptive, Supply-Chain Model for Lightweight Procurement. In: Proceedings of the 15th International Conference on Production Research, Limerick, Ireland, pp. 585–588 (1999)

Ceroni, J.A.: Models for Integration with Parallelism of Distributed Organizations. Ph.D. Dissertation, Purdue University, West Lafayette, IN (1999)

Ceroni, J.A., Nof, S.Y.: Planning Effective Parallelism in Production Operations. Research Memo 97-10, School of IE, Purdue University, W. Lafayette, IN (1997)

Ceroni, J.A., Nof, S.Y.: Planning Integrated Enterprise Production Operations with Parallelism. In: Proceedings of the 15th International Conference on Production Research, Limerick, Ireland, pp. 457–460 (1999)

Ceroni, J.A., Matsui, M., Nof, S.Y.: Communication Based Coordination Modeling in Distributed Manufacturing Systems. International Journal of Production Economics 60-61, 29–34 (1999)

Ceroni, J.A., Nof, S.Y.: Task Parallelism in Distributed Supply Organizations: A Case Study in the Shoe Industry. Production Planning and Control 16(5), 500–513 (2005)

Christensen, C.M.: Exploring the Limits of the Technology S-Curve. Part I: Component Technologies. Production and Operations Management 1(4), 334–357 (1992)

Devadasan, P., Zhong, H., Nof, S.Y.: Collaborative intelligence in knowledge based service planning. Expert Systems with Applications 40, 6778–6787 (2013)

Eberts, R.E., Nof, S.Y.: Tools for Collaborative Work. In: Proceedings of IERC 4, Nashville, TN, pp. 438–441 (1995)

Hirsch, B., Kuhlmann, T., Marciniak, Z.K., Mabow, C.: Information System Concept for the Management of Distributed Production. Computers in Industry 26, 229–241 (1995)

Huang, C.Y., Nof, S.Y.: Viability – A Measure for Agent-Based Manufacturing Systems. In: Proceedings of the 15th International Conference on Production Research, Limerick, Ireland, pp. 1869–1872 (1999)

Huang, C.Y., Nof, S.Y.: The Production Robotics and Integration Software for Manufacturing (PRISM): An Overview. Research Memorandum No. 98-3, School of IE, Purdue University, USA (1998)

Huang, C.Y., Nof, S.Y.: Evaluation of Agent-Based Manufacturing Systems Based on a Parallel Simulator. Computers and IE 43(3), 529–552 (2002)

Huang, C.Y., Ceroni, J.A., Nof, S.Y.: Agility of Networked Enterprises. Parallelism, Error Recovery and Conflict Resolution. Computers in Industry 42, 257–287 (2000)

Karlsson, C., Sköld, M.: The manufacturing extraprise: an emerging production network paradigm. Journal of Manufacturing Technology Management 18, 912–932 (2007)

Karlsson, C.: The development of industrial networks: challenges to operations management in an extraprise. International Journal of Operations & Production Management 23, 44–61 (2003)

Khanna, N., Nof, S.Y.: TIE: Teamwork Integration Evaluation Simulator – A Preliminary User Manual for TIE 1.1. Research Memo 94-21, School of IE, Purdue University, West Lafayette, IN (1994)

Khanna, N., Fortes, J.A.B., Nof, S.Y.: A Formalism to Structure and Parallelize the Integration of Cooperative Engineering Design Tasks. IIE Transactions 30(1), 1–15 (1998)

Kim, C.O., Nof, S.Y.: Coordination and Integration Models for CIM Information. In: Tzafestas, S.G. (ed.) Knowledge Based Systems, pp. 587–602. World Scientific Publishers (1997)

Kiritsis, D.: Product Lifecycle Management and Embedded Information Devices. In: Nof, S.Y. (ed.) John Wiley and Sons Handbook of Automation, ch. 43, pp. 749–765 (2009)

Lin, C., Prassana, V.: Analysis of Cost of Performing Communications Using Various Communication Mechanisms. In: 5th Symposium Frontiers of Massively Parallel Computation, USA, pp. 290–297 (1995)

Lutz, S., Helms, K., Wiendahl, H.P.: Subcontracting in Variable Production Networks. In: Proceedings of the 15th International Conference on Production Research, Limerick, Ireland, pp. 597–600 (1999)

Ming, X.G., Yan, J.Q., Lu, W.F., Ma, D.Z.: Technology Solutions for Collaborative Product Lifecycle Management – Status Review and Future Trend. Concurrent Engineering 13(4), 311–319 (2005)

Newman, B.: An Open Discussion of Knowledge Management. The Knowledge Management Forum (August 2002),
http://www.km-forum.org/what_is.htm

Nof, S.Y.: Integration and Collaboration Models. In: Nof, S.Y. (ed.) Information and Collaboration Models of Integration. Kluwer Academic Publishers, Dordrecht (1994)

Nof, S.Y., Huang, C.Y.: The Production Robotics and Integration Software for Manufacturing (PRISM): An Overview. Research Memorandum No. 98-3, School of IR, Purdue University, West Lafayette, IN (1998)

Nof, S.Y.: Modeling of e-Work with TIE, a teamwork integrator evaluator. Research Memo 99-20 (School of Industrial Engineering, Purdue University) (1999)

Nof, S.Y.: Modeling e-Work and conflict resolution in facility design. Proceedings of the 5th International Conference on Computer Simulation and AI, Mexico City, Mexico (February 2000a)

Nof, S.Y.: Models of e-Work. In: Proceedings of the IFAC/MIM-2000 Symposium, Patras, Greece (July 2000b)
Nof, S.Y.: Design of effective e-Work: Review of models, tools and emerging challenges. Production Planning and Control, Special Issue on e-Work: Models and Cases 15(8), 681 (2003)
Nof, S.Y.: Collaborative control theory for e-Work, e-Production, and e-Service. Annual Reviews in Control 31, 281–292 (2007)
Phillips, C.L.: Intelligent Support for Engineering Collaboration. Ph.D. Dissertation, Purdue University, West Lafayette, IN (1998)
Rodriguez-Clare, A.: Multinationals, Linkages, and Economic Development. The American Economic Review 86(4), 852–873 (1996)
Romer, P.: Idea gaps and object gaps in economic development. Journal of Monetary Economics, 543–573 (1993)
Sahlman, W.A.: The New Economy Is Stronger Than You Think. Harvard Business Review, 99–106 (November-December 1999)
Seok, H., Nof, S.Y.: The HUB-CI initiative for cultural, education and training, and healthcare networks. In: 21st International Conference on Production Research, Stuttgart (August 2011)
Stark, J.: Product Lifecycle Management. Springer, London (2011)
Stephan, P.E.: The Economics of Science. Journal of Economic Literature XXXIV, 1199–1235 (1996)
Teece, D.J., Pisano, G., Shuen, A.: Dynamic capabilities and strategic management. Strategic Management Journal 18, 509–533 (1997)
Warnecke, H.J., Sihn, W., Briel, R.: The Factory of the Future: New Structures and Methods to Enable Transformable Production. In: Handbook of Industrial Engineering: Technology and Operations Management, 3rd edn., ch. 10, pp. 311–323 (2001)
Williams, N.P., Nof, S.Y.: TestLAN: An Approach to Integrated Testing Systems Design. Research Memorandum No. 95-7, School of IE, Purdue University, West Lafayette, IN (1995)
Williams, N.P., Liu, Y., Nof, S.Y.: Analysis of Workflow Protocol Adaptability in TestLAN Production Systems. IIE Transactions 35(10), 965–972 (2003)
Wise, R., Baumgartner, P.: Go Downstream, The New Profit Imperative in Manufacturing. Harvard Business Review, 133–141 (September-October 1999)
Witzerman, J.P., Nof, S.Y.: Integration of Cellular Control Modeling with a Graphic Simulator/Emulator Workstation. International Journal of Production Research 33, 3193–3206 (1995)
Yusuf, Y.Y., Sarhadi, M.: A Framework for Evaluating Manufacturing Enterprise Agility. In: Proceedings of the 15th International Conference on Production Research, Limerick, Ireland, pp. 1555–1558 (1999)

Review and Challenge Questions

1. Consider the portfolio of products and services offered by a given company, and analyze the evolving needs for e-Work processes and solutions that they would experience along their lifecycles.

2. (i) Explain why collaboration enabled by e-Work has become the most important enabling method in the development and maintenance services of products and services. (ii) Illustrate your answer in (i) for (a) digital cameras; (b) engineering design service.
3. (i) Compare several e-Work based solutions in fields such as logistics, manufacturing, or product service. (ii) Comment on their similarities and differences.
4. Discuss the role of knowledge in service and product development, and the role of e-Work tools in it.
5. (i) Analyze the benefits and risks related to virtual enterprises in modern organizations; (ii) Compare those benefits and risks in the others forms of (non-virtual) manufacturing networks.
6. Elaborate on the specific benefits reported by taxonomies in general, and the integration taxonomy in particular.
7. (i) Explain the trade-off between processing and communication times in collaborating systems; (ii) Extrapolate the trade-off to other relevant performance measures in modern organizations.
8. Consider the risks involved in production networks and the remediation methods that companies may implement to control these risks.
9. Establish a simple criterion for determining the weighting factors of the multi-objective collaborative intelligence optimization problem.
10. Establish the future trends confirming that collaboration will continue to be a key factor in product and service development.

Lab Questions

11. Access and explore a PLM service on the Internet. Then explain: (i) Its main functionalities; (ii) The collaborative e-Work inherent in this system; (iii) Any collaboration engineering and CCT principles that can improve its own life cycle development.
12. (i) Explain the CI based optimization model. (ii) Illustrate its usefulness with simple numerical examples and assumptions.

Chapter 7
e-Logistics, e-Production, and e-Supply Networks

Abstract. Logistics is the organization and control of the flow of goods, information, and other resources in a cycle between the points of origin and consumption to meet specific requirements of different entities of a production, manufacturing, supply, or service network, e.g., customers or corporations. Logistics is a primary channel of any supply process that adds the value of time and place utility. Logistics involves the integration of information, transportation, inventory, warehousing, material handling, packaging, and occasionally security. Today, the complexity of production logistics can be modeled, analyzed, visualized, and optimized by plant simulation software. Various research areas and applications of logistics can be found in many references. However, this chapter focuses on material handling, especially in automated material handling, as a set of practical and important problems in the areas of production, manufacturing, and logistics ranging from the physical movement of products to storage and warehousing, and control and protection of products during various processes such as packaging, distribution, disposal, and so forth.

Modern production and manufacturing processes are increasingly exceeding boundaries of single sites, occurring within networks of geographically distributed plants and multiple levels of suppliers (as discussed in Chapter 6). The emerging concepts and paradigms of extended enterprise, as networks of collaborative self-organizing manufacturing enterprises, are becoming a prevailing solution to this transformation. In line with this evolution, a remarkable part of logistics processes are now taking place out of the boundaries of single enterprises and between several networked production and manufacturing sites. Extended enterprises enable the creation of architectures, models, and methodologies to integrate B2B and B2C processes beyond the boundaries of enterprises. The second half of this chapter is then devoted to the basic and emerging concepts in the design, planning, and control of distributed production and supply networks.

S.Y. Nof et al., *Revolutionizing Collaboration through e-Work, e-Business, and e-Service*, Automation, Collaboration, & E-Services 2, DOI: 10.1007/978-3-662-45777-1_7

7.1 Material Handling for e-Work

Material handling refers to the movement, storage, control, and protection of materials, goods and products throughout the process of manufacturing, service, distribution, consumption and disposal. The focus of material handling systems (MHS) is on the development of methods, mechanical equipment, systems and related control mechanisms used to accomplish these tasks. Specifically, MHS is a central part of a *flexible manufacturing system* (FMS), since it interconnects different processes of supplying and taking out raw material, work-pieces, sub-products, parts, and final products. With the increasing applications of automated production process, MHS must respond in concert with the timeliness for all emerging requirements of such processes and systems. Hence, major interests in the design of MHS mainly focus on the *integration*, *automation*, and *collaboration engineering* aspects. We will discuss fundamental concepts, configurations, and architectures of integrated and automated MHS throughout this section.

7.1.1 Concept and Configurations

An integrated, automated material handling system (IAMHS) is an MHS that integrates different types of automated material handling equipment into a single control environment. A pragmatic application of IAMHS is in seaport or airport cargo terminals, which are served by stacker cranes and automated guided vehicle (AGV) systems. The main concern of IAMHS in such applications is how to reduce the waiting time during job transition between two different MHSs in order to increase the throughput. For instance, an AGV must wait after arriving at a load position if a crane is not ready to unload a container to the AGV. If their jobs are not well synchronized, the waiting time will be longer, and as a consequence, the throughput will diminish. Accordingly, effective process integration and collaboration among distributed MHSs are critical for achieving higher efficiency and throughput.

IAMHS is increasingly becoming popular in a variety of industries. In digital picking systems, for instance, IAMHS facilitates the *pick-place* processes through displaying directions on a digital panel installed on the shelves. The IAMHS can also be operated on mobile handheld terminals, which provides several opportunities for warehouse management; the mobile IAMHS is equipped with an RFID (Radio Frequency Identifier) or barcode readers that allow flexible adaptation to changes in the distribution quantity. The wafer fabrication line in semiconductor industry is another application area of IAMHS. A fabrication line consists of more than 300 steps and 500 process tools. Material transportation between tools in a wafer fabrication line is fully automated by the overhead hoist transporter (OHT) system, which is a type of rail-guided vehicle (RGV) system, an AS/RS (Automated Storage/Retrieval System) called the stocker, a lifting system that transfers wafer carriers between different floors, and a mini-environment that is used for a standard interface of machines with the MHS. Fig. 7.1 shows alternative configurations of IAMHSs in semiconductor manufacturing industries (Agrawal and Heragu, 2006):

(a) *Bay-pitch dependent stocker location*: stockers have to be placed on the AGV route for easy access, which causes layout inefficiencies.
(b) *Bay-pitch independent stocker location*: more flexible stocker allocation enabled by OHT technology.
(c) *Bay-pitch independent and multi-bay linking*: similar to (b), except the number of required stockers (typically less; 70% of (b)) and overhead vehicles (typically more; 105% of (b)).
(d) *Spine configuration for fabrication layouts*: several bays connected with a central aisle, where stockers are located at the junction of the main aisle to each bay.

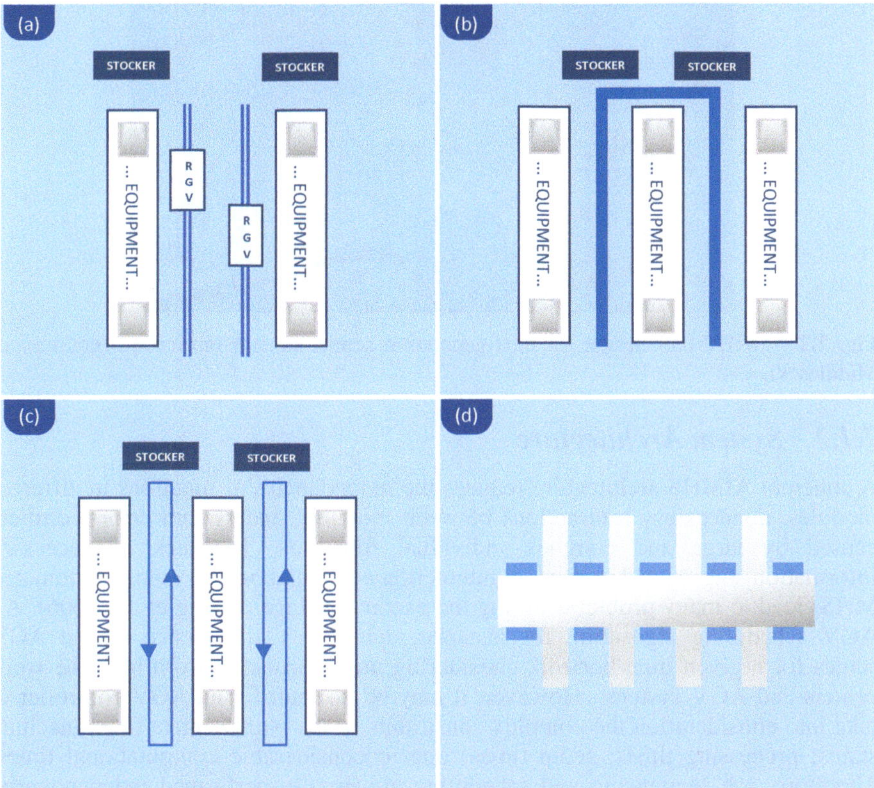

Fig. 7.1 (a) Bay-pitch dependent stocker location; (b) Bay-pitch independent and multi-bay linking; (c) Bay-pitch independent stocker location; (d) Spine configuration for fab layouts.

The next generation IAMHS uses conveyor systems instead of OHT systems and stockers (Fig. 7.2) to guarantee larger-capacity transfers and quick response times for deliveries (More reviews of the IAMHS applications in semiconductor industry are available in (Montoya-Torres, 2006)).

Fig. 7.2 New IAMHS design for next-generation semiconductor fabrication (courtesy of Middlesex).

7.1.2 System Architecture

A coherent AIMHS architecture reduces the redundancies of functions in different modules, unnecessary transactions between modules, and system errors/conflicts caused by large and complex individual functions. The lack of necessary information sharing and inefficient interaction and collaboration among automated MHSs lead to many problems during the execution stage (Chang et al., 2006). An AGV scheduling algorithm, for instance, determines a sequence of the AGV routes for a given time horizon, considering the information from both the work centers and AGV systems. However, it may be difficult for an AGV controller to take into consideration the complex constraints of the work centers (e.g., machine status; processing times; setup times) due to considerable computational times. Therefore, job sequencing and scheduling are usually performed independently, and then connected to the AGV controller using a sequence of protocols. An AGV controller only takes care of requested transfer commands, which specify source and destination locations, priorities, and command trigger times. There is already too much load on the AGV controller in its original tasks, which include path planning for a vehicle, job dispatch for a newly idle vehicle, vehicle dispatch for a new job requested, error recovery, etc. (Egbelu and Tanchoco, 1984). Therefore, the developed AGV scheduling algorithm should be modified based on the structure of the system architecture. One way to carry out this modification is to break up the algorithm for different modules in the system structure.

In production and warehouse systems with large numbers of subsystems, it is critical to develop effective mechanisms to handle interactions and collaborations among those subsystems. A challenging issue, however, is to construct a *scalable* and *modular* system structure that accommodates all different types of MHSs regardless of the size of the system, and the type of business on which the IAMHS is centered. Various architectures exist for the design of an IAMHS, depending on the manufacturing type, system's size, the rate of transactions, and so on. Fig. 7.3 shows an IAMHS architecture, where each automated MHS has its own controller responsible for its own tasks and communication with the material management system (MMS):

- Each AGV controller is responsible for job allocation, path planning, and collision avoidance, communicating with the MMS (e.g., transfer command B in Fig. 7.3), and reporting necessary activities such as vehicle allocation and job completion. The controllers also have to process error recovery routines for robustness of the system control.
- The MMS manages multiple controllers of different automated MHSs, and stores all transactions of the subsystems in the IAMHS database server. It receives transfer commands or short-term scheduling results of processing machines from the scheduling module in a higher-level system in the SCM, Supply Chain Management, (e.g., transfer command A in Fig. 7.3). MMS focuses on efficiencies during the transportation of unit loads.

As shown in Fig. 7.3, the higher level systems of an MMS can be manufacturing execution system (MES), warehouse management system (WMS), and enterprise resource planning (ERP) system. The WMS can be substituted by the MMS if the warehouse is composed of relatively simple systems.

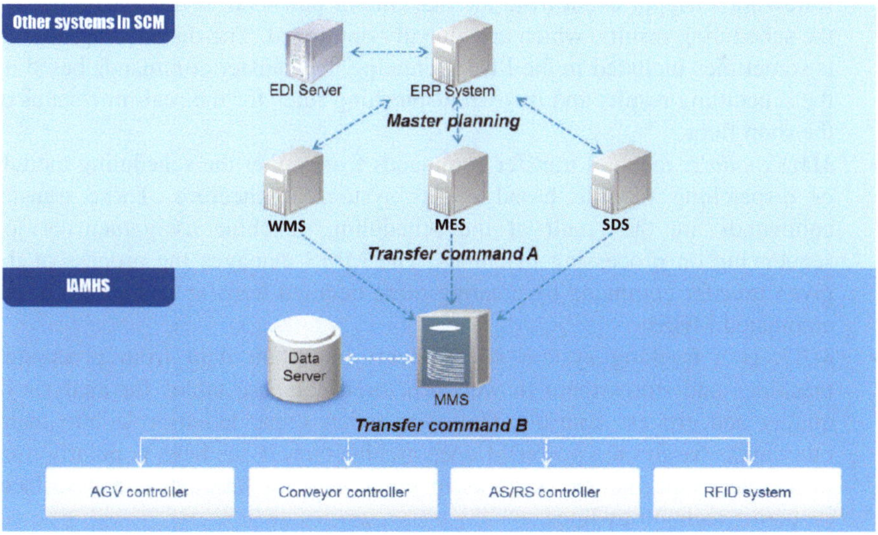

Fig. 7.3 IAMHS in hierarchical system architecture.

Understanding high-level decision-support system in the SCM enhances understanding of the scope of the control tasks performed by the IAMHS. The advanced planning and scheduling (APS) system generally consists of planning and scheduling modules. Sometimes the scheduling module is broken down into scheduling and dispatching modules (SDS). Details of the basic components of the IAMHS architecture are described below.

- *ERP system*: generally includes the planning module; however, the scheduling and dispatching modules can be included in any other systems such as the MES and WMS. In Fig. 7.2, it is assumed that the modules are running in a stand-alone system called the SDS that communicates with the ERP system, MES, and MMS. The planning module makes a long-term production or procurement plan (e.g., production quantities for each product type and time bucket, or production due dates for each product type or product group) based on customer orders, demand forecasting results, and capacity constraints. Detailed resource requirement plans are not specified by the planning module due to the uncertainties and complexities of operations.
- *Scheduling module*: delineates more concrete plans for the shop floor to meet the target production plan from the planning module. It typically optimizes various resource constraints with several objectives such as due-date satisfaction and throughput maximization. Detailed resource requirement plans over time buckets within a time horizon are created by the scheduling module. It sometimes takes into account constraints in the automated MHS for more robust scheduling.
- *Dispatching module*: determines the best unit load for a machine in real time following a trigger event from the machine or unit load. It follows up closely the scheduling results, which are globally optimized. The dispatching module is sometimes included in the MMS, creating the transfer commands based on the scheduling results and its own dispatching rules for the real-time status of the shop floor.
- *MMS module*: receives transfer commands from either the scheduling module or dispatching module based on its system architecture. These transfer commands are the result of the scheduling, machine assignment or job sequencing on processing machines. The MMS manages the process of the given transfer command by creating more detailed transfer commands to the automated MHSs.
- *MES*: is a tracking system that collects important data from processing machines and stores them in well-structured database tables for analysis of quality and process controls. However, it has expanded its role into many other areas based on a powerful open architecture. It has been popularly used in the electronic industry such as in semiconductor fabrication and surface-mounting technology lines.

- *WMS*: is generally used in mid- or large-size warehouse facilities, similar to MES for a production shop floor; it tracks every movement of materials and support operations for material handling in the warehouse. Its focus is on information processing automation. The objective of implementing an ERP system in a company is information sharing for rapid and correct decision-making, and implementation throughout the enterprise by using an integrated database system (Hossain et al., 2002). The whole procedures of order entry, production planning, material procurement, order delivery, and corresponding cash flow are managed by the system.
- *Electronic data interchange (EDI) server*: is responsible for transferring all the data from different intra-/inter-enterprise applications, which allows automated exchange of data between applications. Based on the EDI technology, applications freely exchange purchase orders, invoices, advance ship notices, and other business documents directly from one business system to the other without human support.

The IAMHS takes charge of the final execution of the SCM in an enterprise and also provides useful information as described above.

7.1.3 Material Management System

MMS plays a pivotal role in enabling automation, integration, and collaboration mechanisms in a complex IAMHS. The main functions of an MMS are summarized below:

(1) User interface for monitoring, control, and reporting.
(2) Error detection and recovery for the transfer command.
(3) Transfer command management between different automated MHSs.
(4) Storage and exchange of various data using a database server.
(5) Prioritizing the transfer command.
(6) Optimal routing to get to the destination from a source location via several automated MHSs.
(7) Determining the best destination among several possible automated MHS alternatives.

Functions (6) and (7) are directly related to the best matching (BM) principle of CCT, the collaborative control theory (see Chapter 2). To transfer a unit load, it is necessary to determine its source and destination locations. This task is performed by the dispatching or scheduling module. The dispatching and scheduling system provides a destination group to the MMS.

For example, the execution process of a transfer command is shown in Fig. 7.4. In this example, if a unit load from Machine #2 has just finished its processing and has to be transferred to an AS/RS to be processed next, on one of the machines connected to AGVS #3, there are two candidate AS/RSs connected to AGVS #3: AS/RS #3 and AS/RS #4. The AS/RS group connected to AGVS #3 is named AS Group #3. The MMS receives a transfer command from the dispatching module, specifying the source location as Machine #2 and the final destination as AS Group #3. Since there are two alternative destinations, the MMS may consider the product type of the unit load, the full rate of each AS/RS, the load port status of each AS/RS, the shortest distances from the unit load to the AS/RSs considering current active jobs in each system, and so on. It determines the best AS/RS amongst the two alternatives and triggers a transfer command to AGVS #1, which will first move the load to AS/RS #1 from Machine #2.

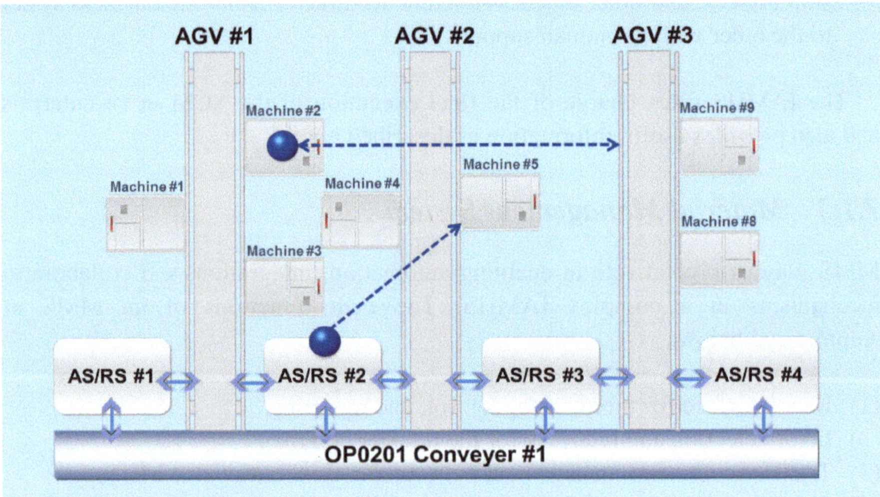

Fig. 7.4 Example of an IAMHS.

The MMS is also responsible for determining the destination subsystem in an automated MHS, such as the load/unload port (or pickup/drop-off port) in an AS/RS; because there are generally multiple load/unload ports with different types and numbers of buffers. The ports may differ, being load only, unloaded only or of unified type. Assume that a unit load in AS/RS #2 in Fig. 7.4 has to be moved to Machine #5, connected to AGVS #2 (solid arrow in Fig. 7.4). First, the unit load has to be moved to one of the output ports in the AS/RS. The AS/RS controller may not know which output port will be the best among the three possible alternatives, since it does not know the next destination of the unit load. The MMS may determine a load port connected to AGVS #2, OP0201 (see Fig. 7.4). In practice, the problems are generally much more complicated than this example due to the larger number of instances in the system (Sun et al., 2005; Jimenez et al., 2002).

Another challenging issue for an MMS is to determine the best route to get to the destination. In a complex IAMHS, there are many possible routes consisting of different automated MHS types. The IAMHS shown in Fig. 7.4 is represented through a graph as shown in Fig. 7.5. The graph can then be encoded in database tables by based on an adjacency (to determine distance) or incidence (to determine connectivity) matrix for computational purposes. Such graphical representation of the IAMHS makes it possible to apply many predefined properties and algorithms of graph theory can be applied in the development of algorithms for the MMS, e.g., in defining the shortest path from a source to a destination location.

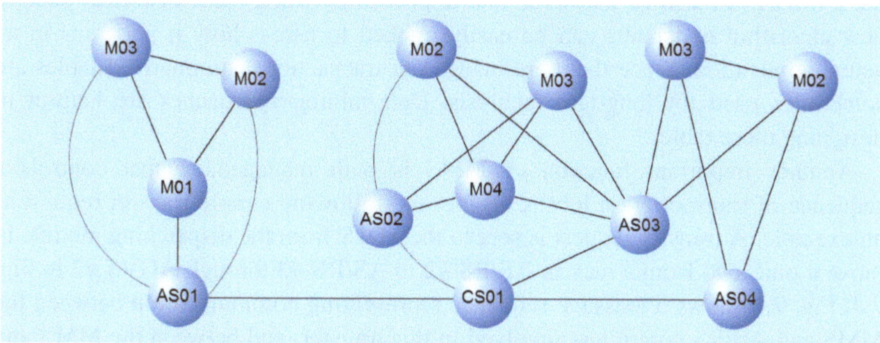

Fig. 7.5 Graphical representation of the IAMHS shown in Fig. 7.4 (Two-digit number in a vertex followed by alphabets represents the index number of machines or automated MHS. For example, M01 is the machine #1 connected to AGVS #1 and CS 01 is the conveyor #1. Acronyms are as follows. M: Machine, AS: AS/RS, CS: Conveyor System).

Transfer commands may be generated randomly over a time horizon, especially when there are significant variations in the number of arrivals during different periods. When the queue size increases on automated MHSs, prioritizing the transfer commands often provides a very useful solution to improve the overall performance of the system (Li et al., 2005).

In an MMS database, two different types of tables may be used; one that stores parameters for control algorithms and status user interfaces (requiring a minimum number of entities to achieve a shorter transaction time when they are queried), and the other stores data of movement histories based on communication messages between component systems of the IAMHS. Advances in information interface technology (IIT) enabled by RFID technology and barcode systems have dramatically improved the accuracy of these historical data. Significant information can be extracted from such historical data, e.g., standard operating time of a machine, processing routes of a unit load over machines in different process stages, and lead time of the unit load from start to finish. Data accuracy has a significant influence on the success of enterprise-level decision support systems explained above.

Accurate data from automated MHSs also enables the implementation of lean manufacturing concept in the shop floor through providing precise measures. For a complicated shop floor, it is often difficult to define a bottleneck stage or performance measures of the bottleneck machines. Lean manufacturing starts from well-defined and accurate performance measures. Many details of the machines can be analyzed by data relating to material movements from machine to machine, examples of which include machine throughput, product lead time, and the work-in-process (WIP) for each processing stage. They may also provide benefits for engineering analysis for the improvement of quality control. The performance of the IAMHS can also be measured and improved by using these historical data. A new algorithm under test can be easily tracked to assess how it performs in an actual application. Since there are many data transactions, summarized tables are sometimes used for long-term analysis. Data-mining approaches are helpful in designing these tables.

Another important function of MMS is path management that controls a sequence of transportation jobs. Consider the following simple transfer request as an example. A transfer request is sent to the MMS from the dispatching module to move a unit load from a rack in AS/RS #2 to AS/RS #3 through AGVS #2 in Fig. 7.4. Fig. 7.6 shows a message sequence representing communication between the MMS and AS/RS controllers involved in this transfer, and between the MMS and AGV controller. A few more messages might be used in actual systems. Although this is a relatively simple transferring task, more than 13 messages are needed to complete the task. First, transfer request #1 is triggered by the MMS. This request message transmits the source location as AS/RS #2, the destination as the unload port of the AS/RS, and the unit load identity to the AS/RS controller #2. If there are other high-level systems such as a WMS or MES, the MMS will send additional messages to these systems. In this case, the status of the unit load possibly needs to be updated from "waiting" to "busy" or "transferring" for the WMS and MES. To send this message to AS/RS controller #2, the MMS has to make at least two major decisions: (1) Selecting the closest idle port to the next destination (i.e., AGVS #2) as the unload port; (2) Determining the destination automated MHS to send the message. After receiving the transfer request from the MMS, AS/RS controller #2 puts the job into its queue, if it is busy; otherwise, it sends the job assign report to the MMS so that it triggers another transfer request command to AGV controller #2. This transfer command could be sent later after the job completion message from the AS/RS controller #2 has been received. However, by sending before the completion message, it can synchronize the transfer activities of two systems and thereby reduce the waiting time of the unit load for the vehicle at the unload port of AS/RS #2.

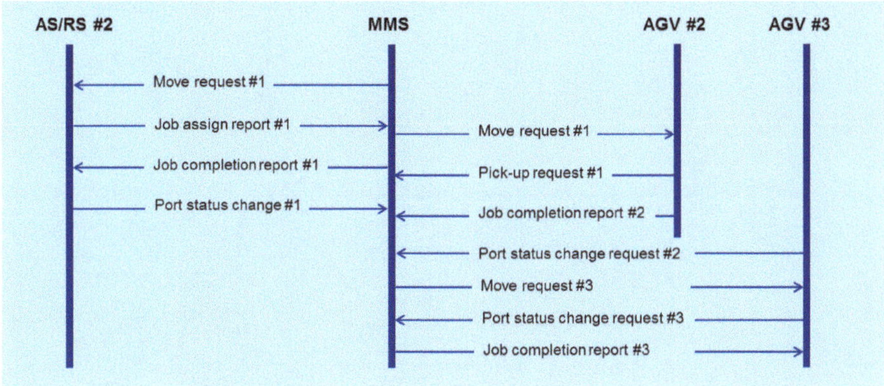

Fig. 7.6 Message sequence for a simple transfer command.

The MMS integrates not only systems but also human operators in the system environment. Hence, a user interface plays a major role in this integration (following the CVC, the Collaborative Visualization and Comprehension, principle of CCT). Operators can monitor the number of automated MHSs using the user interface. In addition, parameters to control an individual automated MHS and an IAMHS can be modified through the user interface. Moreover, another important function is enabled for reporting. Various reports can be queried directly from the database of the MMS.

7.1.4 Material Handling Problems

Material handling problems are mainly concerned with three fundamental questions: Where to locate facilities (e.g., equipment, departments); How to optimally move materials between their source and destination locations. The first question deals with finding the optimal layout of the plant (aka, *facility layout problem*), while the second one is addresses *flow path design problems*. In what follows, the basic problems corresponding to each of the two categories of material handling problems are briefly reviewed.

Facility Layout Problem (FLP)

FLP deals with finding the best arrangement of facilities (e.g., equipment; department; manufacturing cell) needed to produce specific products or services. FLP has been a mainstream research area in the fields of Industrial Engineering and Operations Research. The problem is to determine one (or more) locations and configurations of facilities (multiple alternatives may be needed in multi-period FLP) considering different phases of capacity expansion plans, process designs (Balakrishnan and Cheng, 1998), and loop-based flow path design problems (Asef-Vaziri and Laporte, 2005). Fig. 7.7 summarizes different factors associated with various FLPs, and highlights their relationships through a rough tree representation.

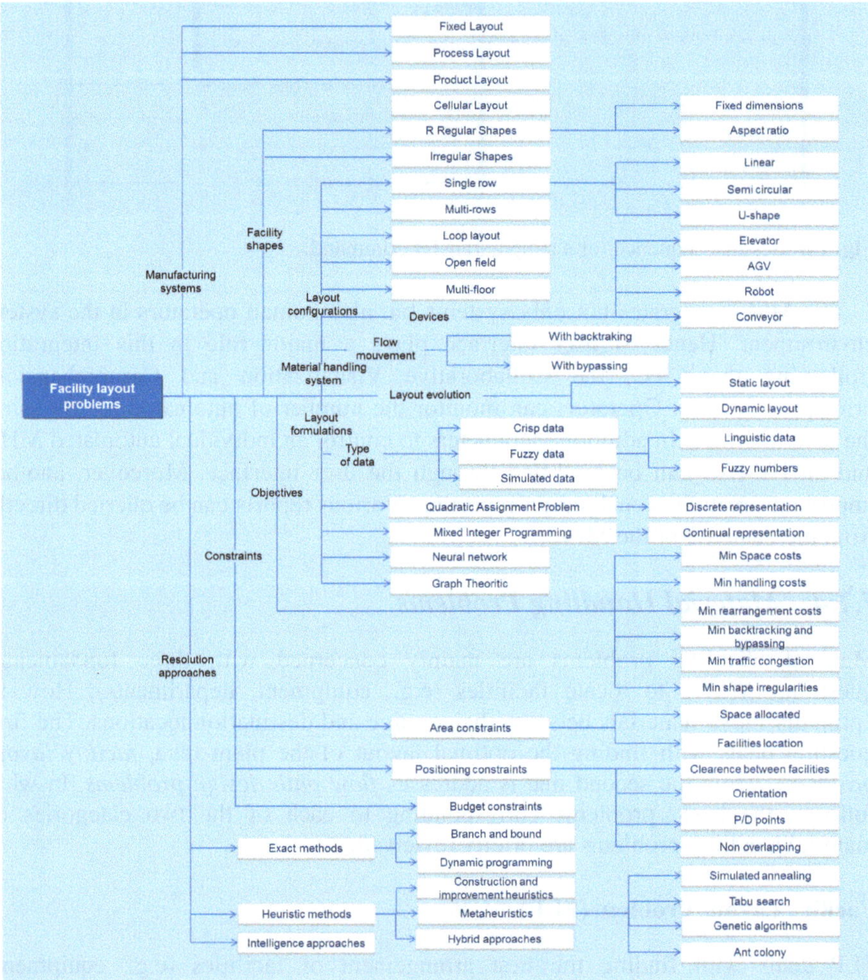

Fig. 7.7 Tree representation of FLPs (after Drira et al., 2007; with permission from Elsevier B.V.).

Major product and process characteristics that influence facility layout decisions include (Drira et al., 2007):

1. *Product type and quantity*: The layout design is influenced by type and variety of products, their interrelationships (i.e., similarities and differences), and the quantity of each type to be produced (We will discuss these product dimensions later on).
2. *Shape and dimension of facilities*: The next set of factors influencing the layout design includes the shape (i.e., regular/irregular) and relative dimensions of facilities. Regular facilities are generally rectangular-shaped, while irregular facilities are mostly polygon-shaped.
3. *MHS*: In some FLPs, it is also required to determine the handling path along with the MHS (e.g., AGV; RGV) that must handle the goods, semi-finished products, raw material, etc. between facilities. This factor is closely related to flow path design, and thus, will be discussed later on.
4. *Multiple floors*: Due to the high land cost in some urban or even industrial regions, besides the *horizontal dimension*, some plants add a *vertical dimension* to locate the facilities on multiple floors enabled by *elevators* (see Fig. 7.8). In these relatively more complicated situations, besides defining the number of required floors, the layout of each floor must be defined with respect to the layout of the other floors.

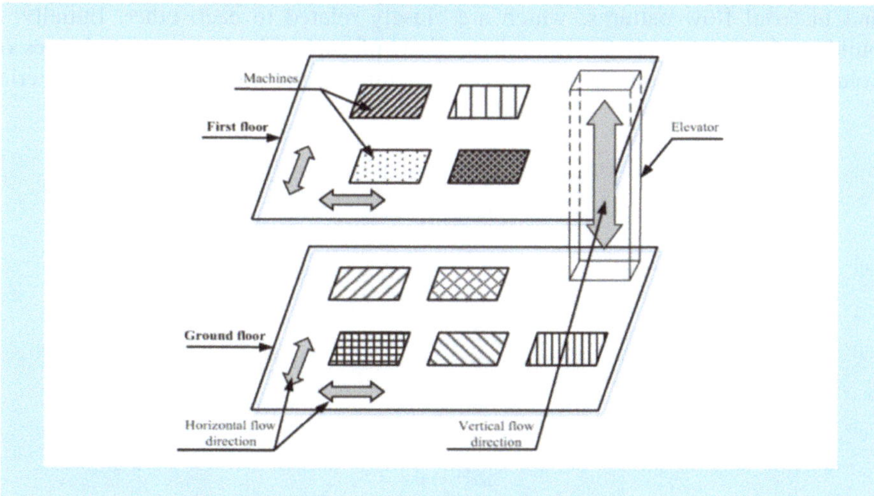

Fig. 7.8 Multi-floor plant.

5. *Backtracking/bypassing*: In flow-line layouts, there are typically two types of undesirable movements that need to be minimized (Fig. 7.9): (1) Backtracking – when a part moves from one facility to a preceding facility in the flow-line arrangement (e.g., for rework); (2) Bypassing – when a part skips a number of facilities in the flow-line arrangement (e.g., parts with different process characteristics).

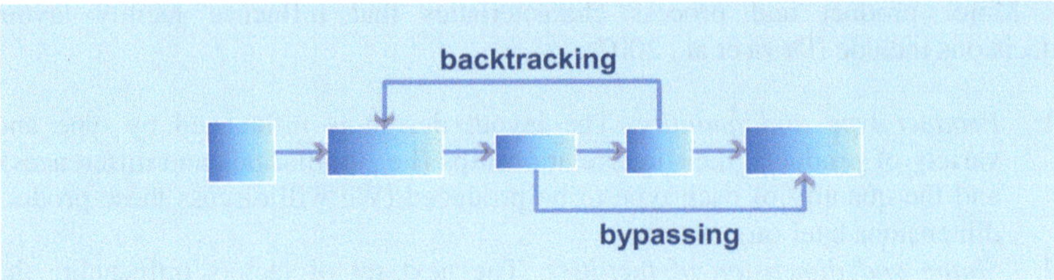

Fig. 7.9 Backtracking and bypassing.

A *layout procedure* helps facility planners identify what actions should be taken in steps to develop a facility layout. One of the most popular layout design procedures follows data flows during the design and uses an activity relationship chart as an important input data of FLP (Muther, 1973). This procedure starts from collecting the input data followed by five next steps: formulating the problem, analysis of the problem, searching for alternative solutions, selection and specification of the solution (Krick, 1962). Ideally, the capacity plan, process design, material flow pattern, and building design should be determined hierarchically (Figs. 7.10 and 7.11). In practice, however, this hierarchical procedure is hardly achieved due to limited project time and other types of uncertainties. The process design and the capacity plan affect the building design and material flow patterns, which are closely related to each other. Initially, a building design including shape, dimensions of each side, locations and sizes of columns in the building, height, and orientation is determined. Then, the material

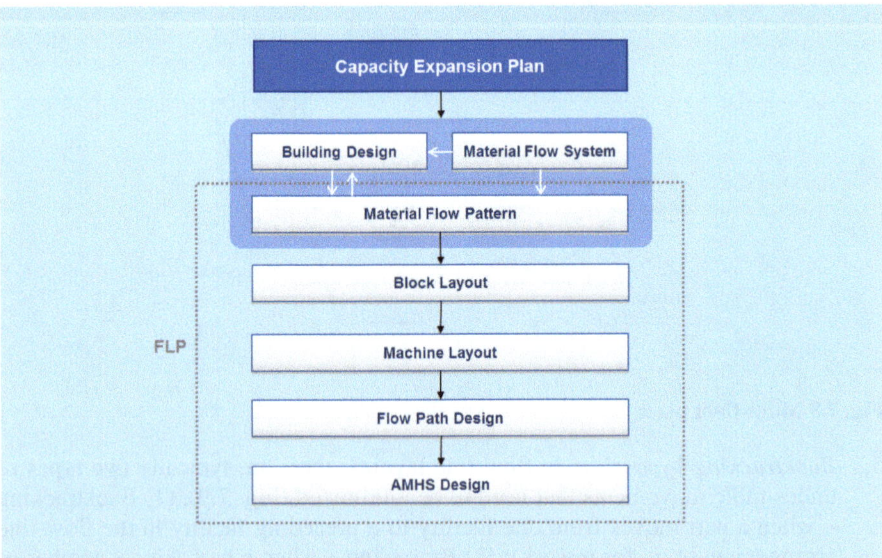

Fig. 7.10 Decision-making procedure for FLP (AMHS: Automated MHS).

flow pattern is continuously improved during the project. Once the material flow pattern is determined, a block layout is developed before a more detailed design of the layout. The machine layout problem is then defined and solved to generate of a more detailed layout.

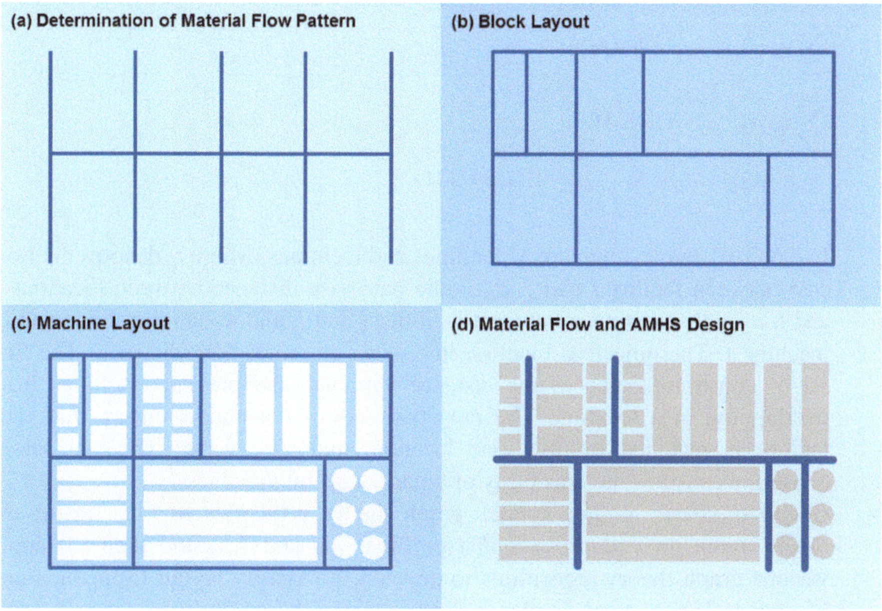

Fig. 7.11 Layout design procedure.

Block layout is one of the most common FLPs, which considers the locations of the facilities in a two dimensional space restricted by the vertical and horizontal lengths. Three most popular approaches addressing this problem are (see Fig. 7.7) QAP (Quadratic Assignment Problem) model, graph theoretic model, and MIP (Mixed-Integer Programming)-based model (Kurosaki et al., 1997;Meller and Gau, 1996;Singh and Sharma, 2006)

- *QAP model*: assumes that all facilities have the same sizes and shapes (Koopmans and Beckmann, 1957). In the original QAP, M facilities are located at M candidate locations with known pair-wise distances. Under this simple assumption, any types of material flow patterns (even 3D flow patterns) can be simply formulated. A basic QAP model has a quadratic objective function with binary variables, which minimizes the material flow cost with respect to the assigned facility locations. The constraints imply that in a feasible layout, there is a one-to-one correspondence between locations and facilities. The formulation of the original QAP is as follows:

$$\min \sum_{i=1}^{M}\sum_{j=1}^{M}\sum_{h=1}^{M-1}\sum_{l=h+1}^{M}\left(f_{ij}+f_{ji}\right)d_{hl}x_{ih}x_{jl}$$

s.t.

$$d_{hl}=\sum_{i=1}^{M}\sum_{t=h+1}^{l-1}w_i x_{it}+\sum_{i=1}^{M}\sum_{j=1}^{M}\sum_{t=h}^{l-1}a_{ij}x_{it}x_{j,t+1}+\frac{1}{2}\sum_{i=1}^{M}w_i x_{ih}+\frac{1}{2}\sum_{i=1}^{M}w_i x_{il}, \qquad 1\le h<l\le M, \quad \text{(QAP)}$$

$$\sum_{i=1}^{M}x_{ih}=1, \qquad 1\le h\le M,$$

$$\sum_{h=1}^{M}x_{ih}=1, \qquad 1\le i\le M,$$

$$x_{ih}\in\{0,1\}, \quad d_{hl}\in R, \qquad 1\le i,h,l\le M.$$

In the QAP above, there are M facilities and locations, where f_{ij} denotes the flow cost between facility i and j, d_{hl} is the pair-wise distance between locations h and l, a_{ij} is the clearance between facilities i and j, and w_i denotes the length of machine i. The objective function minimizes the total flow distance. The first set of constraint balances the distance variables and prevents facilities from overlapping in a location. The next two sets of constraints ensure that each facility is matched to only one location, and vice versa. The last set of constraints ensures the feasibility of binary decision variables.

- *Graph theoretic model*: applies graph theory principles to (1) convert the block layout to a *planar graph* (see Figs. 7.4 and 7.5) and then (2) apply various graph theory algorithms to develop the facility layout (Seppanen and Moore, 1975). A graph is planar if no edges in the graph intersect each other so that it can be represented on a plane. A bounded region by more than two edges form a closure is called a face. If two faces share a common edge, they are adjacent. If each face in a graph has a vertex and all the vertices are connected so that new edges intersect the existing edges only once, the new graph is said to be dual. If a graph is dual its dual graph is also dual. The concept of *dual graph* is important; because in an FLP, if a graph represents a block layout, its dual graph represents the relationships between the blocks. The relationship chart or flow distances between blocks used for developing the layout are given in this format. Another important graph theoretic concept is *critical set*. A non-planar graph can be converted to a planar graph by removing some edges or relationships. The smallest set of edges satisfying the planarity condition (if removed) is called the critical set of the graph. Finding the smallest set from a given flow cost matrix is a major challenge of the graph-theoretic model.
- *MIP-based model*: overcomes a few major limitations of QAP. It does not use predefined discrete locations having equal areas. Instead, it represents the layout space using continuous variables of x and y-axis. Input data of this model include the length and width of available space, area requirements for each facility, upper and lower limits of the length and width of each facility,

material flow frequencies and costs between facility pairs, and minimum and maximum perimeter of the facility. This model can be easily reformulated for the problem with fixed facility lengths and widths instead of the area requirements (Heragu and Kusiak, 1991). The pair-wise distances are obtained from the coordinate system in the layout. Moreover, the non-overlapping constraint is posted through the coordinate system and relative location variables of each facility on x and y-axis. MIP-based formulation is analogous to the TSP, the Traveling Salesman Problem. The objective is similar to the objective of the QAP model. The formulation of the MIP-based FLP is as follows:

$$
\begin{aligned}
&\min \sum_{i=1}^{M}\sum_{j=1}^{M} f_{ij} d_{ij} \\
&\text{s.t.} \\
&e_i - s_i = w_i, \qquad 1 \le i \le M, \\
&\frac{(e_i + s_i)}{2} = x_i, \qquad 1 \le i \le M, \\
&d_{ij} = | x_i - x_j |, \qquad 1 \le i, j \le M, \\
&s_i + w_i + a_{ij} \le s_j + M\left(1 - z_{ij}\right), \qquad 1 \le i, j \le M, \\
&\sum_{i=1}^{M} z_{ij} = 1, \qquad 1 \le j \le M+1, \\
&\sum_{j=1}^{M} z_{ij} = 1, \qquad 1 \le i \le M+1, \\
&s_i,\ e_i,\ x_i \ge 0, \qquad 1 \le i \le M, \\
&z_{ij} \in \{0, 1\}, \qquad 1 \le i \le M+1, 1 \le j \le M.
\end{aligned}
\qquad \text{(TSP)}
$$

In the above model, s_i and e_i denote the starting and ending points of facility i, respectively. z_{ij} is the binary matching variable: 1, if facility j is the right next to facility i; 0, otherwise. The first four sets of constraints are the balance equations that calculate the pair-wise distance between facilities. The next two sets of constraints ensure feasibility of matching. A dummy facility is also added (M + 1) in order for the tour to form a cycle. The rest of the constraints ensure the feasibility of decision variables.

FLP has a variety of instances with different applications. Just to point out, other important and pragmatic classes of FLPs are FLP under uncertainties (in e.g., demand, equipment performance, process design, and product mix), re-layout problem (for dynamic modifications and improvements), and robust FLP (selection of the best layout from a set of candidates based on a *robustness metric*).

Flow Path Design

A flow path design problem is typically the next step in the design of plant after a block or machine layout is determined, covering the selection of flow path type, directions of flow paths, locations of load/unload stations of facilities, and adding cross cuts. The flow and sequence of products/services among different facilities is one of the most significant factors influencing the layout design. A material flow system is defined based on specific requirements of different products/services (Tompkins et al., 2003), and is influences by the process design. The relationship between the facilities is the key to define the material flow system. Hence, this concept is used from the perspective of the flow requirement rather than that of design. Material flow systems are directly influenced by the characteristics of products and processes (see Table 7.1):

- *Product line*: If the process sequences of products or the relationships between facilities can be sequentially defined, there is no *backtracking* required to check the material flow (usually applied for mass production).
- *Product families*: Products in the same group form a specific isolated flow, where there may be some minor inter-facility flows.
- *Process*: Is the opposite to that of the product line, i.e., facilities with similar functionalities are placed close to each other.
- *Fixed product*: In this case, the product does not move (usually related to large products such as airplanes and vessels).

To better understand and analyze these material flow systems in complicated facilities, the P-Q (Product–Quantity) analysis method is usually performed to identify the type of material flow system (Muther, 1973). In practice, since rework and inspection processes sometimes generate ad hoc patterns of material flows, more than one flow systems is typically considered for a facility. The flow patterns are also influenced by frequencies and rates of unit process sequences corresponding to different products/product families.

Table 7.1 Material flow systems, flow patterns and layout types (adapted from Francis and White, 1974; Tompkins et al., 2003; Irani and Huang, 2006).

Material flow system	Conventional flow Pattern	Layout type	Implications
Product line	I-flow, L-flow, U-flow	Product or flow-line layout	Low WIP level; High throughput, Low flexibility
Product family	Manhattan shape flow; Spine flow	Cell layout	High flexibility; Low WIP; Low utilization
Process	Spine flow, Circular flow	Process or functional layout	High utilization and flexibility; High WIP level; Complex control
Fixed product	Custom flow	Fix layout	Long production time; High flexibility

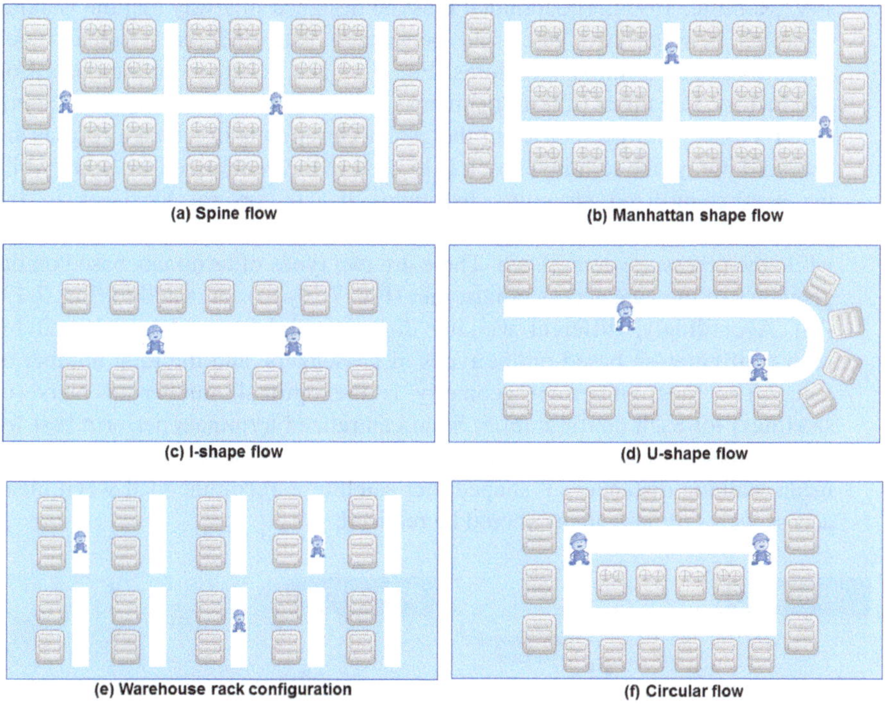

Fig. 7.12 Conventional material flow patterns.

The material flow pattern (or structure) defines overall environments and skeletons of the facility layout. It further defines spatial and directional geometries into the material flow system. Examples of popular material flow patterns used in practical layouts are shown in Fig. 7.12. Material flow patterns may address flow of products/services *between facilities*, *within a facility* or *within a workstation*

(Tompkins et al., 2003). A similar definition called *physical material flow system* also exists, which classifies material flow paths in the existing block layouts (Sinriech, 1995) into three types of *fully connected*, *partitioned*, and *split network*. In terms of design, however, material flow patterns are classified into two types of conventional and unconventional. Conventional flow patterns (Fig. 7.12) point out that in any facility, the locations of entrance and exit are the keys to determine the flow pattern.

In some instances, it is impractical to apply the conventional flow patterns discussed above. We describe several unconventional flow patterns through the following examples.

1. *Airport terminal*: Consider an airport terminal configuration serving three types of passengers: arriving, departing, and transferring (Bandara and Wirasinghe, 1992;Robuste, 1991). The terminal must be designed in a way that the walking distances between two gates and between a gate and the terminal block minimized. The terminal structures (Fig. 7.13 and 7.14) are defined based on two factors: Locations of terminal blocks in which ticking and baggage claims are processed, and geometries of the gate access points in which airplanes are accessed. Only one terminal block is located in the centralized structure while the semi-centralized structure distributes it into several locations in the terminal. For instance, patterns (c) and (i) in Fig. 7.14 are semi-centralized while the others are centralized. There are two types of terminals based on the geometry of the gate access points: pier (Fig. 7.13-(a)) and satellite (Fig. 7.13-(b)). Accordingly, different walking distance models can be evaluated for each configuration based on the types of passengers, and the total number of gates. The most important geometry is the optimal number of piers (or satellites) for each configuration. Semi-centralized terminals perform best for a wide range of transfer passenger rates and number of gates with respect to mean walking distances. Y-shaped pier satellite is recommended when mean access walks of passengers should be reduced.

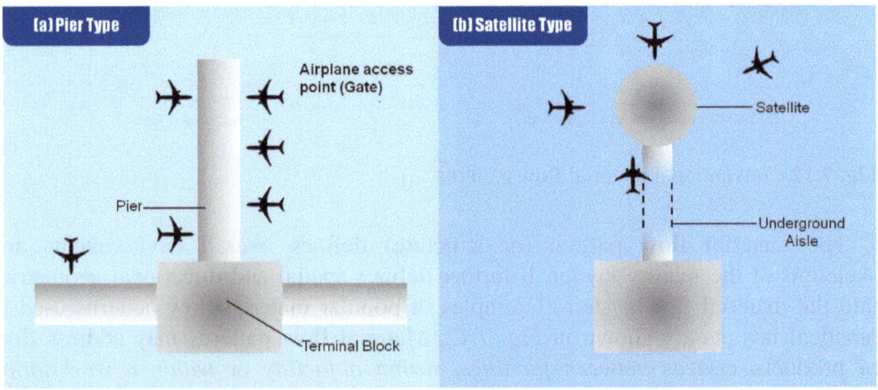

Fig. 7.13 Pier-type vs. satellite-type configuration of airport terminals.

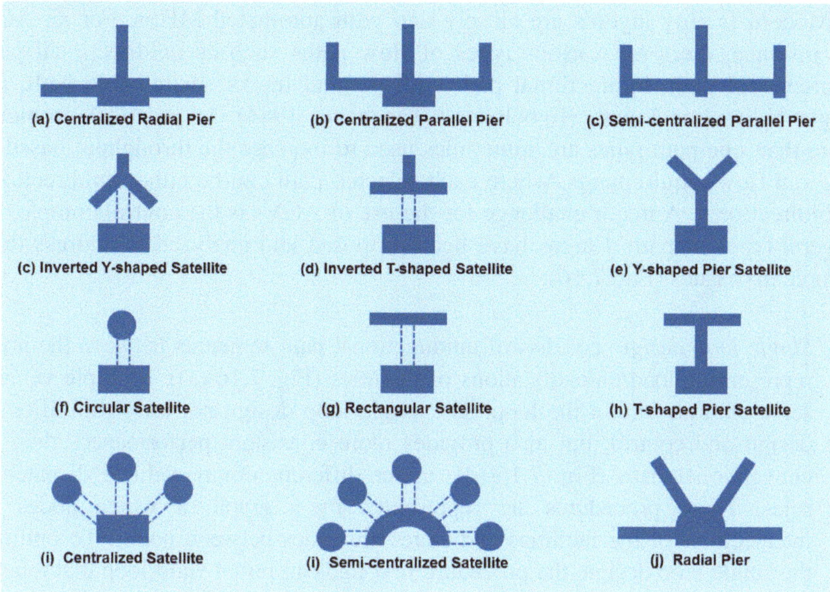

Fig. 7.14 Various airport terminal structures.

2. *Hybrid cellular layout*: This flow pattern (Irani and Huang, 2006) is a combination of cellular and process layout, e.g., cellular layout with reorientation of cells, S-shaped flow-lines in parallel or remainder cells, virtual cellular layout, and cascading flow-line layout. All mentioned layouts basically include a common area between cells. This type of layout can be developed by removing duplicated facilities and also sharing facilities between multiple cells. In the cascading flow-line layouts (Fig. 7.15), for example, the flow line is composed of multiple parallel lines in a cascading layout, where materials flow to the adjacent lines to create common patterns of the process sequences (common-strings). Development of a cascading layout involves two phases. First, the horizontal routings of processes are identified to increase the adjacencies between pairs of flow-lines. Then, process routings are vertically arranged so that machine duplication is minimized.

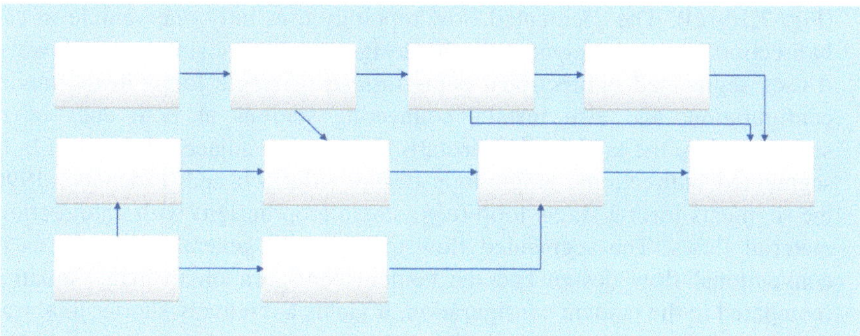

Fig. 7.15 Cascading flow-line structure.

Modern facility layouts are closely tied with automated MHSs. For an AGV, for instance, there are various types of flow paths such as unidirectional path, bidirectional path, bidirectional path with parallel tracks, switch-able path, and single bidirectional path (Egbelu and Tanchoco, 1984). In practical problems, more than one path types are sometimes used to increase the throughput based on material flow requirements, where each selected path can be either unidirectional or bidirectional. A major challenge for the use of AGVs is the control complexity. Several types of path designs have been proposed and analyzed to address these complexity issues (Fig. 7.16):

- *Single loop design*: consists of unidirectional path segments between the nodes representing load/unload stations of facilities (Fig. 7.16-(a)). Multiple vehicles serve stations within the loop. The single loop design not only simplifies the design and control, but also provides more consistent performances than the conventional path (Fig. 7.16-(d)) under different empty vehicle dispatching rules. These procedures are represented by a graph of which nodes are intersections of the facilities and edges are aisles between nodes. To optimize the single loop design, the procedure first finds an initial valid loop that touches at least one edge of all facilities, and then evaluates all possible candidate loops through an enumeration method. The primary assumption of single loop design is that the locations of facilities are known and fixed.
- *Tandem configuration*: divides the flow path into several loops, where only one vehicle services each loop (Fig. 7.16-(b)). This configuration significantly reduces the control complexity (Bozer and Srinivasan, 1991). A single loop has additional pickup/drop-off stations to connect to its adjacent loops. Tandem configuration is less robust than the conventional path design (Farling et al., 2001; Bischak and Stevens, 1995) in terms of handling the changes in several factors such as load/unload time at stations and product mix. In addition, both single loop design and tandem configuration require more vehicles to achieve similar performances as the conventional design, and are vulnerable to vehicle breakdowns.
- *Segmented flow topology*: is similar to tandem configuration in that it divides the material flow path into several non-overlapping discrete flow segments (Fig. 7.16-(c)). The segmented flow topology uses only one vehicle in each bidirectional discrete segment for the design and control simplicity. However, it uses segmented bidirectional paths instead of single loops in the tandem configuration, and also installs connecting stations at both ends of the segment while the tandem flow installs them in any adjacent locations. In the segmented bidirectional single-loop design (Sinriech and Tanchoco, 1996), the segments form a single loop (e.g., single loop design) with unidirectional material flows. The segmented flow topology, in general, outperforms the conventional flow design and the tandem configuration in many situations (compared to the tandem configuration, it yields a relatively shorter total track length, which in turn, reduces the installation cost).

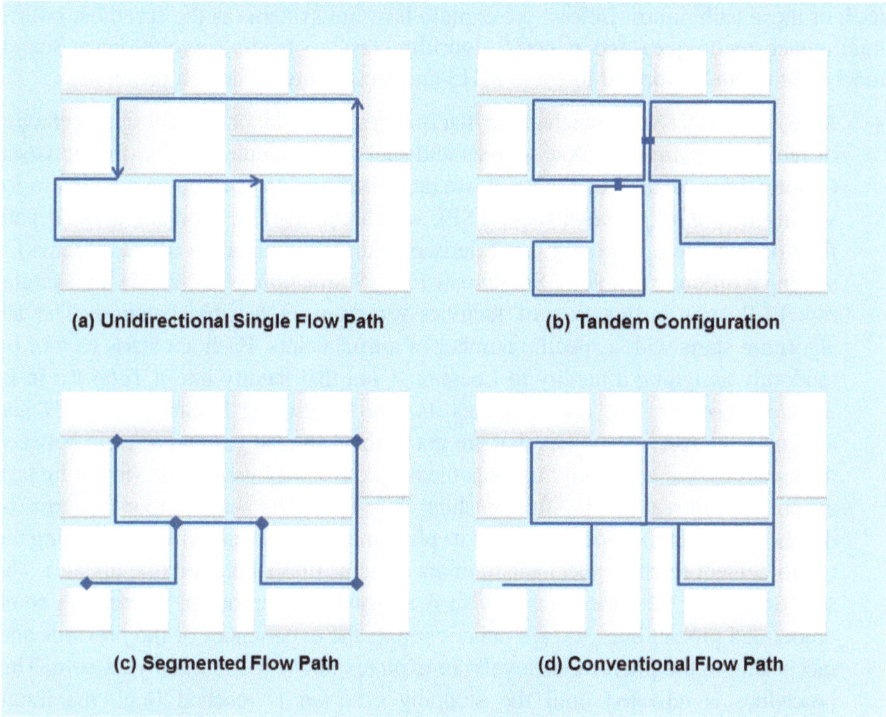

Fig. 7.16 Material flow path designs for automated MHSs.

Design of modern facility layouts is closely tied with automated MHSs. Automated MHS design issues includes optimal selection of the equipment type, location of storage and vehicle path for the design of a new facility, rack design for AS/RSs, and capacity estimation of the system. The design of an automated MHS is usually related closely to the FLP in that they sometimes impose constraints to each other and share performance measures in many areas.

7.1.5 Bio-inspired Solutions

In Chapter 4, we discussed two fundamental bio-inspired techniques based on *natural evolution* and *swarm intelligence* for the design and control of e-Work systems. In general, these techniques have a local beam search structure rooted in AI (Artificial Intelligence), which consists of a state representation, successor function, and evaluation function (Russell and Norvig, 2003). The initial state is the state in which the algorithm starts. The successor function receives a state as a parameter and returns a set of actions and successors. The definition of state together with the successor function determines the overall search space. The evaluation function calculates the fitness of each state, and the corresponding actions and successors. In order to successfully apply these bio-inspired optimization and control techniques, the very first and fundamental step is to identify the analogy of the problem under study with the standard structures of each of these techniques. Below, we explain how ant system (as the foremost swarm intelligence technique) and genetic algorithm (as the leading evolutionary search mechanism) can be applied to solve MHS and facility layout design problems.

- *Ant system*: is a search mechanism that imitates the collaborative foraging behavior of real ants in finding food sources and the shortest route back to the nest (see Chapter 4 for details). In chapter 4, we described how to develop an ant system for solving a *routing problem* (e.g., TSP), which can also be used for optimal path flow design. Here, however, we briefly explain the application of ant systems for solving a different problem: Single-row FLP (Solimanpur et al., 2005). In a single-row FLP, any permutation of facilities represents a feasible solution. The ant algorithm starts with a specific number of artificial ants. Each ant starts its tour by randomly assigning a facility to Location 1 put that facility into a *Tabu list* in its *working memory*. The ant continues its tour by visiting Locations 2 to *M* and assigning the rest of facilities that are not in the Tabu list yet. In each trip between the locations, the ant locally updates the respective *pheromone trail* (imitating trail evaporation in real ants). After finishing their tours, the shortest tour (in terms of distance, time, etc.) is identified and its pheromone trail is intensified, following the reinforcement learning mechanism of ant systems (global pheromone update). The selection of the next locations to visit is a probabilistic process, where with some predefined probabilities the ant either exploits the experiences of the previous ants (according to the pheromone levels) or explores new experiences by its own. This procedure is repeated until the stopping criterion is reached (e.g., maximum number of iterations; no improvement). Fig. 7.17 shows the logic of the ant system used to solve single-row FLP.

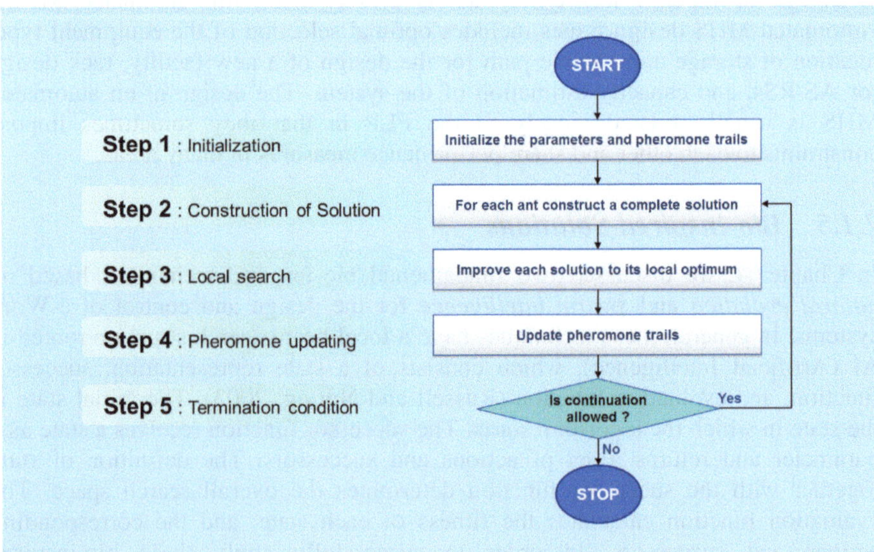

Fig. 7.17 Ant system for solving single-row FLP (adapted from Solimanpur et al., 2005).

- *Genetic algorithm*: is an evolutionary search mechanism based on the natural selection of the fittest individuals (see Chapter 4 for details). The first step in the design of any genetic algorithm is to encode the solution set of the problem as a chromosome, which must guarantee feasibility of the generated solution along with the required computational efficiency. For solving an FLP, the most common encoding scheme is *string representation*, where the entire plant is divided into a number of grids, each representing a specific facility (Azadivar and Wang, 2000). After defining the encoding scheme, the next step is to generate the first population of chromosomes. Then, the fitness of each individual is evaluated, and based on this evaluation the reproduction functions of genetic algorithm are triggered, i.e., crossover, mutation, and selection. The reproduction functions play pivotal roles in improving the computational efficiency of the algorithm and its convergence (see (Azadivar and Wang, 2000) for details). Fig. 7.18 shows the mechanics of the genetic algorithm used for solving FLP.

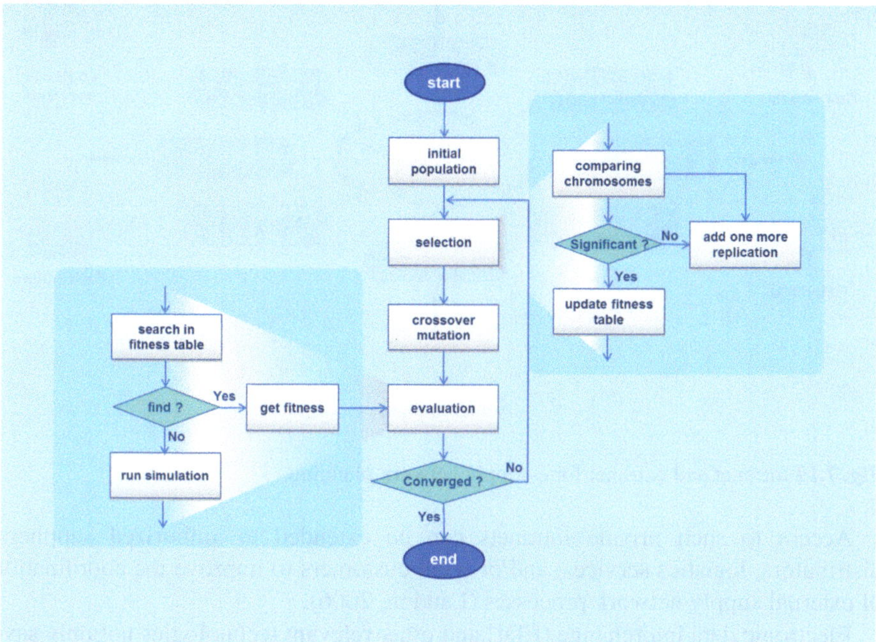

Fig. 7.18 Genetic algorithm integrated with computer simulation for solving FLP.

7.2 e-Supply and e-Production Networks

7.2.1 Supply Network Planning

To effectively design and manage supply networks, it is necessary to coordinate the flow of materials and information both within and between enterprises. The focus of supply network planning process is then on the synchronization of processes and activities, from the raw material suppliers to the final customer. The challenge is to find an integrated solution for the strategic, tactical, and operational activities in order to allow enterprises balance their supply and demand, and the flow of products and services. Information and communication technology (ICT) coupled with collaboration engineering mechanisms play a pivotal role in supply network planning through facilitating the flow of information and enhancing interaction and collaboration among customers, suppliers and third party partners. As show in Fig. 7.19, intranet can be used to integrate information from isolated business processes within the firm to facilitate the management of internal supply network processes.

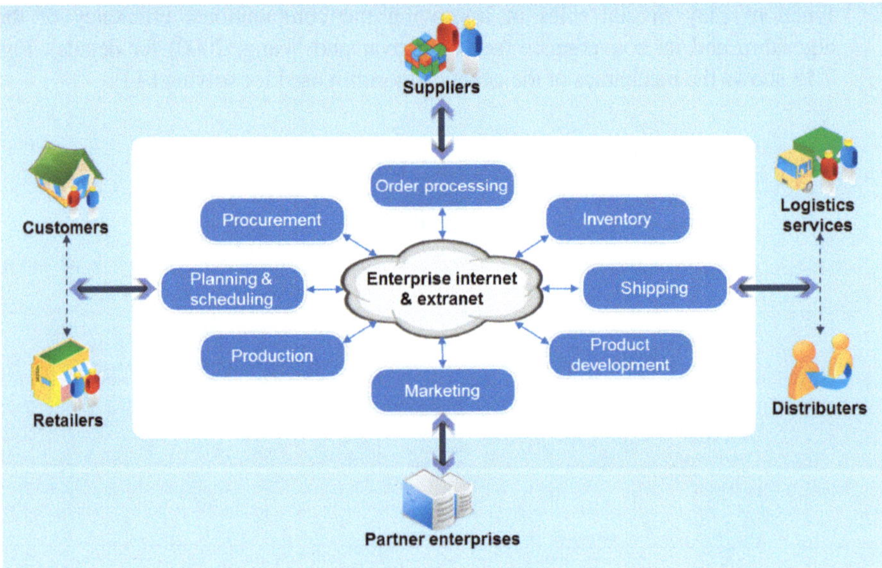

Fig. 7.19 Intranet and extranet for e-supply network planning.

Access to such private intranets can be extended to authorized suppliers, distributors, logistics services, and/or retail customers to improve the coordination of external supply network processes (Laudon, 2006).

Electronic data interchange (EDI) and other relevant technologies not only save money by reducing/eliminating human intervention in data entry processes, but also pave the way for collaboration initiatives between enterprises. A good example is vendor-managed inventory (VMI) where customers send inventory and consumption information to suppliers, who schedule deliveries to keep customer inventory within agreed upon ranges. VMI not only provides benefits to the customer but also increases the demand visibility of suppliers and leads to cost savings and reduced inventory investment. As shown in Fig. 7.20, EDI can significantly improve productivity of a supply network. A typical illustration is the case of Warner-Lambert that increased its products' shelf-fill rate at its retailer

Wal-Mart from 87% to 98% by EDI application, earning the company about US $ 8 million a year in additional sales (Turban et al., 2008).

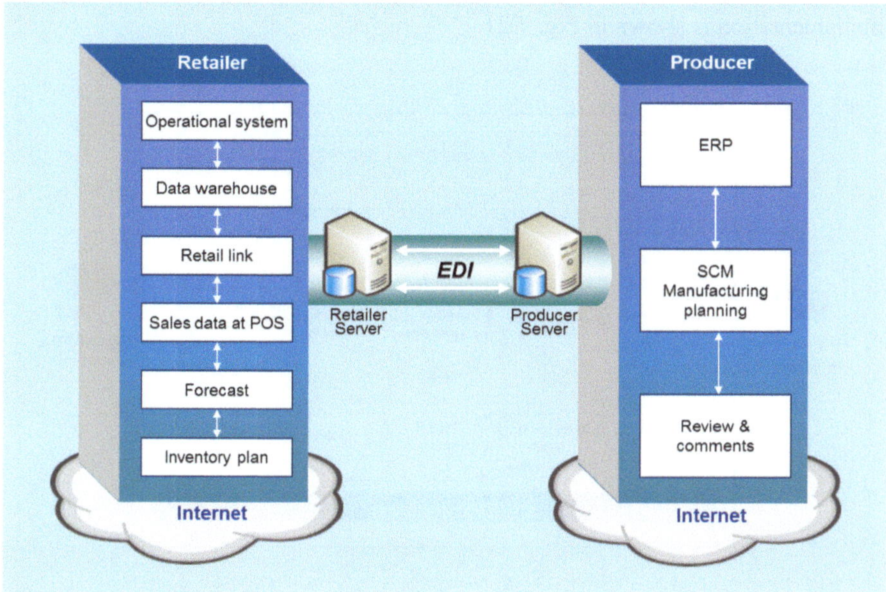

Fig. 7.20 Producer and retailer EDI application.

ICT and collaboration engineering are the keys to the strategic, tactical, and operational analysis required for supply network planning Khator and Nof, 1985, Nof and Moodie, 1989, Nof and Woo, 1991). They enable automation of the acquisition, processing, and communication of raw data related to historic and current supply network operations. This data is utilized by ICT systems for evaluating and disseminating decisions based on some rules. For example, when inventory reaches a certain level, a purchase order for replenishment is automatically generated (Shapiro, 2002). The data from ICT systems can also be used for simulating and optimizing a system (Nof, 1982, 1986, Azavidar and Wang, 2000, Ceroni and Nof, 2002). An example could be where ICT data is used to simulate the impact on a process with different levels of work in process on total productivity.

7.2.2 e-Production Planning and Control

As the rate of technological innovation increases, maintaining a competitive cost structure may rely heavily on production efficiency generated by an effective production planning and control process. For large-scale, global enterprises where traditional MRP-I (Material Resource Planning) and MRP-II (Manufacturing

Resource Planning) approaches are sufficient, new enterprise resource planning (ERP) solutions can be deployed, which not only link all aspects from the bill of materials to suppliers to customer orders, but also utilize different algorithms in order to efficiently solve the scheduling conundrum. A general framework of ERP implementation is shown in Fig. 7.21.

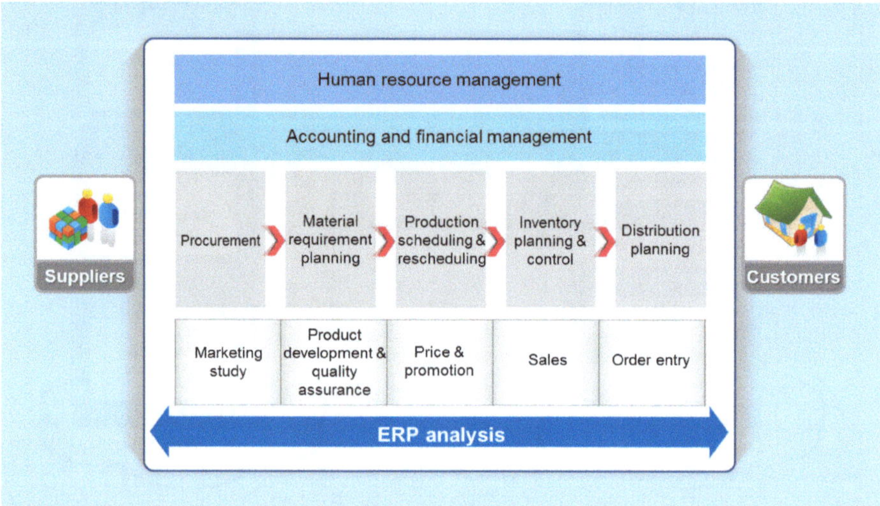

Fig. 21 A general framework of ERP implementation.

Off-the-shelf solutions (e.g., SAP products such as advance planner and optimizer (APO)) provide automatic data classification and retrieval, allowing key measures to be retrieved for strategic, tactical, and operational planning. Typical ERP systems, however, are rigid and hardly adjustable to the complexity in demand requirements and constant innovation of the product portfolio mix. Global enterprises competing in diverse marketplaces may choose to address these issues by building their own large-scale optimization models. With a solid database structure, these models are able to adapt to continuous changes in portfolios and can incorporate external influences on demand, such as market trends in promotions and advertisement. Over the past decade, there has been a shift in focus from business functions to business processes and further into value chains. Nowadays, enterprises focus on the effectiveness of the operation that requires functions to be combined to achieve end-to-end business processes (Anussornnitisarn and Nof, 2003).

7.2.3 e-Logistics Execution Systems

Logistic execution systems (LES) consolidate all logistic functions (e.g., receiving, storage, inventory management, order processing, order preparation, plant management and shipping) into an integrated process. The LES can be designed to communicate with the enterprise network and external entities such as customers and suppliers. The LES may be part of the ERP system; otherwise, it should communicate with the ERP system through interfaces so as to interact with other modules such as finance, purchasing, administration, and production planning. Usually, it also communicates with customers, suppliers, and carriers through EDI or web-based systems with the purpose of sharing logistic information relevant to all parties such as order status, trucks availability, confirmation of order reception, etc.

An LES is usually composed of a WMS, complemented with a labor management system and a transportation management system. Basically, WMS issues, manages, and monitors the tasks related to warehouse operation performance. A WMS improves efficiency and productivity of warehousing operations through barcode and/or RFID technologies. Both of these technologies provide a WMS with instantaneous visibility of warehouse operations, facilitating precise inventory control as well as accurate knowledge of labor and equipment resources availability. The addition of labor and transportation management in a LES has extended the span of WMS functionality to the point of practically embracing every logistic function from receiving to the actual shipping of goods. Emerging LES (such as SAP LES) support functions such as warehouse management and control (manual, semi- and fully-automated), shipment and trade processing, yard management, and connection to Pick-by-Voice, Pick-to-Voice, and RFID technologies.

7.2.4 RFID-Based e-Supply Networks

An emerging technology with great promise is the use of RFID (Radio Frequency Identification) to automate the collection of data. An RFID tag is a small computer chip that can send via radiofrequency a small amount of information to a short distance. The signal is captured by an RFID antenna and then transferred to a computer network for data processing. The general RFID system architectures, applications, frequencies, and standards are shown in Table 7.2. RFID is becoming increasingly prevalent as the price of the technology decreases. In supply network applications, current uses of RFID technology are focused on the location identification of products. These are as varied as identifying trailers and ocean freight trailers in a trailer yard, stopping of shoplifting of small but higher-priced fast-moving consumer goods such as razor blades, as well as an alternative to a WMS. In broader uses, Wal-Mart is discovering RFID technology to increase sales by making sure inventory at the store's loading dock is actually placed on the shelf. We will discuss the details of RFID technologies and applications in the next chapter.

Table 7.2 Key RFID technologies, frequencies, standards, applications, and decision criteria (Courtesy of the Department of Homeland Security).

Frequency	Data & Speed	Read Range	Application	Features	Standards
Low frequency: 125 – 134 KHz	Low read speed, small amounts of data	Very short: inches	Access control, Animal tagging, Inventory control, Car immobilizer	Low tag costs, Small read range, Small data amounts, Low data transfer speed,	ISO 11784/11785 ISO 18000/2
High frequency: 13.553 – 13.567 MHz	Medium read speed, small to med amounts of data	Short to med: 1 to 3 feet	Smart cards, Item or case level tagging	Sufficient data amounts, Most standards in place, Less susceptible to interference	ISO 15963 ISO 14443A ISO 14443B ISO 18000/3 EPC Class 0/1
Ultra high Frequency (1): 433 MHz	Good data speed, med to large amounts of data	Long range: 50 - 300 feet	Active tags, Container seals, Container tracking for DLA	Read speed and range, Costs, Potential interference with certain devices	INCITS 371.2 ISO 18000/7
Ultra high Frequency (2): 900-950 MHz	High read speed, small to med amounts of data	Medium: 2 to 10 feet	Pallet or case level tagging SENTRI/ NEXUS	Better vicinity read range, More susceptible to interference, High data transfer speed High tag cost	EPC Class 0/1 EPG UHF Gen 2 ISO 18000/6
Microwave Frequency (1): 2.45 GHz	High read speed, med amounts of data	Medium to long: 3 to 20 feet	Container or rail car, Toll collection, Pallet level tagging	Long read range, High data transfer speed, High tag costs	ISO 18000/4 INCITS 371.1
Microwave Frequency (2): 5.8 GHz	Very high read speed, high data rates	Long range: 50 -300 feet	Toll tags	Long read range, High data transfer speed, High tag costs, Battery replacement	ISO 18000/5 (rejected)

7.3 Emerging Trends and Challenges

Practically all modern production, logistics, and supply facilities, enterprises, and networks are automated to improve their effectiveness, safety, security, and sustainability. Hence, they can all be identified as e-Logistics, e-Production, and e-Supply systems. In all of them, collaborative e-Work, e-Business, and e-Service are inherent. A major challenge is, therefore, to engineer the collaboration in such systems and enterprises towards augmentation and optimization of their functions (Anussornnitisarn and Nof, 2003, Nof et al., 2006, Nof, 2006).

Integration and automation of material handling systems provide rapid and reliable infrastructures for information systems such as ERP, FMS, WMS, and MES. It enhances accurate data tracking during material handling in shop floors and warehouses. Relying on these accurate data, more intelligent functions can be added to the high-level automated and collaborative systems such as e-Production and e-Procurement planning, scheduling, and dispatching for higher reliability.

The IAMHS has been increasingly implemented in various applications to help complicated material handling operations in large-scale production and warehouse systems. Modern industrial facilities are typically equipped with IAMHSs such as AGV, AS/RS, and conveyor systems. However, these emerging technologies, despite their several advantages, often bring up new challenges to the design and execution of IAMHSs. In particular, the flow path design problem frequently imposes additional constraints on the facility layout, which also turn into bigger problems once the layout is developed. The design and control challenges range from material handling equipment selection to capacity estimation, innovative equipment design, and system design optimization.

There are several potential directions for further augmentation of the level of intelligence and autonomy in material handling systems. While the number of components in the IAMHS continues to increase, communication and long response time along with considerable computational times are the major limitations during the implementations of new control mechanisms. Sensor networks can be used for diagnosis and prognosis of potential errors and conflicts in IAMHSs, to ensure higher reliability and enable closed loop feedback and predictive mechanisms for more precise controls. (In the next chapter we will discuss facility sensor networks in detail).

In this context, RFID technologies can also be seen as a new opportunity to enhance real-time data gathering and communication within a shop floor or even among distributed production/supply networks. Current barriers to implementing RFID technology are cost, effectiveness, and privacy issues. The cost of a RFID tag has declined 90% in the past few years; however, it is still expensive and its use is typically limited to pallets and not on individual cases, products, or boxes. The other cost barrier is the investment in antennas. For companies such as Wal-Mart to utilize RFID technology, they need to install antennas in all of their distribution centers and stores. Moreover, RFID technology has problems sending signals through certain dense materials such as liquids, which limits their use. In addition, some people believe that if the RFID technology improves in terms of the distance a signal can be sent, then people will be able to determine which products are in people's homes, and thoughts of "Big Brother" come to mind. Currently, the technology is not capable of fulfilling these privacy concerns, but the concern may continue to diminish the acceptance of this relevant technology. In the next chapter, we will discuss RFID technologies, networks, and applications in detail.

References

Agrawal, G.K., Heragu, S.S.: A Survey of Automated Material Handling Systems in 300-mm Semiconductor Fabs. IEEE Transactions on Semiconductor Manufacturing 19(1), 112–120 (2006)

Anussornnitisarn, P., Nof, S.Y.: e-Work: the challenge of the next generation ERP systems. Prod. Plan. Control 14(8), 753–765 (2003)

Asef-Vaziri, A., Laporte, G.: Loop based facility planning and material handling. European Journal of Operational Research 164(1), 1–11 (2005)

Azadivar, F., Wang, J.: Facility layout optimization using simulation and genetic algorithms. International Journal of Production Research 38(17), 4369–4383 (2000)

Balakrishnan, J., Cheng, C.H.: Dynamic layout algorithms: a state-of-the-art survey. Omega 26(4), 507–521 (1998)

Bandara, S., Wirasinghe, S.C.: Walking distance minimization for airport terminal configurations. Transportation Research, Part A: General 26(1), 59–74 (1992)

Bischak, D.P., Stevens, K.B.: An evaluation of the tandem configuration automated guided vehicle system. Production Planning and Control 6(5), 438–444 (1995)

Bozer, Y.A., Srinivasan, M.M.: Tandem configurations for Automated Guided Vehicle systems and the analysis of single vehicle loops. IIE Transactions 23(1), 72–82 (1991)

Ceroni, J.A., Nof, S.Y.: A Workflow Model based on Parallelism for Distributed Organizations. Journal of intelligent Manufacturing 13(6), 439–461 (2002)

Chang, T.H., Fu, H.P., Hu, K.Y.: The innovative conveying device application for transferring articles between two-levels of a multi-story building. Int. J. Adv. Manuf. Technol. 28(1-2), 197–204 (2006)

Drira, A., Pierreval, H., Hajri-Gabouj, S.: Facility layout problems: A survey 31(2), 255–267 (2007)

Egbelu, P.J., Tanchoco, J.M.A.: Characterization of Automatic Guided Vehicle Dispatching Rules. International Journal of Production Research 22(3), 359–374 (1984)

Egbelu, P.J., Tanchoco, J.M.A.: Characterization of automated guided vehicle dispatching rules. Int. J. Prod. Res. 22(3), 359–374 (1984)

Farling, B.E., Mosier, C.T., Mahmoodi, F.: Analysis of automated guided vehicle configurations in flexible manufacturing systems. International Journal of Production Research 39(18), 4239–4260 (2001)

Heragu, S.S., Kusiak, A.: Efficient models for the facility layout problem. European Journal of Operational Research 53(1), 1–13 (1991)

Hossain, L., Patrick, J.D., Rashid, M.A.: Enterprise Resource Planning: Global Opportunities and Challenges. Idea Group, Hershey (2002)

Irani, S.A., Huang, H.: Cascading flowlines and layout modules: Practical strategies for machine duplication in facility layouts. International Journal of Flexible Manufacturing Systems 17(2), 119–149 (2006)

Jimenez, J., Kim, B., Fowler, J., Mackulak, G., Choung, Y.I., Kim, D.J.: Operational modeling and simulation of an inter-bay AMHS in semiconductor wafer fabrication. In: Winter Simul. Conf. Proc., vol. 2, pp. 1377–1382 (2002)

Kaspi, M., Tanchoco, J.M.A.: Optimal flow path design of unidirectional AGV systems. International Journal of Production Research 28(6), 1023–1030 (1990)

Khator, S., Nof, S.Y.: A Flow Control Modeling Approach for Computerized Manufacturing Systems. Material Flow 2, 1–10 (1985)

Koopmans, T.C., Beckmann, M.: Assignment problems and the location of economic activities. Econometrica 25, 53–76 (1957)

Krick, E.V.: An Introduction to Engineering and Engineering Design. John Wiley and Sons, New York (1962)

Kurosaki, R., Nagao, N., Komada, H., Watanabe, Y., Yano, H.: AMHS for 300 mm wafer. In: IEEE International Symposium on Semiconductor Manufacturing Conference Proceedings, pp. D13–D16 (1997)

Laudon, K.C., Laudon, J.P.: Management Information Systems, 9th edn. Prentice Hall, Upper Saddle River (2006)

Li, B., Wu, J., Carriker, W., Giddings, R.: Factory throughput improvements through intelligent integrated delivery in semiconductor fabrication facilities. IEEE Trans. Semiconduct. Manuf. 18(1), 222–231 (2005)

Meller, R.D., Gau, K.Y.: The facility layout problem: recent and emerging trends and perspectives. Journal of Manufacturing Systems 15(5), 351–366 (1996)

Montoya-Torres, J.R.: A literature survey on the design approaches and operational issues of automated wafer-transport systems for wafer fabs. Prod. Plan. Control 17(7), 648–663 (2006)

Muther, R.: Systematic Layout Planning, 2nd edn. Cahners Books, Boston (1973)

Nof, S.Y.: On the Structure and Logic of Typical material Flow Systems. International Journal of Production Research 20(5), 575–590 (1982)

Nof, S.Y.: Robotics and Material Flow. Elsevier Science Publishers, Amsterdam (1986)

Nof, S.Y., Moodie, C.L. (eds.): Advanced Information Technology for Industrial Material Flow Systems, Berlin, New York. Springer Series in Computer and Systems Sciences (1989)

Nof, S.Y., Woo, T.C.: Computer Integrated Engineering of Material Flow Systems: Key Research Issues. International Journal of Computer Applications in Technology 4(3), 159–165 (1991)

Nof, S.Y., Morel, G., Monostori, L., Molina, A., Filip, F.: From Plant and Logistics Control to Multi-Enterprise Collaboration. Annual Reviews in Control 30, 55–68 (2006)

Nof, S.Y.: Collaborative e-Work and e-Manufacturing: Challenges for Production and Logistics Managers. J. Intelligent Manufacturing 17(6), 689–701 (2006)

Robuste, F.: Centralized hub-terminal geometric concepts. I. Walking distance. Journal of Transportation Engineering 117(2), 143–158 (1991)

Russell, S., Norvig, P.: Artificial Intelligence: a modern approach. Prentice-Hall, NJ (2003)

Seppanen, J.J., Moore, J.M.: string processing algorithms for plant layout problems. International Journal of Production Research 13(3), 239–254 (1975)

Shapiro, J.F.: Business Process Expansion to Exploit Optimization Models For Supply Chain Planning (2002)

Singh, S.P., Sharma, R.R.K.: A review of different approaches to the facility layout problems. International Journal of Advanced Manufacturing Technology 30(5-6), 425–433 (2006)

Sinriech, D.: Network design models for discrete material flow systems: A literature review. International Journal of advanced technology 10(10), 277–291 (1995)

Sinriech, D., Tanchoco, J.M.A., Herer, Y.T.: The segmented bidirectional single-loop topology for material flow systems. IIE Transactions 28(1), 40–54 (1996)

Solimanpur, M., Vrat, P., Shankar, R.: An ant algorithm for the single row layout problem in flexible manufacturing systems. Computers and Operations Research 32(3), 583–598 (2005)

Sun, D.S., Park, N.S., Lee, Y.J., Jang, Y.C., Ahn, C.S., Lee, T.E.: Integration of lot dispatching and AMHS control in a 300 mm wafer FAB. In: IEEE/SEMI Adv. Semiconduc. Manuf. Conf. Workshop – Adv. Semiconduct. Manuf. Excellence, pp. 270–274 (2005)

Tompkins, J., White, J., Bozer, Y., Tanchoco, J.: Facilities Planning, 3rd edn. John Wiley and Sons (2003)

Turban, E., Leidner, D., Mclean, E., Wetherbe, J.: Information Technology for Management, 6th edn. Wiley, New York (2008)

Review and Challenge Questions

1. (i) Using several examples from this chapter, discuss the role of collaborative e-Work in e-Logistics, e-Production, and e-Supply. (ii) List several functions of engineered collaboration through e-Work in these examples.
2. Why must most advanced, competitive logistics, production, and supply systems and enterprises nowadays be designed and operated as e-Logistics, e-Production, and e-Supply?
3. Describe several configurations of IAMHS and explain their relative advantages and limitations. (Relate your answer to the typical information, communication and control problems.)
4. Explain how the three main tools of e-Work (covered in Chapters 4 and 5) are useful to solve some of the problems indicated in this chapter.
5. Illustrate your answer in Question 4 for the case described in Fig. 7.6.
6. (i) Describe briefly the QAP model and algorithm, and its contribution to solving facility layout problems. (ii) Consider the CCT design principles explained in earlier chapters, and describe how they can address some of the material handling problems beyond the QAP solution.
7. Explain how (i) Collaboration Requirement Planning; (ii) Fault Tolerance by Teaming; (iii) Collaborative Fault Tolerance; and (iv) Emergent Lines of Collaboration and Command can significantly improve the flow systems and patterns listed in Table 7.1.
8. (i) Describe the main steps in bio-inspired optimization techniques; (ii) Explain how they are useful in the design of material handling and facility layout. (iii) Illustrate your answers for the design problem shown in Fig. 7.16.
9. Explain how (i) Conflict and Error Detection Protocol; (ii) Best Matching Protocol; and (iii) Task Requirement Analysis Protocol can significantly improve the design and management of e-Production and e-Supply networks.
10. Describe the role RFID can play in implementing several CCT principles in e-Business and e-Service.
11. Elaborate on some of the emerging trends and challenges in e-Logistics, e-Production, and e-Supply with a focus on collaboration engineering.

Lab Questions

12. Solve a small facility layout planning problem with (i) an ant algorithm; (ii) a genetic algorithm.
13. (i) Find and explore several examples of active and passive RFID applications in e-Logistics, e-Production, or e-Supply. (ii) Explain is each example specific ways by which this technology enables better engineered and augmented collaboration.

Chapter 8
Factory Sensors and RFID Networks

Abstract. In line with the emergence of decentralized e-Systems and the ever-increasing needs for effective interaction and collaboration, a rising challenge is how to enable communication and gather information on the real-time state of distributed agents in an e-Work network. That is, each agent (e.g., machine agent; resource agent; product agent), as an autonomous and intelligent element of a collaborative e-System, must be able to detect and process useful information from its environment and share it with other peers. The supporting technologies for this critical need are sensor and RFID (Radio Frequency Identification) networks. In this chapter, fundamentals of each technology are discussed and some real-life applications are reviewed.

8.1 Sensors

A *sensor* is an instrument that responds to a specific physical stimulus and produces a measurable electrical signal. Sensors have different types ranging from mechanical to electrical, electromechanical, magnetic, optical, chemical or even biological. Indeed, any device that is directly altered in a predictable and measurable manner by changes in a real-life parameter is known as a sensor for that parameter. Sensors play a significant role in any e-System due to the need to gather, process, and share information relevant to specific tasks. Recent advances in micro-device technology, micro-fabrication, chemical processes, and digital signal processing have enabled the development of economically affordable micro-/nano-sensors. Micro-sensors have a variety of applications in many practical areas such as medical and space devices, military equipment, telecommunication, and manufacturing (Luo, 1996; Luo et al., 2002). Compared to conventional sensors, micro-sensors provide some unique advantages such as interfering less with the environment they measure, requiring less manufacturing cost, being used in narrow spaces and harsh environments, etc. Successful design and implementation of micro-sensors depends highly on sensors' capability, cost, and reliability.

S.Y. Nof et al., *Revolutionizing Collaboration through e-Work, e-Business, and e-Service*, Automation, Collaboration, & E-Services 2, DOI: 10.1007/978-3-662-45777-1_8

8.1.1 Sensing Principles

According to their working principles, sensors are technically classified into different types (Table 8.1). Sensors may measure physical phenomena by capturing resistance change, capacitance change, inductance change, thermoelectric effect, piezoelectric effect, photoelectric effect, Hall effect, and so on (Wilson, 2004; Fraden, 2003). Among these effects, majority of sensors utilize the resistance change or *resistivity* of a conductor. Considering uniform current density in the insulator, the resistance R of a conductor of cross-sectional area A can be computed as

$$R = \frac{\rho \cdot l}{A}, \tag{8.1}$$

where l denotes the length of the conductor, and ρ is the electrical resistivity of the material. Resistivity is a measure of the material's ability to oppose electric current. Therefore, the change of resistance can be measured by detecting physical deformation (l or A) of conductive materials or by sensing resistivity (ρ) of conductor. A strain gage, for example, is a sensor that measures resistance by deformation of length or cross-sectional area, and a thermometer is a sensor that measures resistance by examining the resistivity change of a material. Expanding and then simplifying Eq. (8.1) through Taylor series, resistance change can be expressed as

$$\Delta R = \Delta\rho\frac{l}{A} + \Delta l\frac{\rho}{A} - \Delta A\frac{\rho \cdot l}{A^2}. \tag{8.2}$$

Dividing both sides of Eq. (8.2) by the resistance R, the resistance change rate can be expressed as

$$\frac{\Delta R}{R} = \frac{\Delta\rho}{\rho} + \frac{\Delta l}{l} - \frac{\Delta A}{A} = \frac{\Delta\rho}{\rho} + \varepsilon + 2\nu\varepsilon, \tag{8.3}$$

where ε and ν are the strain and the Poisson's ratio of the material, respectively. When the resistivity ρ of a sensing material is close to constant, the resistance can be determined from the values of strain ε and Poisson's ratio ν of the material (e.g., strain gage). When the resistivity of a sensing material is sensitive to the measuring targets and values of ε, ν can be neglected; the resistance can be measured from the resistivity change ($\Delta\rho/\rho$) (e.g., resistance temperature detector (RTD)).

In capacitance-based sensors, the sensor measures the amount of electric charge stored between two plates of capacitors. The capacitance C can be calculated as

$$C = \frac{\varepsilon A}{d}, \tag{8.4}$$

where A is the area of each plate, d is the separation between the plates, and ε is the dielectric constant (or permittivity) of the material between the plates. The dielectric constant for a number of very useful dielectrics changes as a function of the applied electrical field. Thus, capacitance-based sensors utilize capacitance change by measuring the dielectric constant, the area A of, or the separation d between the plates. A capacitive-type torque meter is an example of a capacitance-based sensor.

Inductance-based sensors measure the ratio of the magnetic flux to the current. Linear variable differential transformers (LVDT; Fig. 8.1) and magnetic pick-up sensors are representative inductance-based sensors.

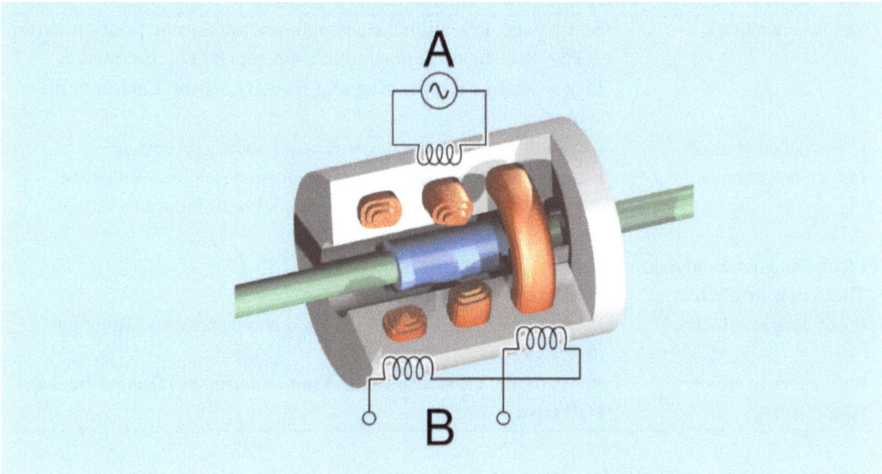

Fig. 8.1 Cutaway view of an LVDT. Current is driven through the primary coil at A, causing an induction current to be generated through the secondary coils at B.

Electromagnetic induction-based sensors are based on Faraday's law of induction, which is involved in the operation of transformers, inductors, and many forms of electrical generators. The law states that *the induced electromotive force in any closed circuit is equal to the time rate of change of the magnetic flux through the circuit.* Quantitatively, the law takes the following form

$$E = -\frac{\mathrm{d}\Phi_B}{\mathrm{d}t}, \tag{8.5}$$

where E is the electromotive force (EMF) and Φ_B is the magnetic flux through the circuit.

In addition to the mentioned types of sensors, thermocouples measure the temperature difference between two points rather than absolute temperature. In classic applications, one of the (cold) junctions is maintained at a reference temperature, while the other end is attached to a probe. Having available a cold junction at a known temperature is simply not convenient for most directly connected control instruments. They incorporate into their circuits an artificial cold junction using some other thermally sensitive device, such as a thermistor or diode, to measure the temperature of the input connections at the instrument, with special care being taken to minimize any temperature gradient between terminals. Hence, the voltage from a known cold junction can be simulated, and the appropriate correction applied. Photodiodes, phototransistors, and photo-interrupters are sensors that use the photoelectric effect. Other types of sensors are listed in Table 8.1.

Table 8.1 Technical classification of sensors according to their working principle.

Sensing Principle	Sensors
Resistance change	Strain gage, Potentiometer, Potentiometric throttle position sensor (TPS), Resistance Temperature Detector (RTD), Thermistor, Piezo-resistive sensor, Magneto-resistive sensor, Photo-resistive sensor
Capacitance change	Capacitive-typetorque meter, Capacitance level sensor
Inductance change	Linear Variable Differential Transformer (LVDT), Inductive angular position sensor (Magnetic pick-up), Inductive torque meter
Electromagnetic induction	Electro Magnetic flow meter
Thermoelectric effect	Thermocouple
Piezoelectric effect	Piezoelectric accelerometer, Sound Navigation And Ranging (SONAR)
Photoelectric effect	Photo diode, Photo transistor, Photo interrupter (Optical encoder)
Hall effect	Hall sensor

8.1.2 Types of Sensors

Sensors are also classified in terms of the physical phenomena they measure (e.g., position, velocity, acceleration, heat, pressure, flow rate, and sound). Below, different types of sensors are briefly described (see Webster, 1998; de Silva, 2008 for more details).

Position Sensors

A position sensor refers to any device that is capable of position measurement, such as limit switches or proximity sensors that detect whether or not something is close to or has reached a limit of travel. Position sensors also include potentiometers that measures rotary or linear position. The linear variable differential transformer (LVDT) is an example of potentiometers that measures linear displacement. Resolvers and optical encoders, on the other hand, measure the rotary position of a rotating shaft. The LVDT and resolver function much like a transformer. The optical encoder produces digital signals that convert motion into a sequence of digital pulses. In fact, there also exist optical encoders for measuring linear motion. Some position sensors are classified according to their measuring techniques. Sonar measures distance with sonic/ultrasonic waves and radar utilizes electronic/radio to detect or measure the distance between two objects. Many other sensors are used to measure position or distance.

Velocity Sensors

Velocity sensors measure *speed* through taking consecutive position measurements at known time intervals and computing the derivative of the position values. A tachometer is an example of a velocity sensor that does this for a rotating shaft. The typical dynamic time constant of a tachometer is in the range 10-100μs. A tachometer is a passive analog sensor that provides an output voltage proportional to the velocity of a shaft. There is no need for an external reference or excitation voltage. Traditionally, tachometers have been used for velocity measurement and control only, but all modern tachometers have quadratic outputs which are used for velocity, position, and direction measurements, making them effectively functional as position sensors.

Acceleration Sensors

An acceleration sensor or accelerometer measures continuous mechanical vibrations; aerodynamic flutter and transitory vibrations (e.g., shock waves, blasts or impacts). Accelerometers are normally mechanically attached or bonded to an object or structure for which acceleration is to be measured. An accelerometer detects acceleration along one axis and is insensitive to motion in orthogonal directions. Strain gages or piezoelectric elements constitute the sensing element of an accelerometer, converting vibration into a voltage signal. The design of an accelerometer is based on the inertial effects associated with a mass connected to a moving object.

Miscellaneous Sensors

Besides position, velocity, and acceleration sensors, as the most popular types of sensors, there are also other miscellaneous types of sensors that measure physical quantities such as force, strain, temperature, pressure, and flow:

- *Force sensors*: are represented by a load cell used to measure a force. The load cell consists of several strain gages connected to a bridge circuit to yield a voltage proportional to the load.
- *Temperature sensors*: are devices that indirectly measure quantities such as pressure, volume, electrical resistance, and strain and then convert the values

using the physical relationship between the quantity and temperature. For example (Cetinkunt, 2007; Alciatore and Histand, 2007).

- A *bimetallic strip* composed of two metal layers with different coefficients of thermal expansion utilizes the difference in the thermal expansion of the two-metal layers.
- A *resistance temperature sensor* constructed of metallic wire wound around a ceramic or glass core and hermetically sealed utilizes the resistance change of the metallic wire with temperature.
- A *thermocouple* constructed by connecting two dissimilar metals in con tact produces a voltage proportional to the temperature of the junction.

- *Flow sensors*: are devices for sensing the rate of fluid flow. A flow sensor is a sensing element used in a flow meter to record the flow of fluids. Some flow sensors have a vane that is pushed by the fluid (e.g., potentiometer), while other flow sensors are based on heat transfer caused by the moving medium.
- *Ultrasonic sensors*: or *transducers* generate high frequency sound waves and evaluate the echo received back. An ultrasonic sensor computes the time interval between sending the signal and receiving the echo to determine the distance to an object. Radar or sonar works on a principal similar to that of the ultrasonic sensor (Fig. 8.2).

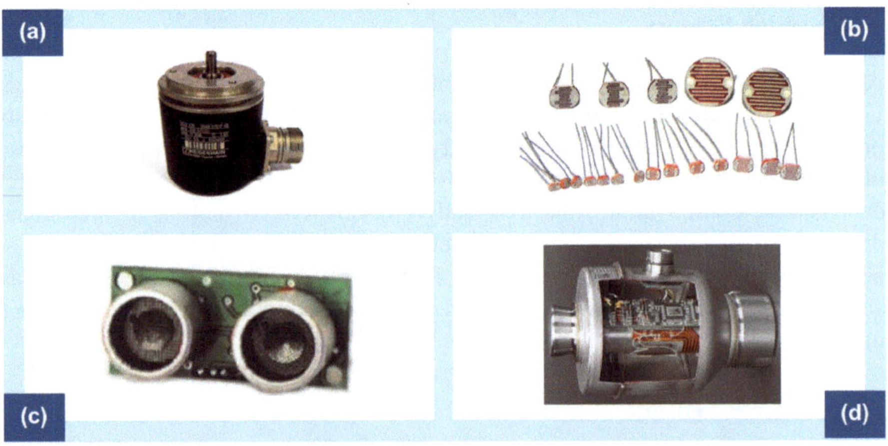

Fig. 8.2 Various sensors; (a) Absolute encoder (b) Photoresistor (c) Sonar (d) Digital load cell cutaway (Courtesy of Society of Robots).

- *Vision sensors*: are typically used embedded in a vision system. A vision system can be used to measure shape, orientation, area, defects, differences between parts, etc. Vision technology has improved significantly over the last decade in that they have become rather standard smart sensing components in most factory automation systems for part inspection and location detection. In general, a vision system consists of a vision camera, an image processing computer, and a lighting system. The basic principle of operation of a vision system is that it forms an image by measuring the light reflected from objects, and the sensor head analyzes the output voltage from the light intensity received. The sensor head consists of an array of photosensitive, photodiodes or charge-coupled devices. Currently, various signal-processing techniques for the reflected signals are applied for many industrial applications to provide accurate outputs (Fig. 8.3).

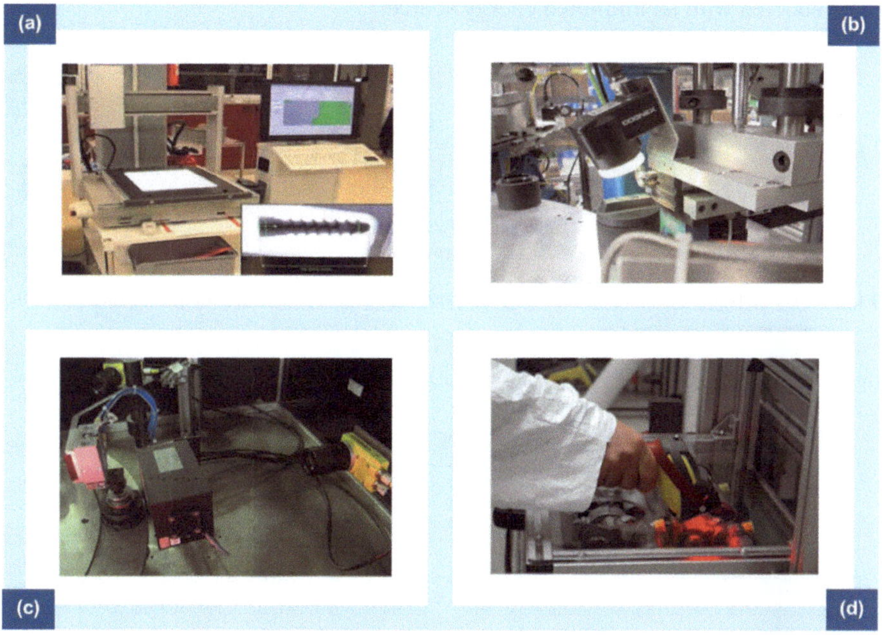

Fig. 8.3 Types of vision sensor applications; (a) Automated low volume/high variety production (b) Vision sensors error-proof oil cap assembly (c) Defect free parts in 360 degrees (d) Inspection of 2D matrix marked codes (Courtesy of Cognex Corporation).

- *Micro-sensors*: are miniature electronic device functioning similar to the existing large-scale sensors. With recent micro-electromechanical system (MEMS) technology, micro-sensors are integrated with signal-processing circuits, analog-to-digital (A/D) converters, programmable memory, and a microprocessor, a so-called smart micro-sensor (Warneke et al., 2001; Yamasaki, 1996). Current smart micro-sensors contain an antenna for radio signal transmission. Wireless micro-sensors are now commercially available and are evolving with more powerful functionalities (Fig. 8.4). A wireless micro-sensor, in general, consists of the following units (Fig. 8.5):
    - *Sensing unit*: is an electrical part detecting the physical variable from the environment.

- *Processing unit*: is a tiny microprocessor performing signal-processing functions, i.e., integrating data and computation required in the processing of information.
- *Power unit*: provides energy source with other units.
- *Communication unit*: consists of a receiver, a transmitter, and an amplifier if needed. Basically, all individual sensor nodes are operated by a limited battery, but a base-station node as a final data collecting center can be modeled with an unlimited energy source.

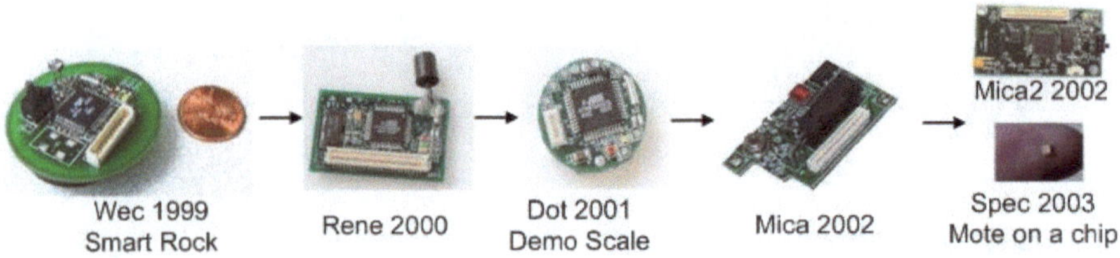

Fig. 8.4 Evolution of smart wireless micro-sensors (Courtesy of Crossbow Technology Inc).

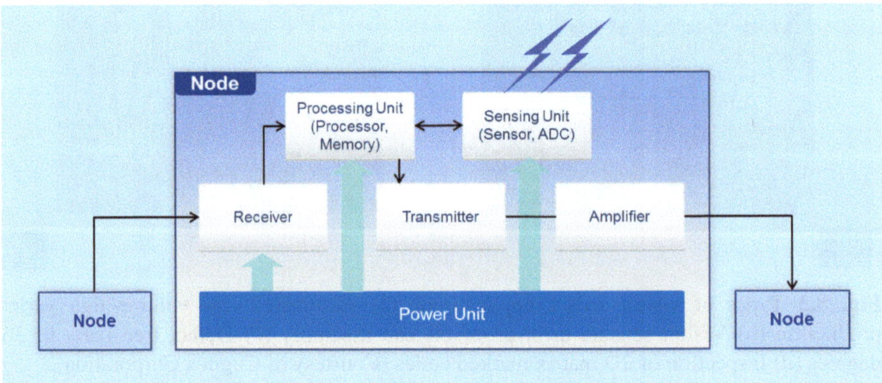

Fig. 8.5 Wireless micro-node model; each node has a sensing module, processing unit, and communication elements.

- *Nano-sensors*: are used, under the micro-scale, in chemical and biological sensory applications to deliver information about *nano-particles*. For example, nano-tubes are used to sense various properties of gaseous molecules (Fig. 8.6). Major challenges of development and commercialization of nano-sensors include reliability and high costs of production. There is, however, a tremendous room to enhance the technology and implement of nano-sensors in various real-life applications.

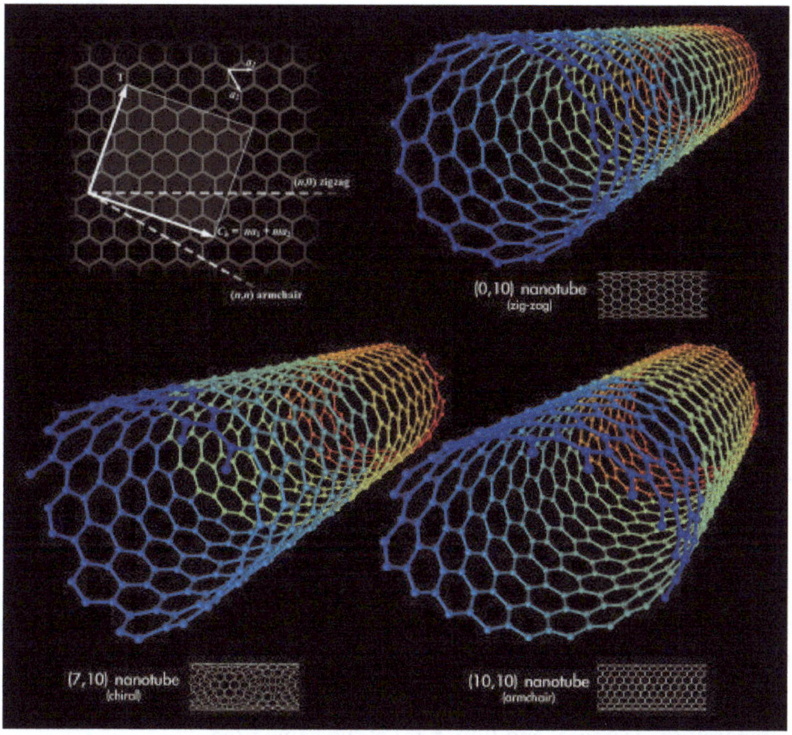

Fig. 8.6 3D model of three types of single-walled carbon nanotubes, like those used to make certain nano-sensors (Created by Michael Ströck on February 1, 2006).

- *RFID* (Radio Frequency Identification) *tags*: facilitates data transmission through portable devices read by RFID readers. The information transmitted through tags typically contains the location or identification along with other specifics of the tagged product (e.g., purchase date, color, and price). Typical RFID tags include microchip with radio antenna, mounted on substrate (Fig. 8.7). RFID tags are configured to respond and receive signals from an RFID transceiver. This allows tags to be read from a distance, unlike other forms of authentication technology. The RFID system has gained wide acceptance in businesses, and is gradually replacing the barcode system. Basic RFIDs consist of an antenna, transceiver and transponder. Antenna emits the radio signals to activate tag and to read as well as write information to it. Reader emits the radio waves, ranging from one to 100 inches, on the basis of used radio frequency and power output. While passing through electronic magnetic zone, RFID tags detect activation signals of readers. Powered by their internal batteries or the reader signals, the tags send radio waves back to the reader. Reader receives the waves, identifies the frequency, generates a unique ID, decodes the data encoded in integrated circuit of tags, and then transmits it to the user.

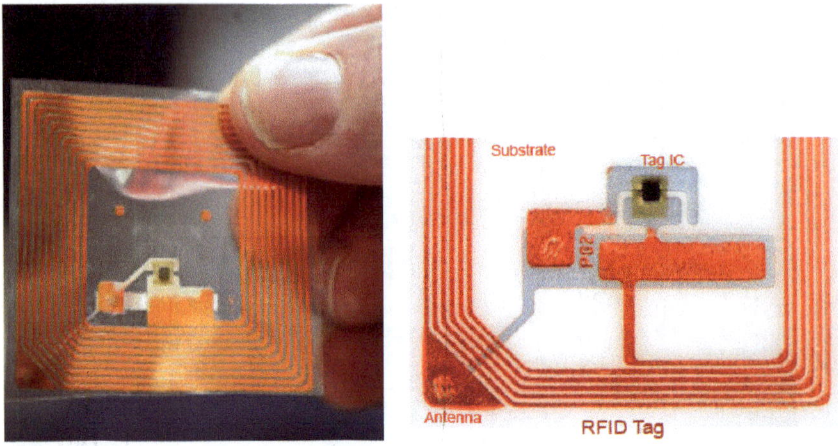

Fig. 8.7 Example of an RFID tag and configuration (Courtesy of Getty Images).

8.2 Sensor Networks

With the growing need for decentralized control along with high fault tolerance and reliability, single-sensor systems increasingly faced various limitations:

- Limited application – when a system should measure several variables (e.g., temperature, pressure, and flow rate) at the same time.
- Low fault tolerance – when the system is prone to a variety of unexpected failures.
- Low accuracy and timeliness – in noisy environments with high likelihood of various uncertain disruptions.

Therefore, single-sensor systems are not suitable when timely, robust, and accurate information is required. Sensor networks, which require replicated sensory information, along with their communication network technologies provide promising solutions to these problems (Jeong and Nof, 2008). Replicated sensor systems are applicable not only because micro-fabrication technology enables the production of various micro-sensors at low manufacturing cost, but also because micro-sensors can be embedded in a system with replicated deployment. These redundantly deployed sensors enable a system to improve accuracy and tolerate sensor failure. Distributed micro-sensor arrays and networks (DMSA/DMSN) are built from collections of spatially scattered micro-sensor nodes. Each node has the ability to measure the local physical variable within its accuracy limit, process the raw sensory data, and cooperate with its neighboring nodes.

Intelligent sensors are those incorporated with dedicated signal-processing functions. The roles of dedicated signal processing functions are to enhance design flexibility, realize new sensing functions, and reduce loads on central processing units and signal transmission lines by distributing information processing to the lower layers of the system (Yamasaki, 1996). A set of micro-sensors deployed close to each other to measure the same physical quantity of interest is called a *cluster*. Sensors in a cluster form a distributed sensor network (DSN), which may be homogenous (sensors with the same type) or heterogeneous (sensors with different types). A DSN can be utilized in a widely distributed sensor system and implemented as a locally concentrated configuration with a high density.

8.2.1 Multi-sensor Data Fusion Methods

In a sensor network, there are three different types of interactions among sensors (Brooks and Iyengar, 1998; Faceli et al., 2004):

(1) *Complementary*: when sensors do not depend on each other directly, but are combined to give a more complete image of the phenomena being studied.
(2) Competitive: when sensors provide independent measurement of the same information regarding a physical phenomenon.
(3) Cooperative, when sensors combine data from independent sensors to derive information that would be unavailable from the individual sensors.

To combine the information collected from the sensors, various multi-sensor data fusion methods can be applied. Multi-sensor data fusion is the process of combining observations from a number of different sensors to provide a robust and complete description of an environment or process of interest. Most recent data fusion methods employ probabilistic descriptions of observations and processes, and apply the Bayes' rule to combine this information (Hugh and Henderson, 2008; Mitchell, 2007).

Bayes' Rule

Bayes' rule is the basis for most data fusion methods. It provides a means to make inferences about an object or environment of interest described by a state x, given an observation z. Based on the rule of conditional probabilities, Bayes' rule is defined through the following conditional probability:

$$P(x \mid z) = \frac{P(z \mid x)P(x)}{P(z)}. \tag{8.6}$$

$P(z \mid x)$ serves the role of a sensor model. The probability is constructed by fixing the value of $x = x$ and then defining what probability density $P(z \mid x = x)$ on x is inferred. The multi-sensory form of Bayes' rule requires the following conditional independence

$$P(z_1, \cdots, z_n \mid x) = P(z_1 \mid x) \cdots P(z_n \mid x) = \prod_{i=1}^{n} P(z_i \mid x_i). \tag{8.7}$$

The recursive form of Bayes' rule is then defined as

$$P(x \mid Z^k) = \frac{P(z_k \mid x)P(x \mid Z^{k-1})}{P(z_k \mid Z^{k-1})}. \tag{8.8}$$

From this equation, one needs to compute and store only the posterior density $P(x \mid Z^{k-1})$, which contains a complete summary of all past information.

Probabilistic Grids

Probabilistic grids apply the Bayesian data fusion technique to solve mapping (Elfes, 1987) and tracking (Stone et al., 1999) problems. Practically, a grid of likelihoods on the states x_{ij} is produced in the form of $P(z = z \mid x_{ij}) = \Lambda(x_{ij})$. It is then trivial to apply the Bayes' rule to update the property value at each grid cell as

$$P^{+}(x_{ij}) = C\Lambda(x_{ij})P(x_{ij}), \quad \forall i, j, \tag{8.9}$$

where C is a normalizing constant obtained by summing posterior probabilities to 1 at node ij only. Computationally, this is a simple point-wise multiplication of two grids. Grid-based fusion is appropriate to situations where the domain size and dimension are modest. In such cases, grid-based methods provide straightforward and effective fusion algorithms. Monte Carlo and particle filtering methods can be considered as grid-based methods, where the grid cells themselves are samples of the underlying probability density for the state.

The Kalman Filter

The Kalman filter is a recursive linear estimator that successively calculates an estimate for a continuous valued state on the basis of periodic observations of the state. The Kalman filter may be considered a specific instance of the recursive Bayesian filter (Mayback, 1979) for the case where the probability densities on states are Gaussian. The Kalman filter algorithm produces estimates that minimize mean-squared estimation error conditioned on a given observation sequence and so is the conditional mean

$$\hat{x}(i \mid j) \triangleq E[x(i) \mid z(1), \cdots, z(j)] \triangleq E[x(i) \mid Z^{j}]. \tag{8.10}$$

The estimate variance is defined as the mean-squared error in this estimate

$$P(i \mid j) \triangleq E[x(i) - \hat{x}(i \mid j)][x(i) - \hat{x}(i \mid j)]^{T} \mid Z^{j}]. \tag{8.11}$$

The estimate of the state at a time k, given all information up to time k, is denoted by $\hat{x}(k \mid k)$. Similarly, the estimate of the state at a time k given only information up to time k–1 is called a one-step-ahead prediction and denoted by $\hat{x}(k \mid k-1)$.

The Kalman filter is appropriate to data fusion problems where the entity of interest is well defined by a continuous parametric state. Thus, it would be useful to estimate position, attitude, and velocity of an object, or the tracking of a simple geometric feature. Kalman filters, however, are inappropriate for estimating properties such as spatial occupancy, discrete labels or processes whose error characteristics are not easily parameterized.

Sequential Monte Carlo Methods

The sequential Monte Carlo (SMC) filtering method is a simulation of the recursive Bayes update equations using sample support values and weights to describe the underlying probability distributions. SMC recursion begins with an a posterior probability density represented by a set of support values and weights $\{x_{k-1}^i, w_{k-1|k-1}^i\}_{i=1}^{N_{k-1}}$ in the form

$$P(x_{k-1} \mid Z^{k-1}) = \sum_{i=1}^{N_{k-1}} w_{k-1}^i \delta(x_{k-1} - x_{k-1}^i). \tag{8.12}$$

Leaving the weights unchanged $w_k^i = w_{k-1}^i$ and allowing the new support value x_k^i to be drawn on the basis of old support value x_{k-1}^i, the prediction becomes

$$P(x_k \mid Z^{k-1}) = \sum_{i=1}^{N_{k-1}} w_{k-1}^i \delta(x_k - x_k^i). \tag{8.13}$$

The SMC observation update step is relatively straightforward and described in (Hugh and Henderson, 2008; Bar-Shamon and Fortmann, 1998). SMC methods are well suited to problems where state-transition models and observation models are highly nonlinear. However, they are inappropriate for problems where the state space is of high dimension. In addition, the number of samples required to model a given density faithfully increases exponentially with state-space dimension.

Interval Calculus

Interval representation of uncertainty has a number of potential advantages over probabilistic techniques. An interval to bound true parameter values provides a good measure of uncertainty in situations where there is a lack of probabilistic information, but in which sensor and parameter error is known to be bounded. In this technique, the uncertainty in a parameter x is simply described by a statement that the true value of the state x is known to be bounded between a and b, i.e., $x \in [a, b]$. There is no other additional probabilistic structure implied. With $a, b, c, d \in R$, interval arithmetic is also possible as

$$d([a,b],[c,d]) = \max(|a-c|,|b-d|). \tag{8.14}$$

Interval calculus methods are sometimes used for detection, but are not generally used in data fusion problems because of the difficulties to get results converged to anything value, and to encode dependencies between variables.

Fuzzy Logic

Fuzzy logic has achieved widespread popularity for representing uncertainty in high-level data fusion tasks. Fuzzy logic provides an ideal tool for inexact reasoning, particularly in rule-based systems. In the conventional logic system, a membership function $\mu_A(x)$ (aka, characteristic function) is defined. Then the fuzzy membership function assigns a value between 0 and 1, indicating the degree of membership of every x to set A. Composition rules for fuzzy sets follow the composition processes for normal crisp sets as

$$\mathrm{A} \cap \mathrm{B} \rightleftharpoons \mu_{A\cap B}(x) = \min[\mu_A(x), \mu_B(x)]. \tag{8.15}$$

$$\mathrm{A} \cup \mathrm{B} \rightleftharpoons \mu_{A\cup B}(x) = \max[\mu_A(x), \mu_B(x)]. \tag{8.16}$$

The relationship between fuzzy set theory and probability is, however, still debated.

Evidential Reasoning

Evidential reasoning methods are qualitatively different from either probabilistic methods or fuzzy set theory. In evidential reasoning, belief mass cannot only be placed on elements and sets, but also sets of sets, while in probability theory a belief mass may be placed on any element $x_r \in \chi$ and on any subset $A \subseteq \chi$. The domain of evidential reasoning is the power set 2χ. Evidential reasoning methods play an important role in discrete data fusion, attribute fusion, and situation assessment, where information may be unknown or ambiguous.

Multisensory fusion methods and their models are summarized in Table 8.2.

Table 8.2 Multi-sensor data fusion methods (adapted from Hugh and Henderson, 2008; Mitchell, 2007)

Approach	Method	Fusion model &rule
Probabilistic modeling	Bayes' Rule	$P(x \mid Z^k) = \dfrac{P(z_k \mid x)P(x \mid Z^{k-1})}{P(z_k \mid Z^{k-1})}$
	Probabilistic Grids	$P^+(x_{ij}) = C\Lambda(x_{ij})P(x_{ij})$
	The Kalman Filter	$P(i \mid j) \triangleq E\{[x(i) - \hat{x}(i \mid j)][x(i) - \hat{x}(i \mid j)]^T \mid Z^j\}$
	Sequential Monte Carlo Methods	$P(x_k \mid Z^k) = C\sum_{i=1}^{N_k} w_{k-1}^i P(z_k = z_k \mid x_k = x_k^i)\delta(x_k - x_k^i)$
Non-probabilistic modeling	Interval Calculus	$d([a,b],[c,d]) = \max(\lvert a-c \rvert, \lvert b-d \rvert)$
	Fuzzy Logic	$\mathrm{A} \cap \mathrm{B} \rightleftharpoons \mu_{A\cap B}(x) = \min[\mu_A(x), \mu_B(x)]$ $\mathrm{A} \cup \mathrm{B} \rightleftharpoons \mu_{A\cup B}(x) = \max[\mu_A(x), \mu_B(x)]$
	Evidential Reasoning	$2^{\mathrm{X}} = \{\{\text{occupied,empty}\}, \cdots \{\text{occupied}\}, \{\text{empty}\}, 0\}$

Multi-sensor fusion is not only the process of combining inputs from sensors with information from other sensors, but also the logical procedure of inducing the optimal output from multiple inputs with one representative format (Luo and Kay, 1989). In the fusion of large-size distributed sensor networks, an additional advantage of multi-sensor integration is the ability to obtain more fault-tolerant information. This fault tolerance is based on redundant sensory information that compensates for faulty or erroneous readings of sensors. There are several types of multi-sensor fusion and integration methods, depending on the types of sensors and their deployment (Iyengar et al., 1995). This topic has received increasing interest in recent years because of the sensibility of networks built with many low-cost micro and nano-sensors. As an example, a recent improvement of the fault-tolerance sensor integration algorithm (FTSIA) (Liu and Nof, 2004; Nof et al., 2009) enables it not only to detect possibly faulty sensors and widely faulty sensors, but also to generate a final data interval estimate from the correct sensors after removing the readings of those faulty sensors.

8.2.2 Design Considerations

Sensor networks are somewhat different from any other networked system in that sensor nodes (especially micro-sensors) are highly prone to failure over time. As sensor nodes weaken or even die, the topology of the active sensor networks changes frequently. Besides, when mobility is introduced into the sensor nodes, maintaining the robustness and discovering topology consistently become even more challenging. Therefore, the algorithms developed for sensor network communication and task administration should be flexible and stable against changes of network topology and work properly under unexpected failure of sensors. DSN systems should be designed with application-specific communication algorithms and TAPs (Task Administration Protocols), because micro-sensors and their networking systems are extremely resource constrained. Therefore, design of application specific protocols with respect to energy consumption and network parameters such as node density, radio transmission range, network coverage, latency, and distribution is another challenging issue. Current network protocols also use broadcasting for communication, while traditional and ad hoc networks use point-to-point communication. Hence, the routing protocols, in general, should be designed by considering crucial sensor network features as follows:

(1) *Fault tolerance*: Over time, sensor nodes may fail or be blocked due to lack of power, physical damage or environmental interference. The failure of sensor nodes, however, should not affect the overall operation of the sensor network. Thus, fault tolerance or reliability is the ability to sustain sensor network functionality despite likely problems.

(2) *Accuracy improvement*: Redundancy of information can reduce overall uncertainty and increase the accuracy with which events are perceived. Since nodes located close to each other are combining information about the same event, fused data improve the quality of the event information.

(3) *Timeliness*: DSN can provide the processing parallelism that may be needed to achieve an effective integration process, either at the actual speed that a single sensor could provide, or at even faster operation speed.
(4) *Network topology*: A large number of nodes deployed throughout the sensory field should be maintained by carefully designed topology because any changes in sensor nodes and their deployments longer lifetime of sensor network systems.
(5) *Lower cost*: Despite the use of redundancy, a distributed micro-sensor system obtains information at lower cost than the equivalent information expected from a single sensor because it does not require the additional cost of functions to obtain the same reliability and accuracy. The current cost of a micro-sensor node, e.g., dust mote (Warneke et al., 2001), is still expensive (US$ 25-172), but it is expected to be less than US$ 1 in the near future, so that sensor networks can be justified.
(6) *Scalability*: The coverage area of a sensor network system depends on the transmission range of each node and the density of the deployed sensors. The density of the deployed nodes should be carefully designed to provide a topology appropriate for the specific application.

A distributed network of micro-sensor arrays (MSA) can yield more accurate and reliable results based on built-in redundancy. Recent developments of flexible and robust protocols with improved fault tolerance will not only meet essential requirements in distributed systems but will also provide advanced features needed in specific applications. While MEMS sensor technology has advanced significantly in recent years, scientists now realize the need for design of effective MEMS sensor communication networks and task administration.

8.2.3 Architectures

A well-designed distributed network of micro-sensor arrays can yield more accurate and reliable results compared to classic single-sensor systems, due to its built-in redundancy. Recent developments of flexible and robust protocols with improved fault tolerance will not only meet essential requirements in distributed systems but will also provide advanced features needed in specific applications. They can produce widely accessible, reliable, and accurate information about physical environments. Various architectures have been proposed and developed to improve the performance of systems and fault-tolerance functionality of complex networks depending on their applications (Wesson et al., 1981; Iyengar et al., 1994; Nadig et al., 1993; Jeong and Nof, 2009).

A network is defined as a general graph $G = (V, \mathrm{L})$, where V is a set of nodes (or vertices) and L is a set of communicating links (or edges) between nodes. For a DSN, a node means an intelligent sensing node consisting of a computational processor and associated sensors, and an edge is the connectivity of nodes. A DSN consists of a set of sensor nodes, a set of cluster-head (CH) nodes, and a communication network interconnecting the nodes (Fig. 8.8). In general, one sensor node communicates with more than one CH, and a set of nodes communicating with a CH is called a cluster. A clustering architecture can increase system capacity and enable better resource allocation (Ghiasi et al., 2002; Lin and Gerla, 1997). Data are integrated in CH by receiving required information from associated sensors of the cluster. In the cluster, CHs can interact not only with other CHs, but also with higher-level CHs or a base station. A number of network configurations have been developed to prolong network lifetime and reduce energy consumption in forwarding data. In order to minimize energy consumption, routing schemes can be broadly classified into two categories: (1) Clustering-based data forwarding scheme (Fig. 8.8-a,b); (2) Multi-hop data forwarding scheme without clustering (Fig. 8.8-b,d).

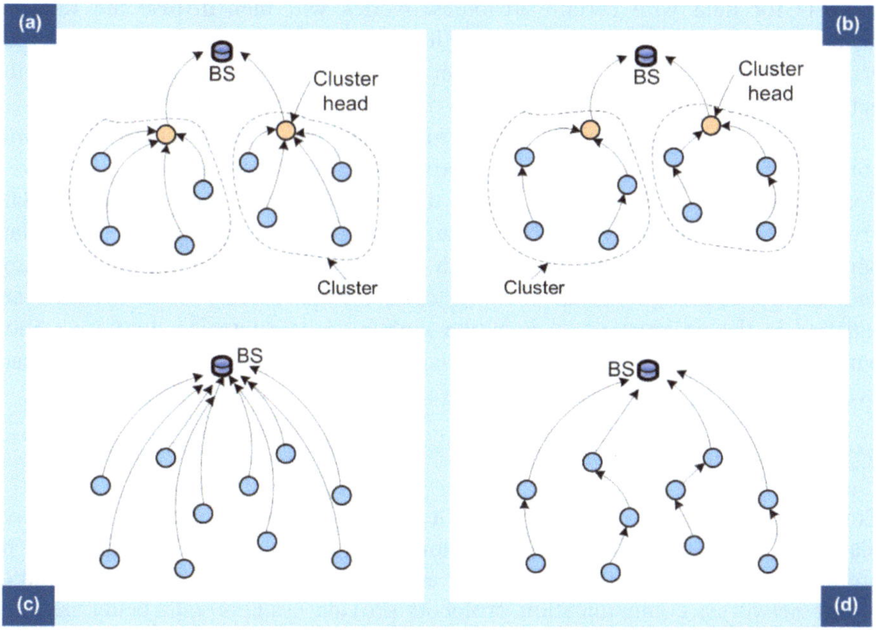

Fig. 8.8 Four different configurations of wireless sensor networks (WSN): (a) Single-hop with clustering; (b) Multi-hop with clustering; (c) Single-hop without clustering; (d) Multi-hops without clustering.

With the advancement of wireless mobile communication technologies, ad hoc wireless sensor networks (AWSNs) have become more popular. With this advancement, the above wired (micro-wired) architectures remain valid only when wireless communication is physically prohibited. Otherwise, wireless architectures are considered superior. The architecture of AWSN is fully flexible and dynamic, that is, a mobile ad hoc network represents a system of wireless nodes that can freely reorganize into temporary networks as needed, allowing nodes to communicate in areas with no existing infrastructure. Thus, interconnection between nodes can be dynamically changed, and the network is set up only for a short period of communication (Ilyas, 2002). Now the AWSN with an optimal ad hoc routing scheme has become an important design concern.

In applications with no given pattern of sensor deployment (e.g., battlefield surveillance or environmental monitoring), the AWSN approach can provide efficient sensor networking. Especially in dynamic network environments such as AWSN, three main distributed services, i.e., lookup service, composition service, and dynamic adaptation service by self-organizing sensor networks, are also studied to control the system (Lim, 2001). In order to route information in an energy-efficient manner, directed diffusion routing protocols based on the localized computation model (Estrin et al., 1999; Intanagonwiwat et al., 2003) have been studied for robust communication. The data consumer will initiate requests for data with certain attributes. Nodes will then diffuse the requests towards producers via a sequence of local interactions. This process sets up gradients in the network which channel the delivery of data. Even though the network status is dynamic, the impact of dynamics can be localized.

A mobile-agent-based DSN (Chau and Geraniotis, 1990) utilizes a formal concept of agent to reduce network bandwidth requirements. A mobile agent is a floating processor migrating from node to node in the DSN and performing data processing autonomously. Each mobile agent carries partially integrated data which will be fused at the final CH with other agents' information. To save time and energy consumption, as soon as certain requirements of a network are satisfied in the progress of its tour, the mobile agent returns to the base station without having to visit other nodes on its route. This logic reduces network load, overcoming network latency, and improves fault-tolerance performance.

8.2.4 Protocols

Communication protocols, as discussed earlier in this book, provide an effective platform for distributed and autonomous agents to interact, collaborate, be productive, and achieve their individual and common goals. In distributed micro-sensor networks, communication protocols provide systems with better network capability and performance by creating efficient paths and accomplishing effective communication between the sensor nodes (Lim, 2001; Heinzelman et al., 2002; Iyengar et al., 1992). The most commons communication protocols for sensor networks are as follows:

- *Point-to-point protocol* (PTP): is the simplest communication protocol and transmits data to only one of its neighbors (Fig. 8.9a). PTP, however, is not an efficient protocol for a DSN, because in case of failure of nodes or links the communication paths are fully disrupted.
- *Flooding protocol* (FP): addresses the information sent out by a sender node to all of its neighbors (Fig. 8.9-b). This disseminates data quickly in a network where bandwidth is not limited and links are not loss-prone. However, since a node always sends data to its neighbors, regardless of

whether or not the neighbor has already received the data from another source, it leads to the implosion problem and wastes resources by sending duplicate copies of data to the same node.

- *Gossiping protocol* (GP): is an alternative to the classic FP (Ween et al., 2000; Hedetniemi et al., 1988), where instead of indiscriminately sending information to all its neighboring nodes, each sensor node only forwards the data to a randomly selected neighbor (Fig. 8.9-c). While the GP distributes information more slowly than FP, it dissipates resources, such as energy, at a relatively lower rate. In addition, it is not as robust relative to link failures as a broadcasting protocol (BP), because a node can only rely on one other node to resend the information for it in the case of link failure.

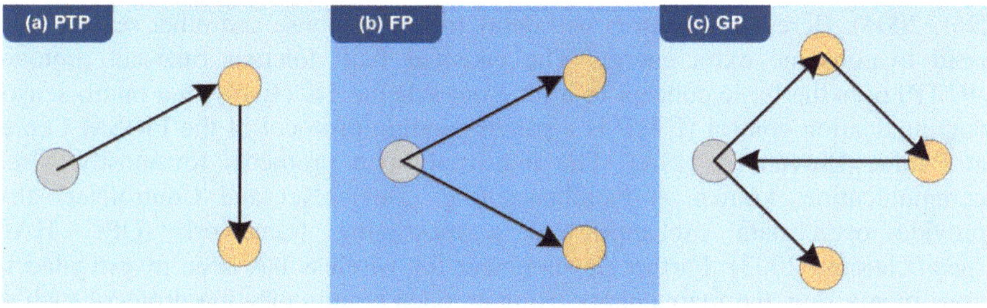

Fig. 8.9 Three basic communication Protocols; (a) Point-to-Point Protocol (PTP), (b) Flooding Protocol (FP), (c) Gossiping Protocol (GP).

A good solution to the implosion and overlap problems is the sensor protocol for information via negotiation (SPIN) (Heinzelman et al., 1999). SPIN nodes negotiate with each other before transmitting data, which helps ensure that only useful transmission of information will be executed. Nodes in the SPIN protocol use three types of messages in three consecutive stages to communicate:

(1) *ADV* (new data advertisement): The protocol begins when a node advertises the new data that is ready to be disseminated. It advertises by sending an ADV message to its neighbors, naming the new data. Upon receiving an ADV, the neighboring node checks to see whether it has already received or requested the advertised data to avoid implosion and overlap.
(2) *REQ* (request for data): If the neighboring node has not received or requested the advertised data, it responds by sending a REQ message for the missing data back to the sender.
(3) *DATA* (data message): The protocol completes when the initiator of the protocol responds to the REQ with a DATA message, containing the missing data.

In a relatively large sensor network, a clustering architecture with a local CH is necessary. In such situations, an alternative solution is the low energy adaptive clustering hierarchy (LEACH), as a clustering-based protocol that utilizes randomized rotation of local cluster base-stations to evenly distribute the energy load of sensors in DSN (Heinzelman et al., 2000).

Energy-minimizing routing protocols have also been developed to extend the lifetime of the sensing nodes in a wireless network. For example, a minimum transmission energy (MTE) routing protocol (Ettus, 1998) chooses intermediate nodes such that the sum of squared distances is minimized by assuming a square-of-distance power loss between two nodes. This protocol, however, results in unbalanced termination of nodes with respect to the entire network. Recently, time-based network protocols have been developed to ensure that, when a task keeps the resources idle for too long, their exclusive service by the resource is disabled. That is, the time-based control protocol is intended to provide a rational collaboration rule among tasks and resources in the networked system (Liu and Nof, 2004). Here, slow sensors will delay timely response and other sensors may need to consume extra energy. The patented fault tolerant time-out protocol (FTTP) uses the basic concept of a time-out scheme effectively in a micro-sensor communication control (FTTP is a patent-pending protocol of the PRISM Center at Purdue University, USA). The industrial open protocols for mostly wired communication, known as Fieldbuses (e.g., DeviceNet and ControlNet) also provide open data exchange and a messaging framework (OPC HAD specifications, 2003). Further development for wireless has been investigated in asset monitoring and maintenance using an open communication protocol such as Zig-Bee (2004) (See Chapter 7 for more detail).

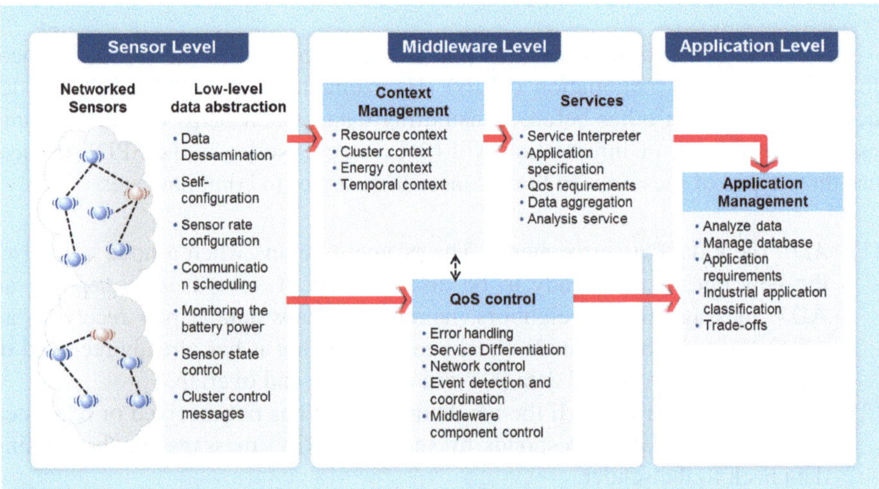

Fig. 8.10 Middleware architecture for facility sensor network applications (MidFSN).

Wireless sensor network (WSN) applications also require a middleware to deliver a general runtime environment that inherits common requirements under application specifications (Jeong and Nof, 2008). In order to provide robust functions to industrial applications under somewhat limited resource constraints and the dynamics of the environment, appropriate middleware is required to contain embedded trade-offs between essential quality-of-service (QoS) requirements from the applications. Typically, a sensor network middleware has a layered architecture that is distributed among the networked sensor systems. Based on classic architectures, a middleware for facility sensor network (MidFSN) can be developed with a layered structure as depicted in Fig. 8.10.

8.2.5 Facility Sensor Networks

A facility sensor network (FSN; Fig. 8.11) refers to a wireless sensor network (WSN) applied in a facility, where wireless sensors continuously monitor the manufacturing processes. Application of wireless devices in real facilities, however, may raise many challenges in terms of accuracy, reliability, and performance. These issues must be resolved before wireless sensor packages are deployed by any facility. Consider, for example, a facility where commercial wireless sensor packages are deployed for process monitoring (Agarwal and Tsoukalas, 2008). In this case, the data transmitted from sensors may be corrupted due to the interferences caused by the several machines and equipment working in the facility. In order to develop a reliable WSN for facility monitoring, it is important to firstly identify the sources of interferences, and then find a way to minimize the influence of those interferences.

In deployment of a FSN, the numbers and locations of individual multi-functional sensors, which are capable of conducting continuous measurements and reporting, must be carefully defined. Typically, the major FSN design criteria are lifetime maximization (through minimizing energy consumptions) (Cheng et al., 2008), QoS (Toumpis and Tassiulas, 2006), and network cost (Xu et al., 2005). However, other criteria may also be involved, which cannot be quantified and measured in a simulated environment. In particular, data interference and noise are major critical issues when wireless sensors are deployed in real facilities. Therefore, proper, application/automation facility-specific deployment strategies must be developed to overcome such problems.

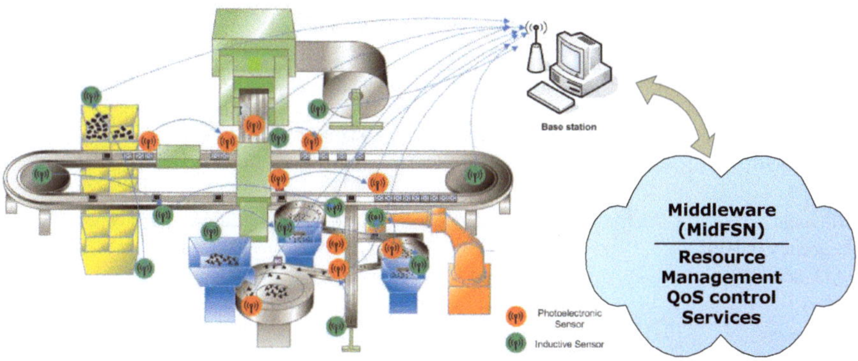

Fig. 8.11 Facility Sensor Network (FSN) for production process monitoring.

The source of interference in an FSN can be identified through statistical analyses. In order to minimize the effect of interference, an interference model must be developed to obtain the optimal deployment strategy that yields the lowest variance (Ko et al., 2010), and possibly define the optimal number and location of sensors. While the deployment model can be used to make a decision at strategic level of FSN design, a protocol needs to be developed to handle the interference at operational level of FSN (Ko et al., 2009). Even though sensor nodes are optimally deployed based on an optimal design strategy, the data are still interfered by electrical noise. Hence, without processing the data during the operation, it is inevitable to make incorrect decisions based on the interfered, distorted data.

In general, considering the reliability of FSNs in highly unpredictable and noisy environments, FSN approaches have much potential to be applied in various real-time automation applications such as electricity and water infrastructural management system and FSN-based certification of manufacturing equipment.

8.3 RFID Networks

RFID is an advanced automatic identification technology, which uses radiofrequency signals to capture data remotely from the tags located in its reading range (Klaus, 2003; Lee et al., 2009). The basic communication principle of RFID, which was firstly introduced in 1948, is the reflection of power. One of the very first applications of RFID technology was identification of friend or foe (IFF), deployed by the British Royal Air Force during World War II. The IFF system allowed radar operators and pilots to distinguish automatically between friendly and enemy aircraft using radiofrequency (RF) signals. The main objective was to prevent friendly fire and to aid the effective interception of enemy aircraft. The radiofrequency is the critical factor for the type of application for which an RFID system is best suited. Basically, the radiofrequencies are classified into four categories of low, high, ultrahigh, and microwave frequencies, as shown in Table 8.3 (See also Table 7.2, Chapter 7).

Table 8.3 Comparison of RFID frequency bands and their respective applications.

Frequency	Approximate read range	Data speed	Cost of tags	Applications
Low frequency (LF) (125 kHz)	< 5 cm (passive)	Low	High	Animal identification Access control
High frequency (HF) (13.56 MHz)	10 cm–1m (passive)	Low to moderate	Medium to low	Smart cards Payment
Ultrahigh frequency (UHF) (433, 868–928 MHz)	3–7m (passive)	Moderate to high	Low	Logistics and supply chain Pallet and case tracking Baggage tracking
Microwave (2.45 and 5.8 GHz)	10–15m (passive) 20–40m (active)	High	High	Electronic toll collection (ETC) Container tracking

A typical RFID system contains several components, including an RFID tag, which is the identification device attached to the item to be tracked, and an RFID reader and antenna, which are devices that can recognize the presence of RFID tags and read the information stored on them. After receiving the information, in order to process the transmission of information between the reader and other applications, RFID middleware is needed, which is software that facilitates the communication between the system and the RFID devices (Fig. 8.12).

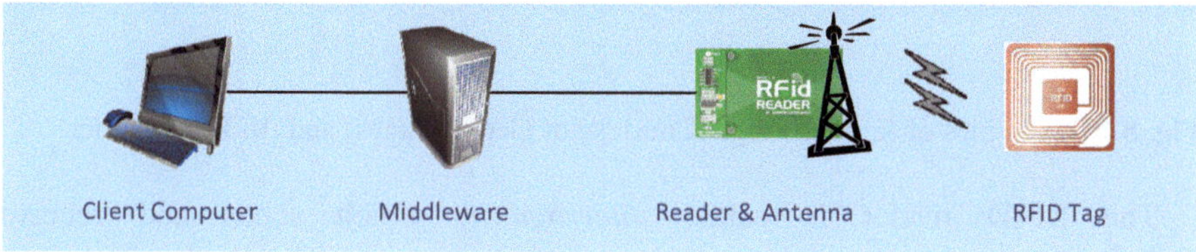

Fig. 8.12 Infrastructure for an RFID system.

RFID tags are devices with identification capabilities that contain the information that can be communicated to a reader from a distance. The tag comprises a simple silicon microchip attached to a small flat aerial which is mounted on a substrate. RFID tags (Fig. 8.13-a) are typically classified into three main types according to the source of energy used to power them:

(1) *Active tags*: use a battery to power the tag transmitter and receiver to broadcast their own signals to readers within the life of batteries. Active RFID tags receive low level signals by tag, but create higher level signals to readers. This allows them to communicate over distances of several meters. This type of RFID is constantly powered, whether in or out of the reader's field. Active tags consist of external sensors for checking humidity, temperature, motion, and other environmental conditions.
(2) *Semi-passive tags*: have built-in batteries to power the chip's circuitry, resist interference, and circumvent the lack of power from the reader signal due to long distance. They are different from active tags in that they only transmit data at the time a response is received.
(3) *Passive tags*: derive their power from the field generated by the reader without having an active transmitter to transfer the information stored, and rely on the energy of radio frequency transferred from the reader. Passive RFID tags require strong signals from the reader, but the signal strength bounced from them is very low.

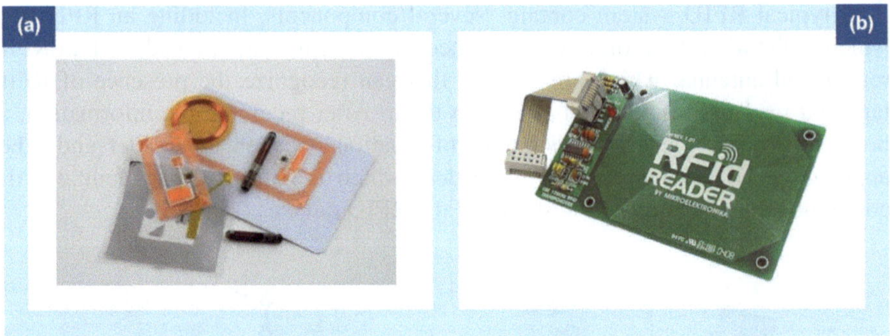

Fig. 8.13 Examples of (a) RFID tags (Courtesy of Getty Images) and (b) RFID reader.

The RFID reader is called *interrogator*, which sends and receives radiofrequency data to and from the tag via antennas. The reader contains a transmitter, a receiver, and a microprocessor (Fig. 8.13-b). The reader unit also contains an antenna as part of the entire system. The antennas broadcast the radiofrequency signals generated by the reader and receive responses from tags within range. The data acquired by the readers is then passed to a host computer, which may run specialist RFID software or middleware to filter the data and route the data to the correct application to be processed into useful information. Middleware refers to software that lies between two interfaces; the readers and the data collection systems (see Fig. 8.12). It is used to collect and filter data from readers. It can transfer those useful data, which are filtered, to the data collection system.

Similar to a radio that can be tuned in various frequencies to connect to different channels, RFID readers and tags need to be tuned in to a same frequency for communication. RFID system uses various frequencies but most common and popularly used frequency is low, high and ultra high frequency. Low frequency is around 125 KHz, high is around 13.56 MHz and ultra high varies between 860-960 MHz. Some applications also make use of microwave frequency of 2.45 GHz. It is imperative to choose right frequency for an application as radio waves work different at various frequencies. RFID tags are used as embedded in the networked system. Accordingly, an RFID network can be a part of sensor network systems.

8.3.1 RFID-Based Automation

RFID is increasingly attracting the attention of many manufacturing industries, as an emerging technology to digitalize the manufacturing information and automate the manufacturing process, especially as the technical problems are slowly being overcome and the cost of RFID is being reduced:

- The worldwide market for RFID technology was US$ 1.49 billion in 2004. In the first 6 months of 2008, 6.8 billion tags were sold for these applications along with 15.3 billion tags for pallets and cases. There is great market demand, which is ever increasing, for various RFID applications. It is predicted that RFID industry figures will increase from US$ 1.95 billion in 2005 to US$ 26.9 billion in 2015 (RNCOS, 2005).

- The rapid increase in use of RFID technology in the retail industry has been driven by major players such as Gillette, Tesco, Wal-Mart, and Metro AG in Germany.
- Wal-Mart, the world's largest retailer, has started deploying RFID applications and implementing new procedures in some of its distribution centers and stores, along with a number of its largest suppliers. The aim of these applications is to reduce out-of-stock occurrences by providing visibility of the location of goods by using RFID tags. Out-of-stock items that are RFID-tagged have been found to be replenished three times faster than before, and the amount of out-of-stock items that have to be manually filled has been cut by 10%.
- Gillette and Tesco implemented an item-level RFID project in the UK. Gillette razor blade cartridges were tagged with RFID tags and Tesco, the retailer, used an RFID reader embedded smart shelf system to search for items in the field, in order to take on-shelf inventories. Metro AG implemented an item-level RFID trial for use in its future stores. RFID tags are attached to each pack of Gillette razor blades, Proctor and Gamble (P&G) shampoos, Kraft cream cheese, and digital versatile disks (DVDs).

These facts indicate how RFID can revamp the retail industry and provide new customer experiences. With better tag and reader technology, declines in the cost of RFID tagging, and the release of information-sharing platforms, it is likely that RFID will be widely adopted across the entire supply network.

RFID enables real-time visibility and increases the processing efficiency of shop-floor manufacturing data. RFID also supports information flow in process-linked applications. Moreover, it can help to minimize the need for reworking, improve efficiency, reduce line stoppages, and replenish just-in-time materials on the production line. RFID can assist in automating assembly-line processes and thus reduce labor and cost, and minimize errors on the plant floor. The integration of RFID with various manufacturing systems is still a challenge to many corporations. As most large retailers will gradually demand the use of RFID in the goods from their suppliers, this creates both pressure and the opportunity for small and medium sized enterprises (SMEs) to adopt this technology in their logistics operations and extend it to the control of their manufacturing processes. RFID can be more cost-effective in bridging the gap between automation and information flow through providing better traceability and reliability on the shop floor (Qiu and Xu, 2004).

Traditional shop-floor control in a production environment, although computerized, still requires manual input of shop-floor data to various systems (e.g., ERP for production planning and scheduling). Such data includes product characteristics, labor, machinery, equipment utilization, and inspection records.

Companies such as Lockheed Martin, Raytheon, Boeing (Poirier and Mccollum, 2006), and Bell Helicopter have installed lean data-capture software and technologies and are in the process of converting barcodes to RFID. Honeywell was using barcodes to collect data related to part histories, inventories, and billing, and sharing data with its clients, and is accelerating its plan to switch from barcodes to RFID. The use of RFID has several advantages over barcodes, as tags that contain microchips that can store as well as transmit dynamic data have a fast response, do not require line of sight, and possess high security. RFID offers greater scope for the automation of data capture for manufacturing process control. In recent years, the cost of RFID tags has continuously decreased (McFarlane et al., 2003; Qiu, 2007) while their data capability has increased, which makes practical applications of RFID technology in manufacturing automation economically feasible.

For manufacturers, it is becoming increasingly important to design and integrate RFID information into various enterprise application software packages and to solve connectivity issues relating to plant floor and warehousing. Real-time manufacturing process automation depends highly on the principle of closed-loop automation that senses, decides, and responds from automation to plant and enterprise operations. For example, a pharmaceuticals manufacturer (Kali Laboratories, 2004) makes use of RFID to trace the route or history taken by an individual product at multiple locations along the production line. This allows the pharmaceuticals manufacturer to easily trace all final products that might have been affected by any production miscarriage. An aerospace company named Nordam Group uses RFID to track its high-cost molds. Through the use of RFID tags, they save the cost of real-time tool tracking. With growing emphasis on real-time responsiveness, manufacturers are seeking to control more effectively the production processes in real time in order to eliminate waste and boost throughput. The desire to extend supply network execution dispatching within a plant makes closed-loop automation an imperative. The inability to achieve true closed-loop manufacturing process automation presents one of the greatest barriers to successful real-time operation strategies. Critical elements that support manufacturing process automation include:

- Dynamic information technology systems to support high-mix, variable environments.
- Dynamic modeling, monitoring, and management of manufacturing resources.
- Lack of integration of manufacturing with the supply network.
- Lack of real-time shop-floor data for predictive analysis and for decision support.
- Lack of a common data model for all operational applications.
- Inability to sense and analyze inputs ranging from the factory floor to the supply network.
- Ineffective integration links between manufacturing process automation and enterprise information technology applications.
- Inability to provide intelligent recommendations and directives to targeted decision points for quick action.

8.3.2 Physical Asset Management

A physical asset management (PAM) program involves the tracking of physical assets (e.g., expensive machinery and equipment) along with returnable items (e.g., containers and pallets). PAM is a major concern in the manufacturing industry. PAM includes engineering activities and adoption of a variety of approaches to maintenance such as reliability-centered maintenance, multi-skills, total productive maintenance, and hazard and operability studies (Hipkin, 2001). Manufacturing enterprises with better PAM systems usually outperform their counterparts in terms of utilization of assets in manufacturing. However, many enterprises today still face significant challenges in relation to tracking the location, quantity, condition, and maintenance and depreciation status of their fixed assets.

A popular traditional PAM approach is to track fixed assets through serially numbered asset tags (e.g., barcodes) for easy and accurate reading. This process is often performed by scanning a barcode on a physical asset identification tag that has been affixed to the asset. However, PAM may also involve a physical asset identification tag with a human-readable number only, which is documented manually. The barcode approach, however, has various limitations such as lack of automation and inability to provide real-time tracking.

To increase the utilization rate of movable equipment, manufacturing enterprises usually adopt a fluid approach to the composition of production lines. Movable equipment is shared across lines for some non-bottleneck procedures. Traditionally, applications for transfer of physical assets required tedious manual operations. Since most manufacturing processes were related to the requested equipment, late arrival of equipment lowered the overall efficiency of the company, which in turn impacted on the company's profit margin. Also, as the transfer records were entered into a database after a long verification process, the timeliness of the information was questionable. Besides, due to inevitable human errors, the database records were incomplete and inaccurate. Common discrepancies included asset locations, utilization, and history of maintenance and repair. As a result, equipment could not be repaired in a timely fashion and some was even lost.

To improve the overall efficiency and increase the visibility of physical assets, an RFID-enabled solution is used to track the movement of physical assets easily when they pass through antenna detection points without interfering with normal operations. RFID antennas should be set up near the door of each production site so that the moving in and out of physical assets would be automatically recorded (Fig. 8.14). Radiofrequency signals are affected by metal because of interference and reflection from it, which affects the readability of RFID tags. Since most equipment is made of metal, this is one of the main challenges of implementing RFID-enabled solutions in a plant. To address this problem, special attention must be paid to the height and angle of the antennas in order to provide optimal results for auto-tracking.

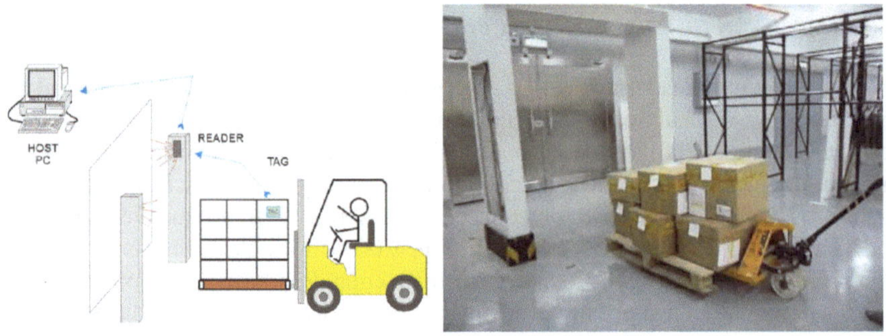

Fig. 8.14 Tracking the movement of equipment by RFID.

Real-Life Application: SAE Magnetics (HK) Ltd.

To evaluate the feasibility of the RFID-enabled PAM system, a case study was carried out in a selected company named SAE Magnetics (HK) Ltd., which is one of the world's largest independent manufacturers of magnetic recording heads for hard-disk drives used in computers and, increasingly, in consumer electronics such as digital video recorders, MP3 players, and even mobile phones. While the bulk of SAE's operations are based in two modern facilities located in mainland China, the company needs to manage large-scale warehouses for its sophisticated logistics. Different kinds of equipment are stored in these warehouses, e.g., semi-products, subassemblies, and equipment parts (collectively known as physical assets). It is normal for movable physical assets to be transferred from one site to another to maximize their use, e.g., production equipment and assets that are costly and advanced. In the past, SAE's warehouses adopted a pen-and-paper approach to record the transfer of physical assets. In time, this became a complicated process that generated piles of forms and involved many approval processes, which ultimately limited the company's operational efficiency.

This new method has enabled equipment or required parts and subassemblies to arrive promptly and has increased the utilization of important and valuable physical assets for the company. This system automatically tracks the transfer of physical assets and greatly reduces the processing time of equipment transfer. With the data transfer between the system and RFID devices being in real time, information about physical assets can be captured. These data are then presented in the RFID-enabled equipment tracking system in a systematic way. The system provides the ability to track and trace the status of equipment, issue equipment transfer orders, and provides reports for asset utilization analysis. As a result, location and utilization of equipment can be easily identified using visual diagrams that help in maintenance planning and repair scheduling.

8.3.3 Warehouse Management

In manufacturing enterprises, warehouse management usually requires complex and accurate systems. Since a single enterprise involves a tremendous amount of stock and multiple vendors at the same time, the warehouse data are usually big. It is also common for a single warehouse to be used to store stock from different vendors so as to enhance utilization rates. Such an approach also increases the complexity of warehouse storage structures. Without an accurate, automated, and comprehensive stock transfer recording procedure, warehouse management becomes a critical challenge. The major challenges that many industrial warehouses management systems involve include

- *Sophisticated manual processes*: Updating inventory lists in the warehouse, checking of replenishment requirements, and entering updated data into the Kerry warehouse management system (KWMS) are all manual processes. Not surprisingly, such paper-based processes are excessively time consuming and susceptible to manual errors.
- *Low warehouse visibility*: Warehouse visibility can be evaluated by measuring the discrepancy between the actual status and the system warehouse inventory status. Since the system is such that warehouse inventory status is only updated after office staffs enter all the latest information into the KWMS, there might be a discrepancy between the actual status and the system warehouse inventory status. As a result, staff might not know the real-time warehouse inventory status from the KWMS. Such information gaps could result in wrong decisions in inbound, relocation, and outbound processes. This is especially important for dialysis solutions since they have a limited period of validity. Extensive warehouse storage time causes product expiry and ultimately financial loss to the company.

Hence, RFID is used to increase the visibility of the warehouse and streamline warehouse operations. Through the implementation of RFID-enabled solutions in warehouse management, instant and accurate inventory records with automatic real-time record update and physical stocktaking can be achieved. An RFID-enabled intelligent forklift provides a higher degree of automation for the warehouse management via RFID (Fig. 8.15). In the RFID-enabled intelligent forklift, two RFID antennas are placed at the front of the forklift, one for location scanning and the other for pallet scanning. A tailor-made application is installed in the mobile computer placed in the forklift, which is used to show the information captured by the antennas and the information stored in the database system. Forklift drivers can view the information on the vehicle-mounted computer screen.

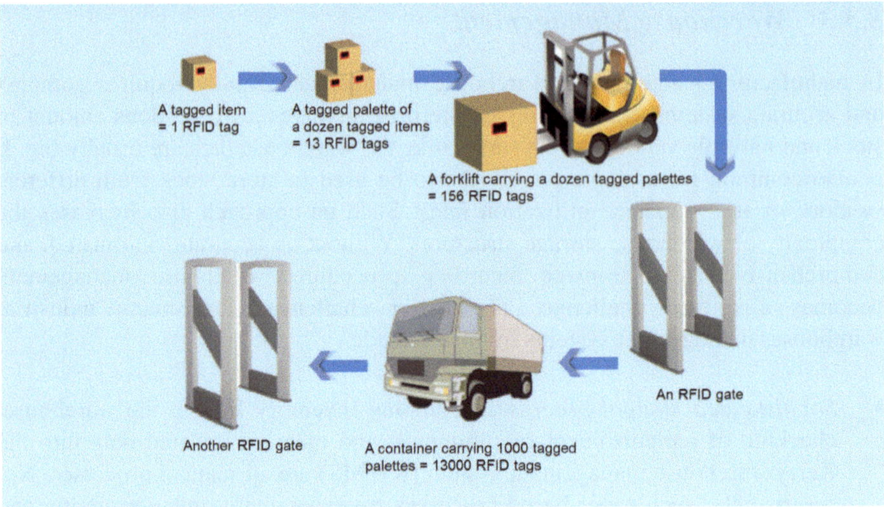

Fig. 8.15 The RFID-enabled intelligent forklift and gateway.

A tag containing product storage information, including particular stock keeping unit (SKU) and lot numbers, is pasted to each pallet. At the same time, RFID tags that store unique location information are placed on the racks. When a driver loads a pallet onto the forklift, the pallet information is scanned automatically and that product information and its planned location are shown on the screen attached to the forklift. After the pallet is placed in the storage position, the forklift scans the location tag on the rack, and completes the stock-in process (Fig. 8.16).

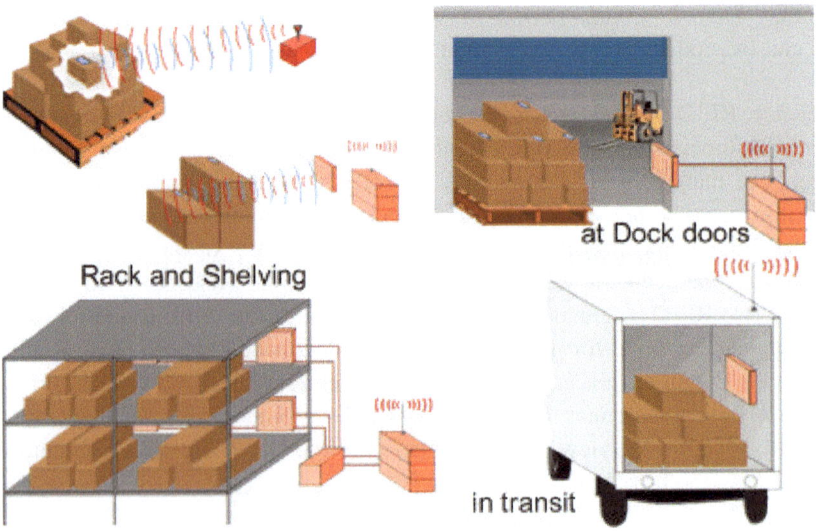

Fig. 8.16 Pallet and location RFID scanning of inventory.

Real-Life Application: Kerry Logistics Ltd.

To realize the capability of the RFID-enabled intelligent forklift, a case study was carried out in a selected company named Kerry Logistics (Hong Kong) Ltd. Kerry Logistics (Hong Kong) Ltd. is part of the diversified international conglomerate, the Kuok Group, and is a subsidiary of the Hong-Kong-listed Kerry Properties Limited. Kerry's Asian logistics business has been built on more than two decades of warehousing and logistics operations in Hong Kong. Its third-party logistics (3PL) business has dramatically accelerated since 1998. Kerry Logistics now encompasses contract logistics, distribution centers, international air and sea freight forwarding, transportation, distribution, and value-added services.

Like most 3PL service providers around the world, Kerry Logistics has recently recognized some inefficiency in the warehouse management of one of its major logistics centers located in Kwai Chung. Because of the complexity of taking care of multiple companies and products at the same time, 3PL service providers require considerable documentation to record the mass of information generated in the course of their work. One warehouse inside the logistics center stores healthcare products, especially dialysis solutions, for renal patients. To reduce human error and streamline the operation flow, an RFID-enabled intelligent forklift was developed.

With such modifications in warehouse operation, numerous benefits can be achieved. Since most paper-based processes are eliminated by RFID-enabled automation, both the speed and accuracy of the stock transaction process are greatly enhanced. For example, warehouse staffs are no longer required to fill in put-away lists and relocation lists, neither do they need to input data manually. Without tedious paper-based location mapping, the overall speed of the inbound process was improved by almost 15%, not to mention the cost saving factors of increased accuracy and reduced manpower. In addition, the RFID-enabled intelligent forklift allows for unstructured storage, which completely eliminates the relocation process. Forklift drivers can freely locate pallets of stock on the racks, and subsequently map their respective location information with only a few clicks. The flexibility of stock positioning is thus improved, resulting in better utilization of warehouse space and further shortened inbound processing time.

8.3.4 Information Interchange

The trend towards the establishment of a global production network in manufacturing industries poses some new challenges to *information interchange*. For example, some components of a product may be produced in one country, some part of the product may be outsourced and made in a second country, and the final assembly of the finished product may be done a third country, before being ultimately sold somewhere else. However, apart from the obvious need for collaboration, the problems of data integration and information sharing within the supply network are challenging. Therefore, an RFID-based intra-supply network information system is much needed in order to streamline the supply network activities and form an intra-supply network. Fig. 8.17 shows an RFID-based architecture, which is basically divided into four platform tiers. The system consists of the application of RFID tags to generate the data within the supply network. Data from the RFID tags is scanned by readers and synchronized with the internal information systems. The RFID-based architecture enables the different partners in the supply network to share real-time information.

(1) *Data acquisition tier*: The data from the operational level of the supply network are captured by RFID technology. RFID readers are installed in the warehouses of the manufacturers and retailers. Tags are placed either at the pallet level or the item level, depending on the stage in the supply network. At the level of the individual item of merchandise it is difficult to achieve 100% reading accuracy with RFID tags. However, it is already technically feasible to track RFID tags at the pallet or carton level. As the raw materials progress through the stages of WIP to the finished goods, the tagging level changes from the pallet level to the carton level and finally to the item level.
(2) *Information systems tier*: This tier integrates the RFID-based information systems with the internal information systems of the different supply network partners, such as ERP systems. The most successful new business models are probably those that can integrate information technology into all activities of the enterprise-wide value chain.
(3) *Intra-supply network data integration platform*: The supply network information is shared and integrated with the different supply network partners. On this platform, an integrated information system is built to enable the supply network partners to download the information. The RFID gateway can act as a check-in and check-out system and synchronize the internal information systems without delay. Supply network partners can select the information to be shared, which will be stored in a centralized database.
(4) *Web-based platform*: On this platform, a web-based information-sharing portal is built so that supply network partners can retrieve related information. Intra-supply network partners can use the web browser to search for required information. The production status, the supplier's inventory level, the delivery status, and retail sales data can be displayed on the web-based platform. The above platforms help an intra-supply network company to obtain data on their operations and provide an interface permitting them to share information. To facilitate the intra-supply network collaboration, different optimization and control technologies such as intelligent agent and data mining can be used to analyze these data.

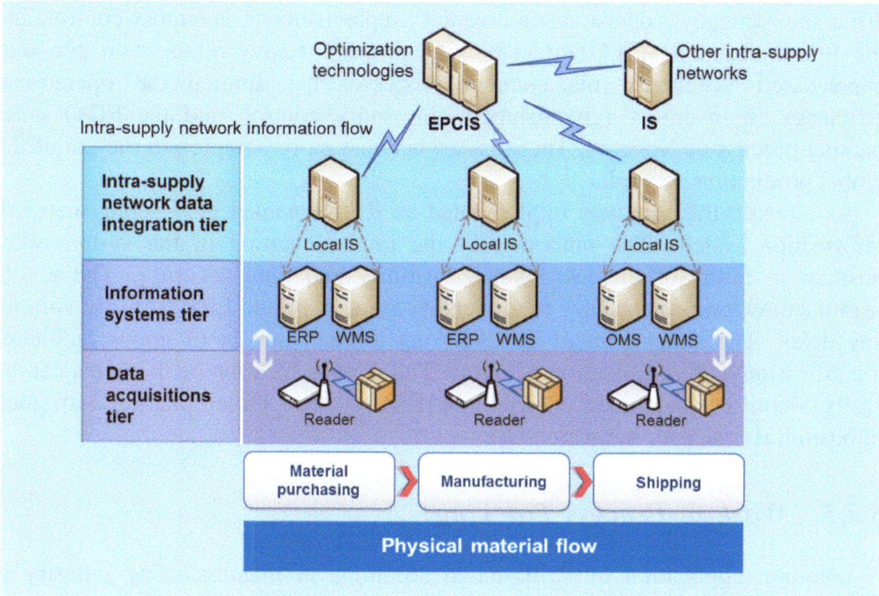

Fig. 8.17 Architecture of RFID-based intra-supply network information system (WMS: warehouse management system; IS: Information System; OMS: Office Management System).

Real-Life Application: Novetex (Textile Industry)

To realize the capability of the RFID-based intra-supply network information system, a case study was conducted in a reference site in the textile industry named Novetex. Novetex Spinners Limited was established in 1976 and, employing over 2000 people, is now recognized as the world's largest single-site woolen spinner. As part of the Novel Group, Novetex has its headquarters in Hong Kong and a factory located in Zhuhai, Southern China. The factory is composed of four spinning mills, a dyeing mill, and 51 production lines, resulting in an annual capacity of 7500 t of high quality yarn. Novetex has recognized the need to be a truly global supplier and has invested in offices and agents worldwide to offer customers the most efficient service possible. The trend towards the establishment of a global production network in the textile industry poses some new challenges to Novetex. For example, fibers might be produced in one country, spun to yield yarns in a second country, woven into fabrics in a third, and sewn into clothing in yet another one, before being ultimately sold somewhere else.

Novetex has traditionally used pen-and-paper-based warehouse management processes. The recording of the inventory, storage, order picking, packaging, and stocktaking processes were heavily dependent on the efficiency of warehouse operators, who used their own judgment and perceptions to decide how, when, and where to store the goods. This heavy reliance on paperwork and on human input to update the inventory data inevitably leads to data inaccuracy. This in turn may affect the company's decisions on inventory replenishment, inventory control, and the first-in first-out (FIFO) order-picking process. Heavy reliance on pen-and-paper-based warehouse management processes has limited the operational efficiency in inventory replenishment, inventory control, and the FIFO order picking process of Novetex. These issues become more complex in the era of the global production network.

As a result, the company implemented an RFID enabled intra-supply network information system. The outcomes of the implementation of the system were positive in terms of increased stock visibility and data accuracy. The textile manufacturer can synchronize the inventory status with the ERP software without any delay. The attachment of an RFID tag to each bundle of goods facilitates identification and visualization of items. The storage location of the items can be easily identified and retrieved from the ERP system. Order picking and stock relocation is easier for the warehouse.

8.3.5 Work-in-Process Tracking

A common application of RFID-based solutions in manufacturing industry is tracking of assembled items, i.e., work-in-process (WIP). The efficiency of the manufacturing process can be increased while entry errors and manpower are saved. For WIP tracking, the RFID tags are attached to products as they are being assembled or created along the production line (Datamonitor, 2005). The status of the product can be updated as it progresses along the production line via RFID readers placed above and below its path (Brown, 2007). The history or route taken by an individual product can be ascertained from data stored on tags attached to the item, by installing RFID readers at single or multiple locations along the production line (Hedgepeth, 2006). Accordingly, the location and flow history of any item can be recorded in the manufacturer's database.

Real-Life Application: Nypro Tool Hong Kong Ltd.

To realize the advantages of RFID-enabled WIP tracking, a case study was conducted in a precision mold manufacturing company named Nypro Tool Hong Kong Ltd. Nypro Tool Hong Kong Ltd. is a precision injection mold manufacturer which provides high-quality molding tools globally. Since the company provides products with a wide variety, WIP items need to be transported to different workshops for various processes such as machining, heat treatment, quality checking, etc. Each of the items is accompanied by its unique drawing so that operators can refer to its required operations. Traditionally, Nypro used a barcode system to monitor the flow of WIP items within the plants. For every processing operation of the WIP items, the status of the process must be captured manually so that management can check follow-up information for WIP through the database system. However, the manual process induced a lot of human errors and is time consuming. The application of RFID in WIP tracking was believed to be able to automate the tracking process and reduce the number of errors. With the RFID technology, the management of the company can accurately plan job allocation during the production process through automatic and accurate capture and tracking of the information related to the flow, location, and processing status of the WIP items. In other words, visibility of WIP information can be significantly enhanced and this makes manufacturing process automation possible.

8.4 Emerging Trends and Challenges

In this chapter, various types of sensors and sensor networks, their properties, benefits, limitations, and real-life applications were reviewed. With the growing advances of micro-miniaturization along with the ever-increasing needs for decentralization, intelligent, and automated, and collaborative control, emerging e-Criteria such as latency, fault tolerance, scalability, and connectivity have become the foremost issues in the design of wireless micro-/nano-sensors and RFID networks. Application of sensor network technologies is moving forward to more practical areas (Fig. 8.18); but, at the same time, facing various challenges such as:

- *Heterogeneity*: Integration of multiple types of sensors (e.g., seismic, acoustic, optical) sensors in one network platform and evaluation of the overall coverage of the system present several interesting challenges. When sensor nodes should be shared by multiple applications with differing goals, protocols must efficiently serve multiple applications simultaneously. Hence, to deploy heterogeneous sensors and develop their network application, several research questions must be taken into account regarding (1) Optimal resource utilization; (2) Handling and processing heterogeneous data; (3) Meeting QoS requirements while minimizing energy usage.
- *Security*: Since the sensor network may operate in a hostile environment, network techniques that support low-latency, survivability, and security are required. Indeed, low probability of communication detection is needed since some sensors may be envisioned for use behind enemy lines. For the same reasons, the network should be protected against intrusion and spoofing. The security challenges, in general, involve the following research questions: (1) The amount and type of security needed; (2) The way data can be authenticated; (3) The way misbehaving nodes can be prevented from providing false data; (4) The optimal trade-off between energy and security such that the level of network security can be easily adapted.
- *Integration with other networks*: In some applications, sensor networks must be able to interact and interface with other types of networks such as Wi-Fi, cellular networks, or Internet. Hence, to find the best way to interface these networks will be a big challenge. Sensor network protocols should support the protocols of the other networks; otherwise sensors could have dual network interface capabilities.

- *QoS*: QoS needs support from media access control (MAC), routing, and transport layers. Most existing ad hoc routing protocols do not support QoS. Besides currently used metrics such as shortest path or minimum hop, bandwidth, delay, jitter, and packet loss (reliability or data delivery ratio) are other important QoS metrics that must be considered in the design of a sensor network. Hence, mechanisms of current ad hoc routing protocols should allow for route selection based on both QoS requirements and availability. QoS considerations are necessary to ensure the support of end-to-end QoS requirements.

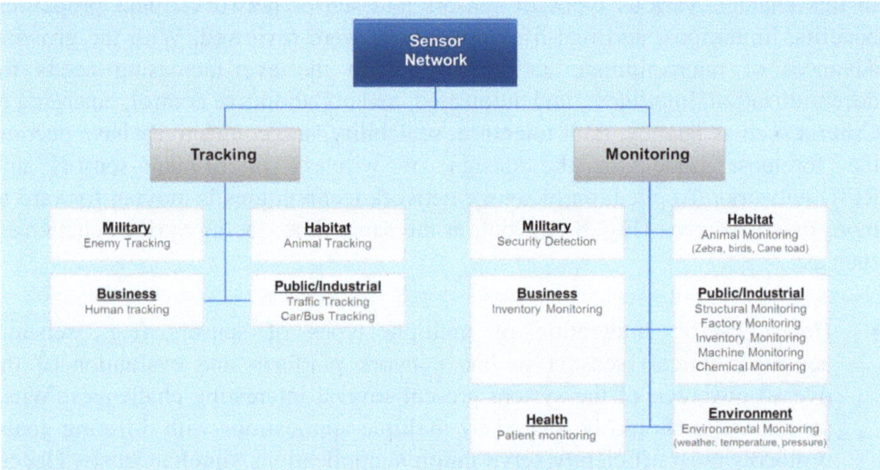

Fig. 8.18 Emerging applications of sensors and sensor networks (after Yick et al., 2008; with permission from Elsevier B.V.).

RFID, as a special class of wireless sensors, is becoming a promising technological solution adopted by various industries for real-time capturing of data from the movement of materials, people, equipment, and assets in production value chains. Asset tracking is one of the most popular applications of RFID. Companies are using RFID tags on the products that might get stolen or misplaced. RFID technology can be used for tracking parts and working in process for reduction of defects, managing production of various versions and increasing output. The technology has also been useful in the closed looped supply networks for years. More and more companies are turning to this technology for tracking shipments among the supply network allies. Not just manufacturers but retailers also are using this RFID technology for proper placement of their products and improvements in the supply chain. RFID also plays an important role in the access and security control. The 13.56 MHz RFID systems provide long range readings to the users. The best part is that RFID is convenient to handle and requires low maintenance at the same time.

Other significant challenges concerning the emerging RFID technologies are privacy and security. Emerging privacy and security applications of RFID technology include (Juels, 2006) automated toll-payment transponders (small plaques positioned in windshield corners), libraries (automated book checkout and inventory control), passports (automated passenger check-in/check-out), human implantation (indexing of medical records), smart appliances (e.g., intelligent dishwasher that automatically sets the wash cycle according to the loaded dishes), shopping (e.g., checkout of the entire shopping cart at once and perhaps automatic charging of the shopper's RFID-enabled payment device), and medication compliance (e.g., medication verification; home navigation for the elderly and cognitively impaired people).

Present trends point towards the fast growth of RFID in the near future. With around 600 million RFID tags sold in 2005, the value of market including systems, services, and hardware has drastically grown by almost factor of 10 between 2006 till now. It is expected that total number of RFID tags delivered in the year 2016 will be around 450 times as compared to the ones delivered in the year 2006. Commercial applications using RFID technology, e.g., logistics, transport, supply network supervision, processing, manufacturing, medicine, and access control, are also likely to grow by leaps and bounds. This smart technology, however, will influence consumer sectors and government as well. Barcodes and RFID will coexist for years to come, although the latter is expected to replace the former in many sectors. Combined with the growing demand for RFID technology for automatic capturing, tracking, and processing of the huge amount of data available in the production of goods and services, digital manufacturing and service industry are going to revolutionize the way in which the supply chain is managed. It will also greatly change the behavior of global producers and consumers.

References

Agarwal, V., Tsoukalas, L.H.: An experimental measurement of power consumption in facility sensor networks. In: Advanced Manufacturing Summit VII, West Lafayette, IN (2008)

Alciatore, D.G., Histand, M.B.: Introduction to Mechatronics and Measurement Systems. McGraw-Hill, New York (2007)

Bar-Shalom, Y., Fortmann, T.E.: Tracking and Data Association. Academic, New York (1998)

Brooks, R., Iyengar, S.: Multi-Sensor Fusion. Prentice Hall, New York (1998)

Brown, D.E.: RFID Implementation. McCraw-Hill, New York (2007)

Cetinkunt, S.: Mechatronics. Wiley, New York (2007)

Chau, Y.A., Geraniotis, E.: Multisensor correlation and quantization in distributed detection systems. In: Proc. 29th IEEE Conf. Decis. Control, vol. 5, pp. 2692–2697 (1990)

Cheng, Z., Perillo, M., Heinzelman, W.B.: General network lifetime and cost models for evaluating sensor network deployment strategies. IEEE Transactions on Mobile Computing 7(4), 484–497 (2008)

Datamonitor: RFID in Manufacturing: The Race to Radio-Tag is Heating Up in Manufacturing, Datamonitor, New York (2005)

de Silva, C.W. (ed.): Mechatronic Systems: Devices, Design, Control, Operation and Monitoring. CRC, Boca Raton (2008)

Elfes, A.: Sonar-based real-world mapping and navigation. IEEE Trans. Robot. Autom. 3(3), 249–265 (1987)

Estrin, D., Heidemann, J., Govindan, R., Kumar, S.: Next century challenges: Scalable coordination in sensor networks. In: Proc. 5th Annu. Int. Conf. Mobile Comput. Netw., MobiCOM 1999 (1999)

Ettus, M.: System capacity, latency, and power consumption in multihop-routed SS-CDMA wireless networks. In: Radio Wirel. Conf. (RAWCON 1998), pp. 55–58 (1998)

Faceli, K., Andre de Carvalho, C.P.L.F., Rezende, S.O.: Combining intelligent techniques for sensor fusion. Appl. Intell. 20, 199–213 (2004)

Fraden, J.: Handbook of Modern Sensors: Physics, Designs, and Applications, 3rd edn. Springer, Heidelberg (2003)

Ghiasi, S., Srivastava, A., Yang, X., Sarrafzadeh, M.: Optimal energy aware clustering in sensor networks. Sensors 2, 258–269 (2002)

Hedetniemi, S.M., Hedetniemi, S.T., Liestman, A.L.: A survey of gossipingand broadcasting in communication networks. Networks 18, 319–349 (1988)

Hedgepeth, W.O.: RFID Metrics: Decision Making Tools for Today's Supply Chains. CRC, Boca Raton (2006)

Heinzelman, W.R., Kulik, J., Balakrishnan, H.: Adaptive protocols for information dissemination in wireless sensor networks. In: Proc 5th Annu. ACM/IEEE Int. Conf. Mobile Comput. Netw. (MobiCom 1999), pp. 174–185 (1999)

Heinzelman, W.R., Chandrakasan, A., Balakrishnan, H.: Energy-efficient communication protocol for wireless microsensor networks. In: Proc. 33rd Annu. Hawaii Int. Conf. Syst. Sci., pp. 3005–3014 (2000)

Heinzelman, W.B., Chandrakasan, A.P., Balakrishnan, H.: An application-specific protocol architecture for wireless microsensor networks. IEEE Trans. Wirel. Commun. 1(4), 660–670 (2002)

Hipkin, I.: Knowledge and IS implementation: case studies in physical asset management. Int. J. Oper. Prod. Manage. 21, 1358–1380 (2001)

Hugh, D.-W., Henderson, T.C.: Multisensor data fusion. In: Siciliano, B., Khatib, O. (eds.) Springer Handbook of Robotics. Springer, Heidelberg (2008)

Ilyas, M.: The Handbook of Ad Hoc Wireless Networks. CRC, Boca Raton (2002)

Intanagonwiwat, C., Govindan, R., Estrin, D., Heidemann, J., Silva, F.: Directed diffusion for wireless sensor networking. IEEE/ACM Trans. Netw. 11(1), 2–16 (2003)

Iyengar, S.S., Sharma, M.B., Kashyap, R.L.: Information routing and reliability issues in distributed sensor networks. IEEE Trans. Signal Process. 40(12), 3012–3021 (1992)

Iyengar, S.S., Jayasimha, D.N., Nadig, D.: A versatile architecture for the distributed sensor integration problem. IEEE Trans. Comput. 43(2), 175–185 (1994)

Iyengar, S.S., Prasad, L., Min, H.: Advances In Distributed Sensor Integration: Application and Theory. In: Environmental and Intelligent Manufacturing Systems, vol. 7. Prentice Hall, Upper Saddle River (1995)

Jeong, W., Nof, S.Y.: Performance evaluation of wireless sensor network protocols for industrial applications. J. Intell. Manuf. 19(3), 335–345 (2008)

Jeong, W., Nof, S.Y.: A collaborative sensor network middleware for automated production systems. Comput. Ind. Eng. 57(1), 106–113 (2009)

Juels, A.: RFID Security and Privacy: A Research Survey. IEEE Journal on Selected Areas in Communications 24(2), 381–394 (2006)

Kali Laboratories: RFID Implementation (2004), `http://www.icegen.net/implement.htm`

Klaus, F.: RFID Handbook, 2nd edn. Wiley, New York (2003)

Ko, H.S., Lim, H., Jeong, W., Nof, S.Y.: Monitoring and learning facilities by sensor network deployment and protocol. In: Proceedings of the 20th International Conference on Production Research (2009)

Ko, H.S., Lim, H., Jeong, W., Nof, S.Y.: A statistical analysis of interference and effective deployment strategies for facility-specific wireless sensor networks. Computers in Industry 61, 472–479 (2010)

Lee, W.B., Cheung, C.F., Kwok, S.K.: Springer Handbook of Automation: Digital manufacturing and RFID-based automation, ch. 49. Springer (2009)

Lim, A.: Distributed services for information dissemination in self-organizing sensor networks. J. Franklin Inst. 338(6), 707–727 (2001)

Lin, C.R., Gerla, M.: Adaptive clustering for mobile wireless networks. IEEE J. Sel. Areas Commun. 15(7), 1265–1275 (1997)

Liu, Y., Nof, S.Y.: Distributed micro flow-sensor arrays and networks: Design of architecture and communication protocols. Int. J. Prod. Res. 42(15), 3101–3115 (2004)

Liu, Y., Nof, S.Y.: Distributed micro flow-sensor arrays and networks: Design of architecture and communication protocols. Int. J. Prod. Res. 42(15), 3101–3115 (2004)

Luo, R.C., Kay, M.G.: Multisensor integration and fusion in intelligent systems. IEEE Trans. Syst. Man Cybern. 19(5), 901–931 (1989)

Luo, R.C.: Sensor technologies and microsensor issues for mechatronics systems. IEEE/ASME Trans. Mechatron. 1(1), 39–49 (1996)

Luo, R.C., Yih, C., Su, K.L.: Multisensor fusion and integration: approaches, applications, and future research directions. IEEE Sens. J. 2(2), 107–119 (2002)

Mayback, P.S.: Stochastic Models, Estimation and Control, vol. I. Academic, New York (1979)

McFarlane, D., Sarma, S., Chirn, J., Wong, C., Ashton, K.: Auto-ID systems and intelligent manufacturing control. Eng. Appl. Artif. Intell. 16, 365–376 (2003)

Mitchell, H.B.: Multi-Sensor Data Fusion. Springer, Heidelberg (2007)

Nadig, D., Iyengar, S.S., Jayasimha, D.N.: A new architecture for distributed sensor integration. In: Proc. IEEE Southeastcon 1993E (1993)

Nof, S.Y., Liu, Y., Jeong, W.: Fault-tolerant time-out communication protocol and sensor apparatus for using same, U.S Patent No. 7,636,038 (2009)

OPC HAD specifications, Version 1.20.1.00 edition (2003), `http://www.opcfoundation.org`

Poirier, C., Mccollum, D.: RFID Strategic Implementation and ROI: A Practical Roadmap to Success. Book News, Portland (2006)

Qiu, R., Xu, Q.: Standardized shop floor automation: An integration perspective. In: Proc. 14th Int. Conf. Flexible Automation and Intelligent Manufacturing, pp. 1004–1012. NRC Research Press (2004)

Qiu, R.: A service-oriented integration framework for semiconductor manufacturing systems. Int. J. Manuf. Technol. Manage. 10, 177–191 (2007)

RNCOS: RFID Industry – A Market Update (2005), `http://www.rncos.com/Report/COM16.htm`

Stone, L.D., Barlow, C.A., Corwin, T.L.: Bayesian Multiple Target Tracking. Artech House, Norwood (1999)

Toumpis, S., Tassiulas, L.: Optimal deployment of large wireless sensor networks. IEEE Transactions on Information Theory 52(7), 2935–2953 (2006)

Warneke, B., Last, M., Liebowitz, B., Pister, K.S.J.: Smart dust: communicating with a cubic millimeter computer. Computer 34(1), 44–51 (2001)

Webster, J.G. (ed.): The Measurement, Instrumentation and Sensors Handbook. CRC, Boca Raton (1998)

Ween, A.S., Tomecko, N., Gossink, D.E.: Communications architectures and capability improvement evaluation methodology. In: 21st Century Military Commun. Conf. Proc. (MILCOM 2000), vol. 2, pp. 988–993 (2000)

Wesson, R., Hayes-Roth, F., Burge, J.W., Stasz, C., Sunshine, C.A.: Network structures for distributed situation assessment. IEEE Trans. Syst. Man Cybern. 11(1), 5–23 (1981)

Wilson, J.S.: Sensor Technology Handbook. Newnes Elsevier, Amsterdam (2004)

Xu, K., Wang, Q., Hassanein, H., Takahara, G.: Optimal wireless sensor networks (WSNs) deployment: minimum cost with lifetime constraint. In: Proceedings of the IEEE International Conference on Wireless and Mobile Computing, Networking and Communications, WiMob 2005 (2005)

Yamasaki, H. (ed.): Intelligent Sensors. Handbook of Sensors and Actuators, vol. 3. Elsevier, Amsterdam (1996)

Yick, J., Mukherjee, B., Ghosal, D.: Wireless sensor network survey. Computer Networks 52(12), 2292–2330 (2008)

ZigBee Alliance. Network Specification, Version 1.0 (2004)

Review and Challenge Questions

1. (i) What are the advantages of sensor networks in comparison with single sensor systems? (ii) Justify the application of sensor networks with respect to the following CCT design principles:
   (a) e-Work parallelism
   (b) Error prevention and conflict resolution
   (c) Collaborative fault-tolerance
   (d) Emergent lines of collaboration and command
2. (i) Describe, along with examples, three types of interactions among sensors in a sensor network. (ii) Compare this classification with the classification of different types of collaboration discussed in Chapter 2 (i.e., mandatory, optional, and concurrent collaboration).
3. Explain briefly the unique features, advantages, limitations, and practical applications of the seven multi-sensor data fusion methods discussed in this chapter.
4. Analyze the relationships between the seven design considerations for sensor networks and the emerging e-Criteria discussed in Chapters 1 (Section 3) and 4 (Section 2). You may prepare a simple comparison table for this purpose.

5. (i) Describe different WSN (Wireless Sensor Network) architectures and protocols, their relative advantages and limitations. (ii) Illustrate cases where each of those architectures is advantageous.
6. Describe the role of sensor network middleware and protocols in factories and other facilities. Are they active?
7. (i) How do sensor networks enable applications of CCT? (ii) How can CCT improve industrial communication and sensor networks?
8. Justify, with examples, the use of different types of RFID tags.
9. (i) What are some key technologies of RFID, and how do they relate to production with computers and communication? (ii) How can RFID technologies enable digital manufacturing?
10. Illustrate your answers to Question 9 with a few case studies.
11. Explain some real-life applications of RFIDs in transportation, inventory management and control, maintenance, healthcare, and home/office automation.
12. What are the potential privacy and security challenges of WSN and RFID technologies?

Lab Questions

13. With the increasing popularity and availability of sensors and RFIDs, the amount and rate of new generates data also increases, resulting in data overload or "flooding". Admittedly, the data and their accuracy can yield high values, but typically for only a "very limited time". Explain the roles of software agent and data mining concepts in resolving the data flooding challenge.
14. Reverse logistics refers to the process of moving goods from their final destination for the purpose of remanufacturing, refurbishing, or proper disposal. In this process, a basic challenge is to gather precise and real-time information on the End-of-Life (EoL) products. Select a real-life reverse logistics example, and prepare a short proposal on how RFID technologies can potentially enhance this process along its different stages.
15. Consider a typical three-echelon supply network. Explain with examples:
    (i) How can RFID tags reduce backorder (backlogging) cost?
    (ii) How can RFID tags reduce shrinkage (e.g., losses) in inventory, thus avoid stock shortages?
    (iii) Do RFID tags increase or decrease the setup cost?
    (iv) Do RFID tags change the probability distribution of demand?

Chapter 9
e-Service Industry

Abstract. e-Service refers to the provision and support of services remotely via information and communication components and services, and over electronic networks. Such services are not limited to the scope of businesses and organizations, and with the global approach to service-orientation, all e-Systems even e-Manufacturing systems are increasingly shifting towards the development and implementation of sound service practices over e-Networks. To achieve this end, this chapter first reviews the state-of-the-art developments in various e-Service industries, their commons characteristics, and emerging challenges. Then, techniques for service performance monitoring, evaluation, and optimization are discussed. Following the preliminary definitions, advanced service design, optimization, and control techniques are introduced, including Task Administration Protocols (TAP) for real-time e-Service design and monitoring, collaborative e-Service portal design, and protocols for diagnosis and prognosis of errors and conflicts in e-Service networks.

9.1 e-Work in Service Industry

The service industry, also referred to as *tertiary sector of industry*, involves industry sectors that provide services (intangible products) to customers, either other businesses or end consumers, as opposed to primary and secondary industries. Examples of service industries include banking, cleaning, advertising, healthcare, insurance, communications, entertainment, security, storage, wholesale and retail trade, government services, etc. Advances in ICT (Information and Communication Technologies), automation, and in particular collaboration engineering have revolutionized service industries, which nowadays provide services virtually through communication networks, enabling integration of distributed resources and interaction with the customers.

Services provided via electronic networks such as the Internet, intranets, or extranets, without being limited to service organizations, are called e-Service. In line with the evolution of businesses to distributed and interconnected networks of SMEs (Small-to-Medium-size Enterprises), e-Services are no longer restricted to the "tertiary sector of industry", and are contributing ever more to manufacturing

S.Y. Nof et al., *Revolutionizing Collaboration through e-Work, e-Business, and e-Service*,
Automation, Collaboration, & E-Services 2, DOI: 10.1007/978-3-662-45777-1_9

and supply of both physical and digital goods. Indeed, there are several value creating e-Services related to manufacturing and logistics, e.g., shipping, repairs and maintenance, tooling services, design services, recycling services, and so on.

Various application domains are influenced by e-Services, among them: e-Business, e-Commerce, e-Government, e-Learning, e-Design, e-Manufacturing, ERP (Enterprise Resource Planning), SCM (Supply Chain Management), CRM (Customer Relationship Management), e-Operations, e-Logistics, and e-Maintenance. All these applications require support from the theoretical

Table 9.1 CCT design principles and applications to e-Service design.

Principle*	Definition	Applications
CRP	Advance, adaptive, and real-time planning for necessary collaboration	Collaborative user requirements analysis / document repository / Interactive resource visualization
EWP	Distributed planning of integrated execution methods	Task assignment, administration ability / interactive, collaborative work environment / smart search engine
KISS	Enhance e-Work by eliminating unnecessary complexity and improving automation	Support human users in face of complexity / minimize needs for retraining
EPCR	Error and conflict diagnosis, resolution, and prognosis	Error and conflict detection and resolution / emergency crisis management / authentication control of knowledge
CFT	Better fault tolerance by collaborative teaming	Collaboration to tolerate faults / intelligent, selective posting protocols / magnetic tag cloud-to-cluster
AD	Conditions for individual agents to J/L/R in a CNEs (Collaborative Network of Enterprises)	Knowledge exploration and management / asynchronous interactive collaboration / incentive-based project collaboration
ELOCC & BM	Better collaborative decision making under dynamic requirements	Cross-cultural interaction / BM protocols / intuitive interaction techniques
CVC	Integration of interactive visualization with analytic tools to manage complexity	Enhance human-machine interactions by integration of computer graphics, interaction, visualization, analytics, perception, and cognition methods

* CRP: Collaboration Requirement Planning
* EWP: e-Work Parallelism
* KISS: Keep It Simple, System
* EPCR: Error Prevention and Conflict Resolution
* CFT: Collaborative Fault Tolerance
* AD: Association and Dissociation
* ELOCC: Emergent Lines of Collaboration and Command
* BM: Best Matching
* CVC: Collaborative Visualization and Comprehension

foundations (see Chapter 1) and design principles (see Chapter 2) of collaborative e-Work in order to provide users with an environment with a large amount of information and various support functions (Table 9.1).

Emerging technologies, transitions in businesses and economies, competitive and dynamic nature of markets, governmental and organizational policies, and rising demands for mass customization are increasingly transforming service industries. In this emerging competitive environment, those enterprises will survive that are (1) resilient and flexible to dynamic changes, (2) capable of providing quality and highly customized products/services, (3) lean and cost-effective, and (4) proactive to uncertainties, conflicts, and complexities, in a collaborative environment. The wide variety of functional challenges in service industries and emerging solutions by collaborative e-Work and e-Service are illustrated in Figure 9.1. The most significant challenges in service industries and the most promising emerging solutions are summarized below (see Nof et al., 2006).

1. *Management of complexity, change, and uncertainty*: Similar to the other industrial sectors, service industries are subject to uncertainties caused by internal (e.g., processes) and external (e.g., market) sources. Moreover, with the increasing demand for collaboration and the transition of independent enterprises to CNEs, increasing complexity and globalization have become other significant challenges for the service industries. These challenges necessitate considerable levels of agility, flexibility, and resilience addressed through effective collaboration. The emerging concept of *digital enterprise*, as networks of digital models, methods and tools (e.g., simulation and visualization), and continuous data management, which map the processes to digital structures, is a primary solution to the aforementioned challenges. The main enablers are advances in ICT, methods and tools of AI (e.g., machine learning), and intelligent DSSs, which enhance real-time service performance monitoring, analysis, and optimization, thus alleviate and control complexities, changes, and uncertainties. These issues are further elaborated in Section 9.3.
2. *Scalability, cost-effectiveness, coordination, and resource allocation*: A key concern is lack of tools and platforms for evaluation and validation of the intelligent DSSs in real-world problems (see Chapter 5). It is necessary to test the design concepts in real-world service through *emulation*, taking into account basic requirements, in order to generate proof-of-concepts that are more accurate and have higher QoS (Quality of Service). QoS in service automation involves issues of affordability, sustainability (i.e., economic, social, and environmental aspects), and dependability (see Erbe, 2009). Another design concern is better understanding of the scalability and robustness criteria, in order to address issues of emergency, contingency, and self-repair/self-organization. Section 9.4 of this chapter is devoted to design of TAPs (Task Administration Protocol) for dealing with the aforementioned challenges in service industries.

3. *Customization, and automation reliability and dependability*: Interoperability of distributed intelligent systems (e.g., data/information processing, communication), and reliable and dependable integration, control, and maintenance of technical management systems are other challenging issues in design of e-Service systems. Development of distributed embedded systems must address the temporal predictability of the underlying infrastructures (both software and hardware) and flexibility in face of dynamic and emerging applications and services. The increasing use of networked control systems influences the dependability of the systems, depending on the setup and design principles of the network. Collaborative enterprise systems, e.g., e-Service Portal (ESP), provide promising solutions to these challenges, as further discussed in Section 9.5 of this chapter.
4. *Sustainability, large-scale infrastructure, and security*: Design of e-Service systems must be compatible the legal, ethical, strategic, and operational decisions and policies made by the enterprise or the society. The major issues are concerned with sustainability (e.g., asset/ product life cycle management) and security (e.g., ISO 28000: 2005 Supply Chain Certification). The considerable complexities in large-scale infrastructures, however, may cause mismatches, errors, and conflicts during the information exchange, allocation of resources, and task execution among highly distributed and autonomous agents. These shortcomings, especially in critical infrastructures (e.g., energy, transportation, communication), may in turn cause huge disruptions and impose considerable losses to the system. Section 9.6 of this chapter discusses detailed techniques for error prevention and conflict resolution in e-Service systems.

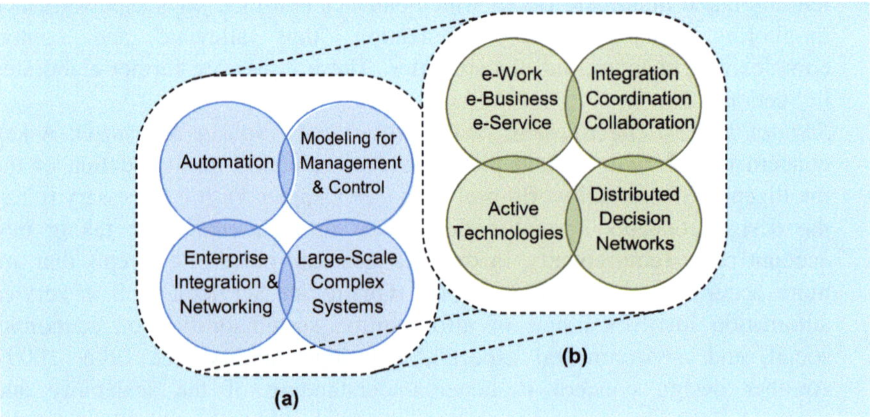

Fig. 9.1 Scope of (a) functional challenges/solutions, and (b) emerging trends for solutions.

There are three fundamental dimensions (or information workspaces) in the design of e-Service systems:

(1) *Domain knowledge (or content space)*: involves the contents created/shared/used by the users.
(2) *Interaction/access (or communication space)*: enables users to interactively access/ manipulate/visualize the contents.
(3) *Collaboration (or coordination space):* includes optimized and rationalized services for interactions and knowledge sharing, and workflows to support collaboration.

The first two dimensions, i.e., knowledge creation and sharing through the Web, are quite well established and addressed by various e-Service systems, e.g., Wikipedia, Adobe Acrobat Connect. Collaboration, however, is the emerging dimension in design of e-Service systems, *partially* addressed by some recent studies, e.g., the Hub technology initiative at Purdue University (Qiao et al., 2006; Lundstrom et al., 2008), briefly introduced in Chapter 6. The emerging challenges regarding this dimension are increased levels of complexity, scalability, efficiency, and real-time capability, which in turn make control of domain knowledge and interactions more complicated. Hence, design principles of CCT and collaborative e-Work are worthy of investigation in design of e-Service systems.

As shown in Table 9.1, the initial step in design of collaborative e-Service systems is to identify the *collaboration requirements* of users. This identification requires methodologies for *modeling* processes and requirements in e-Services, i.e., *describing* and *specifying* the e-Service system at a high *level of abstraction*. In this context, *process* refers to operations or events, taking up resources (e.g., time, space, expertise) to produce an outcome or change the properties of an object. The purpose of process modeling in design of a system, in general, is to (1) trace what actually happens during a process (*descriptive*), and (2) define desired processes and how they should/could/might be performed (*prescriptive*). The foremost methodologies for process modeling in e-Work and e-Service systems are (applications in e-Service systems are further discussed in Sections 9.3 to 9.6):

1. *Petri nets*: is a mathematical modeling language for description of distributed systems, and logic specification and analysis of reachability and performance. A typical Petri net is composed of *positions (factors)*, *transactions (processes)*, *arcs*, and *tokens*. A simplified Petri net for modeling TAP is presented in Chapter 1 (see Figure 1.10). Their applications include distributed computing, manufacturing, control, communication networks, and transportation.
2. *Reference architecture*: or a reference model, is a software architecture, which provides structures and relations between the elements as *template solutions* for an architecture in a particular domain. Several reference

architectures have been introduced and discussed in Chapter 1. See, for example: the HMS architecture (Figure 1.20), the IBM® service-oriented modeling and architecture (Figure 1.12), the integrated enterprise system architecture (Figure 1.14), the extended products in dynamic enterprises (EXPIDE) architecture (Figure 1.16), the OQIS reference architecture (Figure 1.17), and the active middleware architecture (Figure 1.19).

3. *Unified Modeling Language (UML)*: also referred to as object oriented programming, is a process modeling paradigm based on *discrete reusable code blocks* taking variables, performing specific functions, and returning values. UML is a standardized (ISO/IEC 19501:2005) and general-purpose graphical modeling language for specifying, visualizing, constructing, and documenting models of systems, considering their structure and design. UML clarifies specific requirements, and enables users for better understanding. UML is widely accepted as a standard for modeling object-oriented programs and hardware/software products. It is composed of six types of diagrams: *class*, *use case*, *activity*, *sequence*, *state*, and *object*. Figure 9.2 shows application of use case diagrams in e-Service modeling.

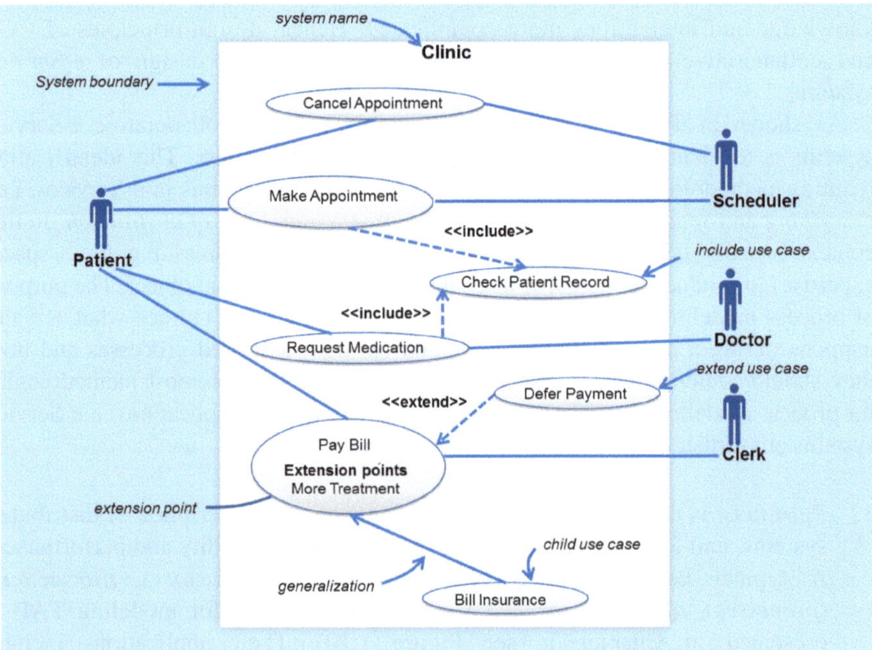

Fig. 9.2 Application of use case diagram in process modeling of a clinic.

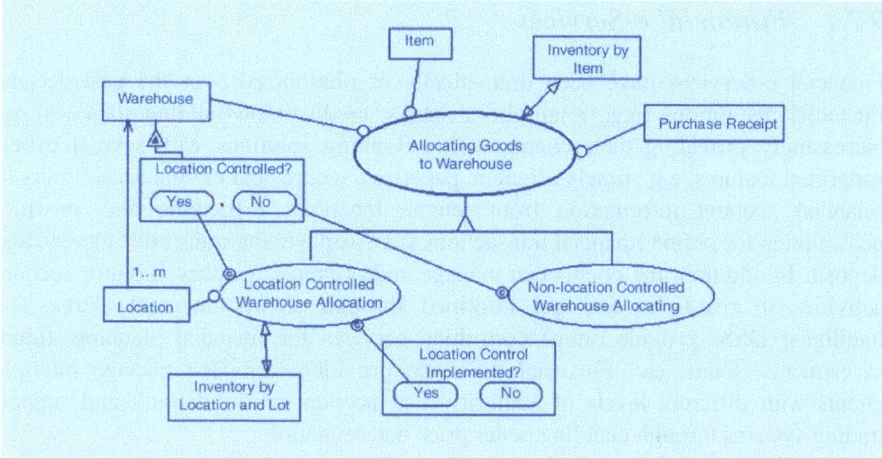

Fig. 9.3 Application of the OPM diagram in modeling a warehouse (after Soffer et al., 2003; with permission from Elsevier B.V.).

(4) *Object-Process Modeling (OPM)*: is a modeling language (Dori, 2002), combining a minimal set of *building blocks* with dual graphic-textual diagrams. It resolves *model multiplicity conflicts* that arise due to the coexistence of a number of objects from different classes. The OPM diagram uses three types of entities (see Figure 9.3): (1) *objects*, i.e., existing entities (rectangles), (2) *processes* that transform objects by either changing their states or creating or consuming them (ellipses), and (3) *states*, i.e., situations or status of the objects (rounded rectangles). Since processes are separated from objects, the OPM diagram resolves the model multiplicity conflict.

9.2 Tasks and Features of Service Industries

Advances in ICT, automation, and collaboration engineering have made e-Service systems emerging and evolving solutions enhancing diverse domains of knowledge and decision support tools using sophisticated interactions and access technologies. Such e-Service systems have dramatically improved the ability of service providers in delivering variety of services to their clients, e.g., e-Commerce, e-Logistics, e-Purchasing, e-Government, e-Finance, engineering design, construction, news and travel services, communication, maintenance, financial planning, and many other services and service industries. In the following, various examples of emerging e-Service industries, their features, functionalities, and goals are presented.

9.2.1 Financial e-Services

Financial e-Services have been dramatically revolutionized over the past decade. Financial institutions (e.g., retail/virtual banks, credit union/building societies) are increasingly providing their clients with e-Banking solutions with several cyber-supported features, e.g., timely efficient, paperless, secure, and mobile accessibility to financial account information from remote locations. e-Banking also provides possibilities for online financial transactions such as payment, transfer of money, and deposit. In addition, the clients can manage and organize finances, monitor account activities in real-time, and get informed through smart statement alerts. The intelligent DSSs provide online consulting services for financial planning, future investments, loans, etc. Financial e-Service providers can also manage multiple clients with different levels of authority (i.e., account consolidation), and support trading systems through enabling better price determination.

Globalization of financial services is among the major drivers of transformation and integration of financial services. Financial e-Services are increasingly influencing and merging various industries (e.g., portal and telecommunication companies, banks, etc.). Figure 9.4 depicts recent developments in financial e-Services. In the presented model, electronic enablers refer to e-Service provision and support companies. Financial products are the (customized) services provided to the clients by the financial e-Service provider (i.e., financial institution). Moreover, portals represent the links between access devices and financial e-Service providers, supported by the aggregators for higher quality of services (e.g., service comparison).

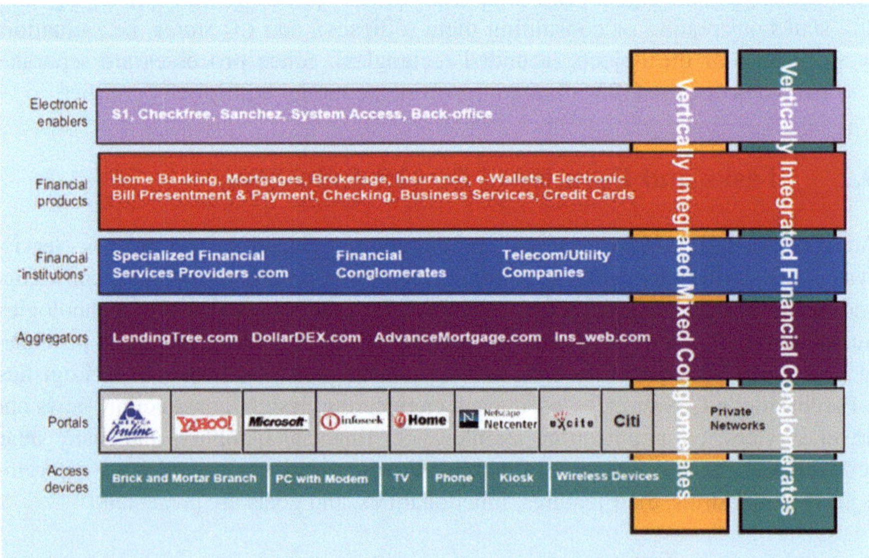

Fig. 9.4 Emerging financial e-Services (after Claessens et al., 2002).

9.2.2 *Power e-Systems*

Power generation and transmission systems are facing challenges regarding the security and efficiency of services. Power failures occasionally affect large grids and diminish the efficiency and productivity of power plants. Power network reliability is also an issue of high importance. The reliability of a power system mostly depends on the quality of the automated and manual decisions. For instance, the power system backbone, electric power transmission, and its operational reliability are highly affected by e-Service enabling automated control. Figure 9.5 illustrates the framework for securing the operations of an electrical power system network, in terms of active power balancing through both automatic and manual interventions. The objective is to determine the constraints for a given control law such that the control system has the authority in taking proper actions. Constraints appear in the form of regulating/ balancing power reserves that have to be planned and reserved in advance so that the area control error, the instantaneous difference between the actual and the reference value for the power interchange of a control area, is kept within reasonable bounds. Balancing power is accessible in the form of *ancillary services*, as provided by power-generating

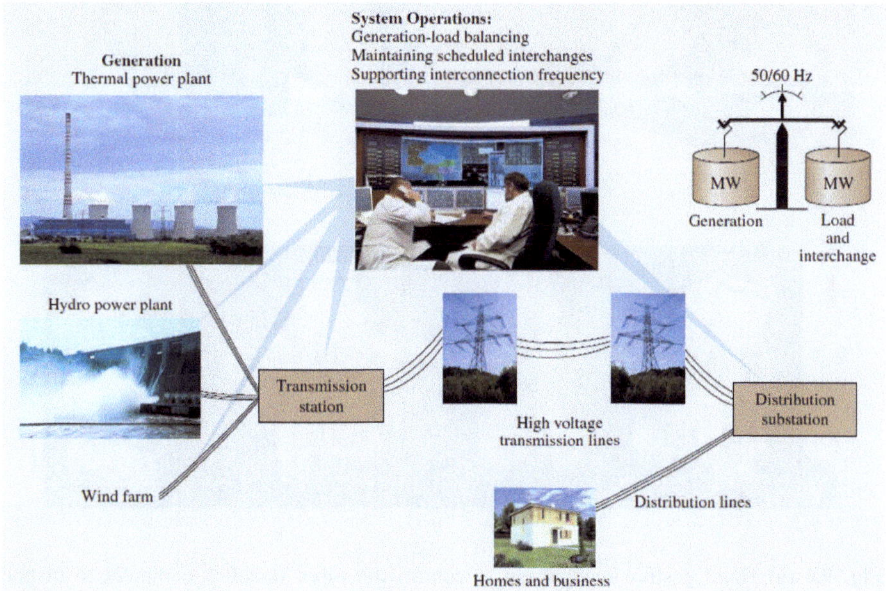

Fig. 9.5 Securing and controlling the operation of an electrical power system network (after Horacek, 2009).

companies and traded in the market. Balancing power reserve activations are variables manipulated by a system operator responsible for power balancing in a geographical area. Power reserves must be carefully planned and purchased so that the system operator is able to use them when required.

9.2.3 Healthcare e-Services

Emerging healthcare systems demand for automated and collaborative e-Services that provide customized, cost-effective, and timely efficient service solutions. The scope of healthcare e-Services varies from automation of operating rooms to management and control of the entire supply chain. An example of CSS (Collaboration Support System) in intelligent operating rooms has been discussed in Chapter 2. Healthcare e-Service is one of the most demanding areas for robotics, and medical robots (e.g., robotic nurse, robotic companion, nano-robots

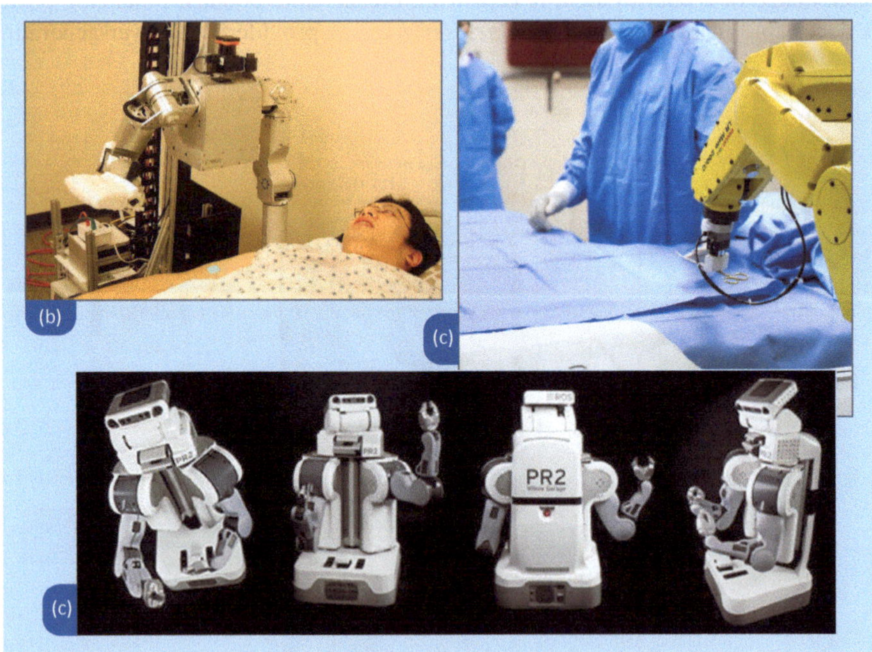

Fig. 9.6 (a) Hand gesture recognition to control the robot or tell a computer to display medical images of the patient during an operation (Wachs and Kölsch, 2011); (b) Robotic nurse assistant enhanced with a direct physical interface for control by human nurses (c) Assistive mobile manipulation for elderly people at home (Courtesy of Healthcare Robotics Lab, Georgia Tech).

for neurosurgery) are becoming more popular, reshaping the entire healthcare systems (Figure 9.6). Some other emerging e-Service areas in healthcare systems are collaborative clinical information systems (e.g., medicine inventory) and data warehousing, electronic medical record data evaluation, technology and information flow for disease diagnosis and management, purchasing and invoice matching, and healthcare delivery integration.

9.2.4 Transportation e-Services

Design and control of next generation transportation e-Services need to address the emerging challenges in public transport, e.g., accidents, road congestion, and environmental pollutions, relying on advances in computerization of vehicles, and wireless and satellite-based communication technologies. Intelligent transportation systems (ITS) are promising solutions that improve road/traffic awareness based in real-time information, safety, and efficiency, while reducing traffic congestion in public transportation. ITS is also becoming more popular in cargo and distribution control of goods, which enhance real-time monitoring and control of distribution operations, transportation time, departures and arrivals, delays, position of the containers, and potential risks. The principal concerns in ITS design are real-time monitoring and alert systems solutions. Major elements / features of the ITS e-Services are (see Figure 9.7):

- *Live tracking*: is performed using GPS, RFID, Wi-Fi, and/or WSN, enhanced with ICT and Internet applications. The information is sometimes stored in local memories in GPS-enabled devices (e.g., smart phones) in the vehicle.
- *Map-based visualization*: applies Google Maps API or Google Earth API for visualized tracking and monitoring of the vehicle.
- *Position storage*: sends the acquired position to a data access layer (Figure 9.7), if an error occurs during sending the information to the server. The information is stored locally and will be retransmitted when the connection is restored.
- *Travel management*: controls the transport status, which may be: 1) open, 2) in progress, or 3) finalized. This is perfumed using the *virtual fence* technology.
- *Alerts and Virtual Fencing*: virtual fence, also known as *geo-fence*, is an emerging technology, which compares the vehicle's position compared to a predetermined zone or a point of interest. Various types of alerts may be triggered based on the system's requirement, e.g., delays or robberies (security), or package status alerts (e.g., temperature, humidity, etc.) for product quality protection.
- *Automation*: is improved through 1) reducing the human intervention, 2) providing e-Service providers with in-house ITS, 3) allowing drivers to focus on safety, 4) avoiding the need for additional driver training, 5) reducing or

eliminating the daily time spent by workforce on transport/cargo tracking, 6) minimizing the risks of accident/incidents, and the corresponding losses, where the driver is unable to report the position manually (security applications).

- *Package tracking, loading/unloading records*: is supported by RFID technologies.

The ITSs are being further investigated for development of self-driving cars (also called autonomous/robot cars). The aforementioned ITS technologies along with AI (Artificial Intelligence), sensor networks, and computer vision enable self-driving cars to sense the environment and navigate autonomously. This emerging technology (e.g., Jeong and Nof, 2009a,b,c; Ko et al., 2010) will enable further automation of cargo and public transportation systems in the near future.

Another example of emerging transportation e-Service is e-Logbooks (or simply, e-Logs), which eliminates paperwork and minimizes audit risk. Figure 9.8 illustrates Boeing's e-Log for operational performance monitoring of aircrafts and better integration of information in the air and on the ground. e-Logs improves productivity of aircraft maintenance through real-time provision of information via three main interfaces: 1) pilot, 2) laptop, and 3) maintenance control. It can also be used to avoid logbook losses and increase security in critical situations (e.g., hurricanes).

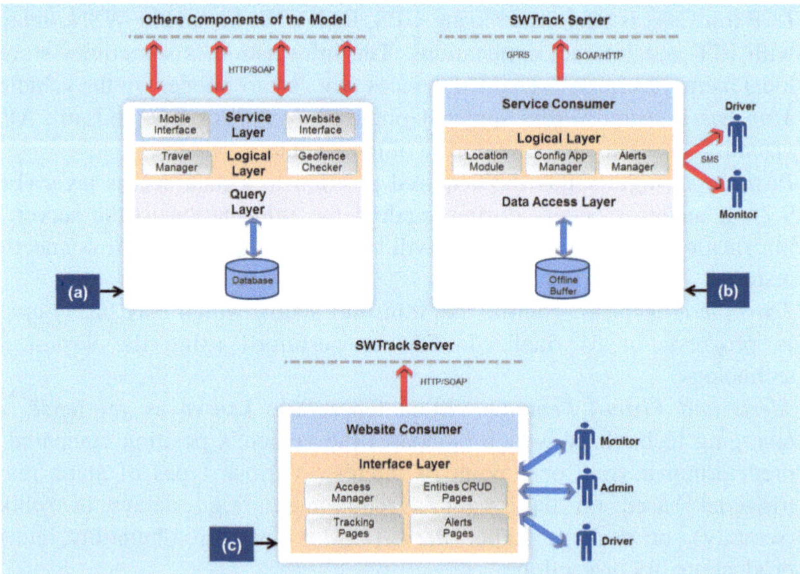

Fig. 9.7 An ITS architecture for transport monitoring; (a) Server architecture (information storage, integration of mobile devices, server, and interfaces, and virtual fence checker); (b) Mobile client architecture (interaction with server, information collection and transmission, local storage; alert manager); (c) Website architecture (interaction between server, mobile client, and decision-maker; interface layer) (after Oliveira et al., 2013; with permission from Elsevier B.V.).

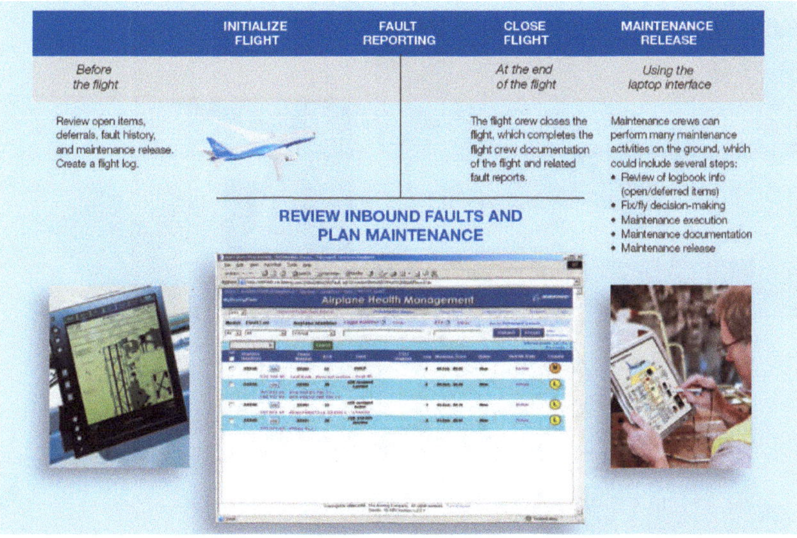

Fig. 9.8 e-Log providing real-time fleet information; 1) *Pilot interface*: follows the flight deck standards and fits into the flight deck environment; 2) *Laptop interface*: maximizes the productivity of engineers and mechanics using task-driven user interfaces; 3) *Maintenance control interface*: enhances monitoring/control of logbook status of the entire fleet (Courtesy of Boeing).

9.2.5 Space Exploration e-Services

Space exploration programs worldwide are actively exploring the space and planetary bodies in the solar system. Automation and robotics are increasingly playing significant roles in mitigating the risks to the lives of astronauts while conducting space missions, varying from low-Earth orbit to Earth's moon and beyond throughout the entire solar system. Some applications are installing space exploration infrastructure e.g., Earth-orbiting satellites, space stations, and satellites that operate for extended periods in orbit around other planets. Examples include transporting scientific instruments and equipment to space destinations for deployment and operation, performing mundane or repetitive tasks that allow humans to focus on deliberative, cognitive or otherwise complex tasks, and creation, delivery, and operation of scientific, commercial, and military assets in space (Figure 9.9-a). The NASA Space Shuttle Orbiter with Canada's Remote Manipulator System has been assisting astronauts with successful on-orbit servicing tasks such as satellite retrieval, repair, and rescue, e.g., servicing of the Hubble Space Telescope involving critical parts replacement and instrument upgrades (Figure 9.9-b). Other relevant e-Service examples include the assembly of space structures, post-assembly inspection (Figure 9.9-c), and assistance to astronauts performing extravehicular activity.

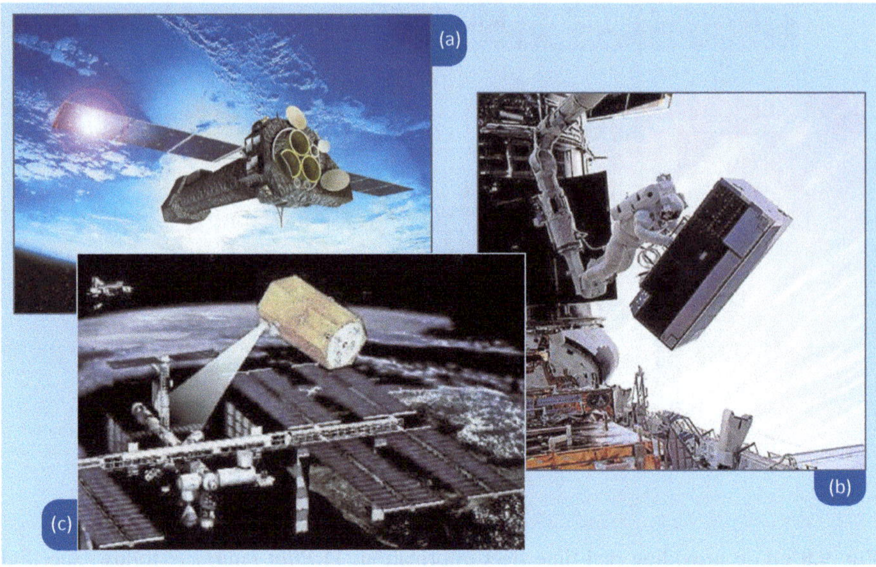

Fig. 9.9 (a) The European Space Agency's XMM-Newton as the most sensitive x-ray telescope enhanced with high-technology design using 170 wafer-thin cylindrical mirrors spread over three telescopes (Courtesy of ESA); (b) Astronaut servicing the Hubble Space Telescope while secured by foot restraints to the end of the Canadarm; (c) Robotic inspection in space.

9.2.6 Cleaning Automation e-Services

Various automation solutions have been developed so far for supporting cleaning services. Common examples are automatic floor cleaning robots and floor vacuum cleaners for household use (Figure 9.10-a). More advanced applications include professional floor cleaning, airplane washing, ship cleaning, facade cleaning, autonomous power supply, sensor systems, environment modeling, and path planning in dynamic environments. Automatic cleaning systems are usually extremely complex robot systems that operate autonomously in unstructured environments or outdoor areas. Some examples of automatic cleaning systems for floors, facades, swimming pools, ventilation ducts, and sewer lines are shown in Figure 9.10 to highlight the potential of cleaning automation. Such cleaning e-Services, however, will not be mass produced until their cost effectiveness, performance, efficiency, and total attendant costs make them superior to manual cleaning. Development of service robots' cognitive capabilities, environment modeling sensor systems, and multimodal user interfaces are the major challenges regarding cleaning robots in the professional sector.

Fig. 9.10 (a) iRobot Roomba; sold ten thousand times over every year at prices below US $300; (b) Hefter cleaning robot ST82R; (c) WEDA B680 pool cleaning robot; (d) Fraunhofer IFF cleaning robot in Leipzig; (e) Fraunhofer IFF Filius roof cleaning robot.

9.2.7 Library e-Services

Automation of library services originated from the standardization of card catalogs over 130 years ago to the creation of the machine-readable cataloging communications format, bibliographic utilities, and integrated library systems. In line with advances of Internet and digital era, characterized by the proliferation of content in electronic format, library e-Services became available publically, e.g., OpenURL linking (Figure 9.11), meta-searching, and electronic resource management systems. Nowadays, libraries are further customized, reacting to personal requirements of the users through quick, simple, and effective discovery and delivery of results. Examples include Internet heavyweights such as Google, which have developed new series of Library 2.0-based discovery-to-delivery applications. Such emerging applications deliver up-to-date user experiences and allow libraries to retain their back-office systems (e.g., acquisitions, cataloging, and circulation) and add to or replace them as needed.

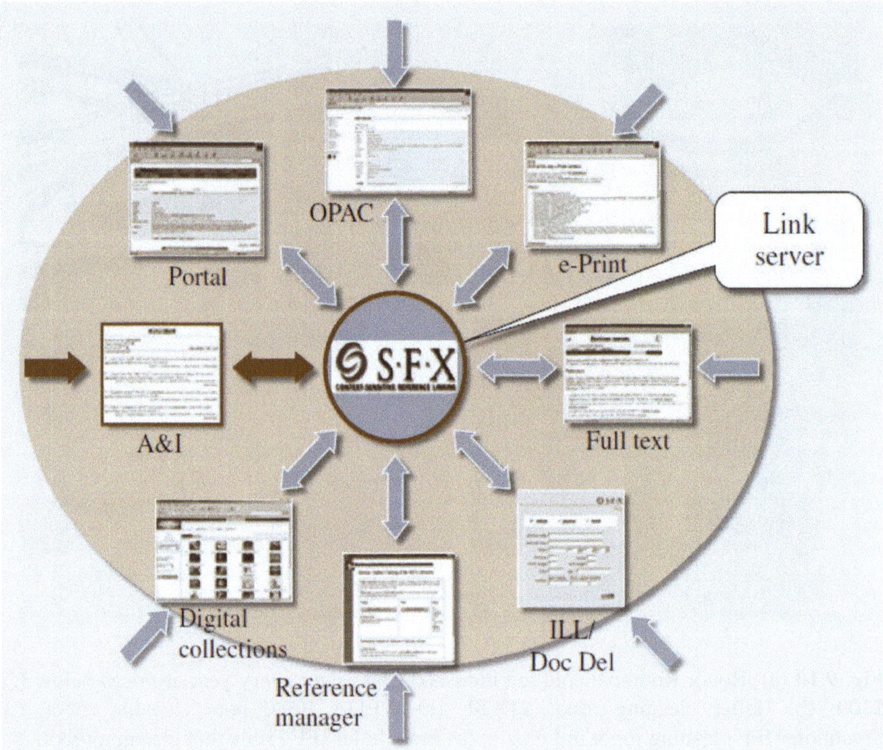

Fig. 9.11 SFX and OpenURL linking.

9.3 Service Performance

Upon designing and starting to operate a system, or a subsystem, services come into play. Performance of a system, however, varies over time during its *life cycle*. Various strategies are known for dealing with deterioration of the system performance, each with certain costs and improvements in the QoS. Prominent strategies are continuous maintenance, overhaul, and system upgrade, as illustrated in Figure 9.12. The optimal strategy then needs to be identified, taking into account the economic situation of the service enterprise and the time constraints (as two major driving forces in service automation), along with other critical e-Criteria, depending on the requirements, plans, and available resources. Implementation of those strategies requires effective collaboration among distributed service agents as well as strong ICT infrastructure to enhance and rationalize their performance.

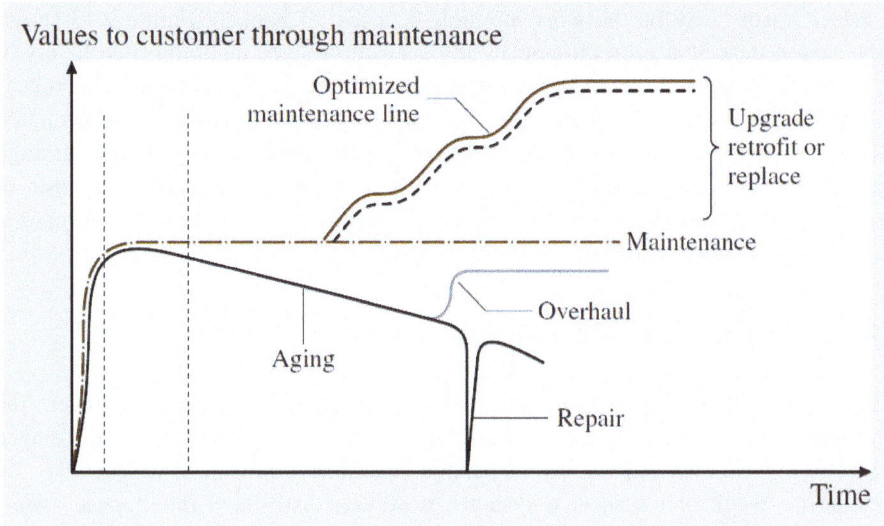

Fig. 9.12 Various strategies for maintenance of a system during its life cycle (after Ola, 2004).

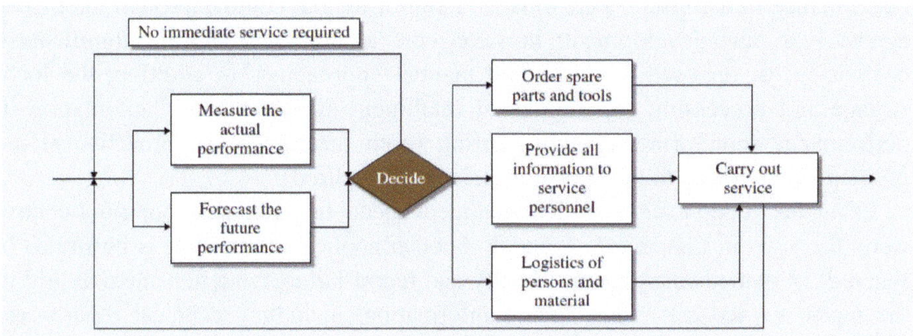

Fig. 9.13 Major steps in a service system (after Pinnekamp, 2009).

The ideal situation for a service enterprise is when the system and all its components are available 24/7, and operating with the optimum performance. Nevertheless, outage due to scheduled maintenance, overhaul, repair, etc. is inevitable and intrinsic in any system. Each cause of outage, however, imposes different costs and has different influences on the system's performance. For instance, unexpected breakdowns due to some errors or conflicts between some service agents may be the worst possible scenario to happen. Figure 9.13 shows the major steps that are common in any service system, including two layers of information exchange and physical transport. The purpose of such architecture is to monitor the actual performance of the system, forecast possible variations in the system's performance and enable real-time interaction and collaboration through adequate communication channels, provide knowledge regarding the use of different resources (e.g., service agents), and optimize access to data/information and logistics for effective and proactive actions.

9.3.1 Performance Monitoring

In order to keep the performance of a system at an acceptable level, the performance of all components must be monitored. Traditionally, performance monitoring was carried out by inspection operators under certain schedules. In modern e-Work and e-Service systems, however, distributed intelligence agents (e.g., sensors, holons, RFIDs, shop floor control devices) monitor and report the performance of system's components in real-time, or even act autonomously when needed (see Chapter 8). The intelligent agents incorporate advances in ICT and provide valuable information/knowledge for prediction of the system's performance in the future. Performance monitoring and control through intelligent agents is a new development; however, the associated costs are significantly decreasing, as opposed to traditional manual approaches. In addition, the local storage and processing capabilities of intelligent agents enhance automation in performance monitoring through alerting (both reactively and proactively) the problems and threats to the decision-makers (if required).

Examples of application of such intelligent agents in automated shop floor control were discussed in Chapters 1, 2, and 4. Another application example is illustrated in Figure 9.14, where smart agents identify and report failures in a transmission grid to the repair service crew. The failure information, including technical reports and history of the component is transferred through GPS (Global Positioning System). Application of agent-based systems in real-time performance monitoring enhances collaborative control in e-Service industry through provision of necessary information/ knowledge from remote locations, enabling cost-effective service automation, and improving decision-making. It also enables *mixed reality concepts*, where the information and decisions flow dynamically and in a bidirectional manner between the real and the virtual systems. The major challenges, however, is adaptability of the networked agents (e.g., sensors) to practical applications, communication, robustness to various interferences, and fault tolerance, which require active middle structures. These issues are addressed in detail in Chapter 8 about sensor and RFID networks (see Figures 8.12 and 8.17).

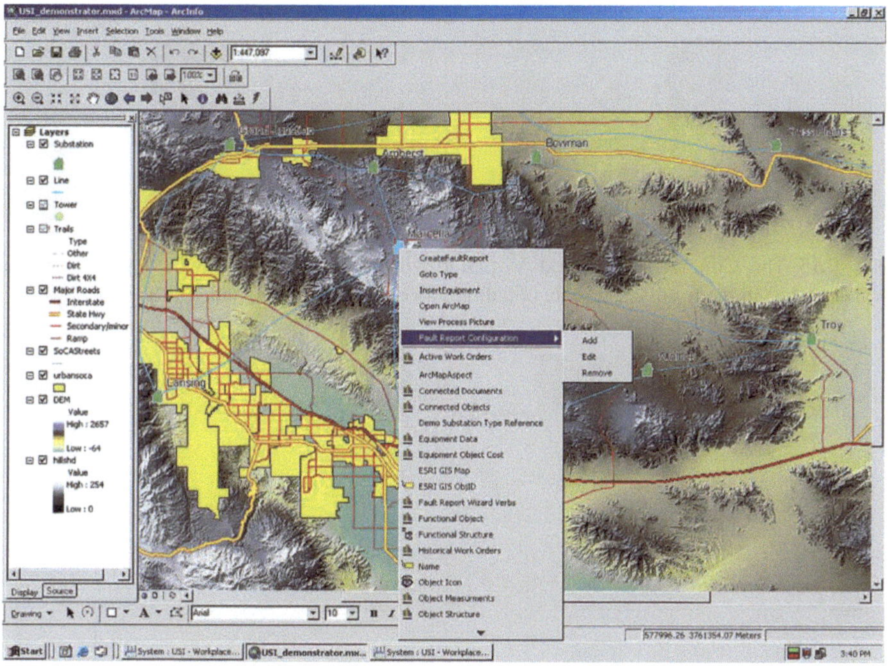

Fig. 9.14 The user interface for geographical information regarding faults in a transmission grid (after Bugge et al., 2004).

An emerging area in automated real-time monitoring of e-Work systems is *e-Maintenance*, where the assets are monitored and the actions are synchronized through wireless distributed agents (Figure 9.16). Necessary considerations in design of e-Maintenance systems are affordability, powerful human-machine interaction and collaboration frameworks (to make use of the experiences and expertise of human operators), and e-Training. *Tele-service* is another emerging concept, which enables human operators to monitor the system from remote locations. Tele-service enables remote accessibility to the system and improves agility and responsiveness, reduces conflicts, and minimizes the costs (as opposed to traditional wired connections). In e-Maintenance, it minimizes trouble-shooting costs, enhances flexibility, the installation and start-up of components, and facilitates the transfer of new software versions, provision of replacement parts, and ordering spare parts on time (Erbe, 2009).

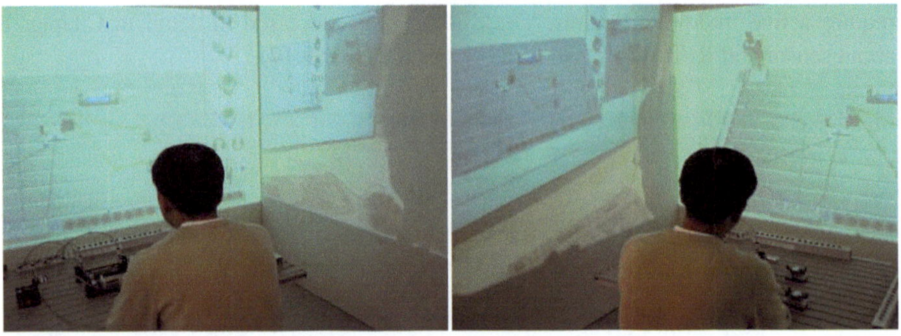

Fig. 9.15 The Computer Animated Virtual Environment workspace with real and virtual workbenches in the front and remote collaborators onto the side canvases (after Erbe, 2009).

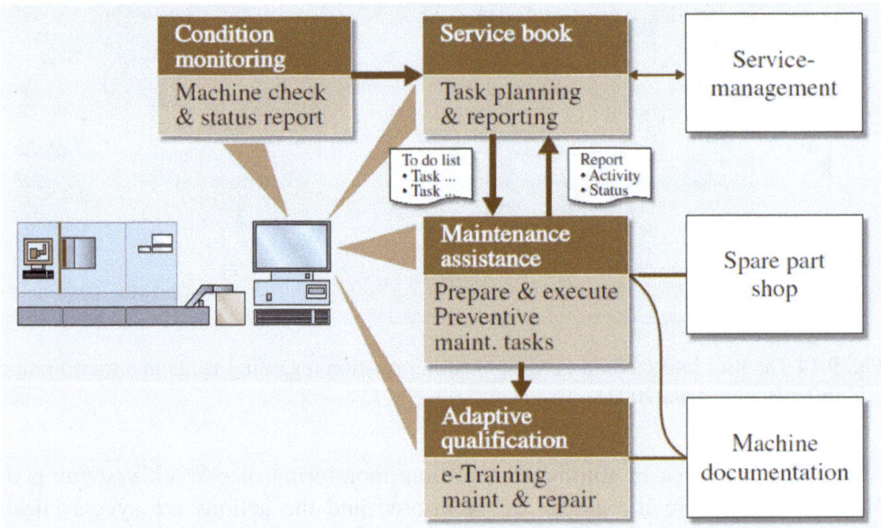

Fig. 9.16 e-Maintenance system architecture (after Berger and Hohwieler, 2003).

9.3.2 Performance Analysis and Optimization

Upon receiving the real-time information of distributed components, the system's performance can be evaluated and the required actions are determined by knowledge-based systems. As discussed in Chapter 1, knowledge-based systems enable the mining, refining, constructing, and extracting the information received from various agents, and transform it into knowledge through knowledge discovery in databases. The accumulated knowledge improves the decisions over time and enables automation in diagnostics and prognostics of the faults. Implementation of the decisions, in turn, requires effective logistics support to ensure availability of the required resources and hardware for performing the corresponding actions. The actions are determined with respect to the critical KPIs (Key Performance Indicator), depending on the requirements, plans, and policies of the e-Service system. Figure 9.17, for instance, illustrates a conceptual model for evaluation of e-Government portals with respect to a certain set of KPIs, e.g., response time, adoption rate (i.e., the ratio of successful transactions), customer satisfaction, and efficiency. Several models exist for performance evaluation of the e-Services based on multiple KPIs (e.g., balanced scorecard). The integrated AHP-DEA framework, presented in Chapter 3, is a legitimate and promising approach for evaluation service systems based on multiple KPIs (see Figure 3.6).

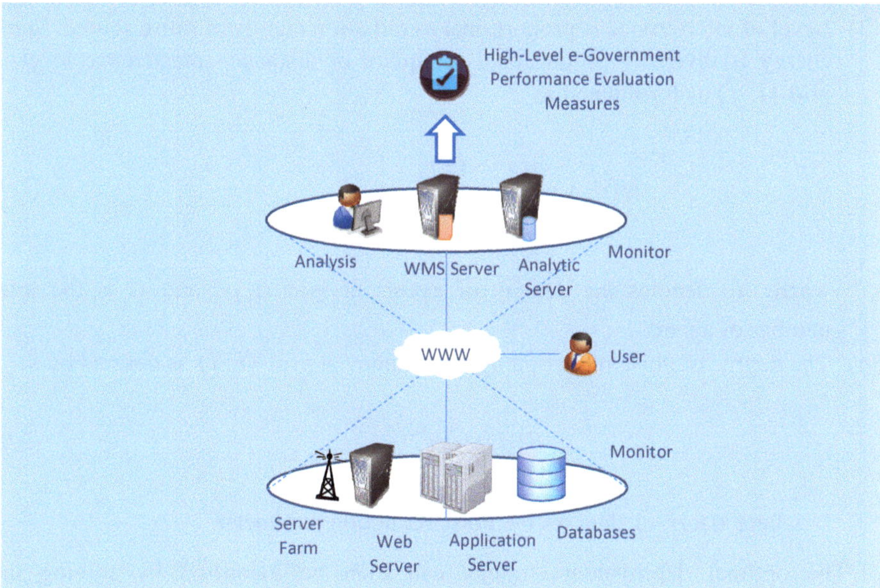

Fig. 9.17 A framework for performance evaluation of e-Government systems (WMS: Web Map Service).

As discussed in Chapter 8, the information received from the distrusted agents (e.g., sensors, RFIDs), in practice, is subject unpredictable interferences. Such noises in information may lead to inaccurate decisions, and the corresponding deviation may even be intensified as the number of the system's layers increases. In order to mitigate the effects of these noises, statistical analyses need to be performed to identify the sources of interference. The appropriate configuration of the network topology (e.g., FSN or RFID network) can then be defined through finding the optimal agent deployment strategy. The optimal strategy yields the lowest variance among various possible strategies as well as the optimal number of agents. In the deployment optimization procedure, it is assumed that agents are clustered into specific grids. The noises interfere with signal transmission by any agent node, independently. Thus, the total variance in each grid segment is the number of agents multiplied by interference effect to the grid segment. Besides, agents must be redundantly deployed to each grid segment in order to reduce the variance and to obtain more accurate, reliable data under noisy environment. Other assumptions of the deployment optimization procedure are as follows:

(1) Level of interference on each agent in the same grid is identical.
(2) Interference by a noise source affects every agent node, individually. Thus, the level of noise is additive and total interference effect in each grid is proportional to the number of agents in the grid.
(3) Level of interference is proportional to radiation energy of noise source. Since energy is inversely proportional to square of distance, interference level at grid (i, j) is formulated as

$$N(d_{ij}) = \frac{\Omega d_{ij}}{(i-2)^2 + (j-2)^2}, \tag{1.9}$$

where d_{ij} denotes the density of agents at grid (i, j), and Ω is the total number of agents.
(4) The degree of noise reduction by redundancy at grid (i, j) is defined by

$$F(d_{ij}) = \alpha^{\Omega d_{ij}}, \tag{2.9}$$

where $0 < \alpha < 1$ denotes the noise reduction parameter.

The optimal deployment strategy can then be identified by solving the following optimization problem:

$$\begin{aligned} \min \; & \sum_{i=1}^{k}\sum_{j=1}^{m} N(d_{ij})F(d_{ij}), \\ \text{s.t.} \; & \sum_{i=1}^{k}\sum_{j=1}^{m} d_{ij} \le 1, \\ & \Omega d_{ij} \ge 2, \quad \forall i, j = 1, \ldots, \Omega. \end{aligned} \tag{3.9}$$

The model above minimizes the total variance. The first constraint implies that the total number of agents deployed must not exceed the given number of agents. The second set of constraints ensures existence of at least two agents in each grid for redundancy. Implications of the aforementioned model in wireless sensor and RFID networks are further discussed in Chapter 8.

9.4 Task Administration Protocols for e-Service

In modern e-Service systems, due to the increased complexity and collaboration requirements, geographical distribution, ease of development, etc., service resources are usually distributed, networked, and controlled in a decentralized manner. Hence, coordination protocols are required in such environments to enable efficient processing of several *tasks* (e.g., service requests, intermediate processes) through interaction and collaboration among the distributed resources and agents. A task (in distributed systems) is formally defined as *an input that comes to an agent in the systems, and then is processed by the agent that has the required skill and capability*. The fulfillment of each single task, however, extremely depends on how individual agents coordinate with each other. Therefore, an effective control mechanism is needed to coordinate and harmonize tasks by exchanging information and decisions among the participating agents. The set of tasks in a distributed collaboration network can be described according to a task data model. According to ISO 14258 (1996), a task data model identifies three main groups of data elements for the tasks: primary (e.g., product, shop floor, planning and control), secondary (e.g., responsibility), and tertiary (e.g., resource). In addition to ISO 14258 (1996), these data elements comprise four information classes for the user to view (Table 9.2).

Table 9.2 Information classes of task data model (adapted from Anussornnitisarn et al., 2005).

Information class	Information contained
Product	- Product features and its processes or services - Bill of materials, operations, and services - Part geometries and properties - Process plans
Manufacturing operation, service planning and control	- Data for handling orders and services - Customer and production orders - Supplier's logistic data of materials and tools - Capacity constraints - 3rd party logistic data
Shop floor	- Data on current operations (production/service) - Shop schedule - Utilization of resources - Work procedures of operations and tools
General	- Data on company standards and guidelines - Equipment - Description of organization

Coordination protocols provide the agents in the system with the rules and the interaction procedures on how to cooperate and coordinate more effectively. In this section, TAP (Task Administration Protocol) that was introduced in the previous chapters is further elaborated for e-Service industry applications. TAP is capable of handling two major critical situations that other coordination protocols are unable to address. First, in some cases, the dependencies among the tasks/resources/agents may be dynamic (variable or conditional dependence). Moreover, complex situations may require decisions beyond simple coordination, e.g., tasks close to the deadline, dynamic priorities, and preemption. Thus, the systems need to be operated under the TAPs, as relatively more intelligent protocols capable of indentifying the current situation in the system and taking actions to deal with it. TAPs are designed for effective task allocation and administration in distributed task environments. The three general elements in task administration are *task*, *resource*, and *time*.

In TAP, resource allocation decisions follow a bidding process, which can be applied on the contract net protocol framework or similar logics. The task agents and resource agents communicate under certain coordination protocols to exchange information and decisions. Each resource agent tries to provide services to task agents, even though it may have to compromise its individual goal. Each agent, however, still tries to maintain its viability at a satisfactory level. A key issue in this type of cooperative multi-agent system resource allocation is that whether the system, as a whole, achieves its goal of efficient resource allocation, while agents make independent decisions to maximize their own criteria based on limited information of the entire system. The agent-based technology and Internet offer the structure of a network of flexibly interconnected capabilities, as shown in Figure 9.18. The objective is to achieve dynamic utilization of existing resources,

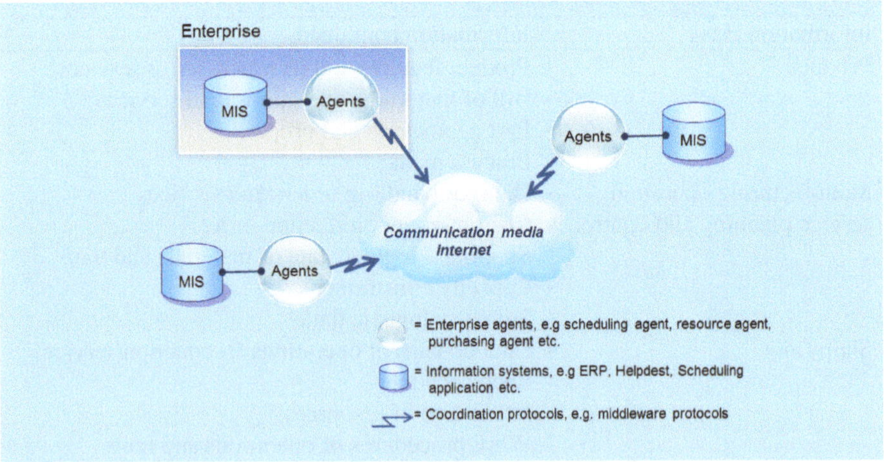

Fig. 9.18 Distributed network of enterprise with agent-based technology.

better customer service, and lower cost. TAPs are composed of three component protocols (see also Figure 1.10, Chapter 1): (1) Task Requirement Analysis Protocol (TRAP), (2) Shared Resource Allocation Protocol (SRAP), and (3) Synchronization and Time-Out Protocol (STOP). Details of each sub-protocol are presented in the following sub-sections (see also Ko and Nof, 2008, 2010).

9.4.1 Task Requirement Analysis Protocol (TRAP)

The primary step in administration of a task administration is its allocation to appropriate resources. TRAP is then activated to handle the event. This sub-protocol finds the dependence between tasks, their priorities, and precedence relationships. Upon arrival of a new task t_i to an agent, the agent calculates the priority of t_i and the previous tasks in the system at time τ through a dynamic priority evaluation function. If the priority of t_i is less than the priority of previous tasks still in the system, it is just added to the end of the task queue following the FIFO (First-In First-Out) discipline. Otherwise, the tasks in the queue are resorted with respect to their priorities. The task currently being processed may also need to be preempted by an urgent task with higher priority. Each task t_i has certain specifications, e.g., requirements, type, quantity, prerequisites, co-requisites, post-requisites, due date, and estimated cost. In general, a task is defined as follows:

$$t_i = < tp_i, qt_i, dd_i, ct_i, pr_i(\tau) >, \quad (4.9)$$

where,

t_i	Task *i*	tp_i	Type of task *i*
qt_i	Quantity of task *i*	dd_i	Due date of task *i*
ct_i	Cost of task *i*	$pr_i(\tau)$	Priority of task *i* at time τ

The dynamic priority of each task t_i at time τ is calculated using the following formula:

$$pr_i(\tau) = \underbrace{w_1 s_i}_{(1)} + \underbrace{w_2 \frac{qt_i(ct_i - p_i)}{\sum_i qt_j(ct_j - p_j)}}_{(2)} + \underbrace{w_3 \left(1 - \frac{dd_i - \tau}{\sum_j dd_j - \tau}\right)}_{(3)} + \underbrace{w_4 \left(1 - \frac{qt_0^\tau}{qt_0}\right)}_{(4)} + \underbrace{w_5 \sum_k (1 - pl_k)}_{(5)}, \quad (5.9)$$

where,

w_n	Weights ($0 \le w_n \le 1$; $\sum_n w_n = 1$)	j, 0	Index of other tasks/current task
s_i	1, if $tp_i = tp_0$; 0, otherwise	p_i	Price of task i
qt_0^{τ}	Remaining quantity of t_0 at time τ	k	Index of tasks preempted by task i
pl_k	The penalty of interruption of task k by task i ($0 \le pt_k \le 1$)		

The implications of the five terms in Eq. (5.9) are as follows:

(1) If t_i is the same type as the current task (t_0), its priority is increased, due to setup time minimization/elimination. In some cases, however, the opposite is required.
(2) The priority of t_i is proportional to its quantity and profit.
(3) A task relatively closer to its deadline has higher priority.
(4) If the current task (t_0) is not finished yet, but almost done, it has a relatively higher priority.
(5) Priority decreases t_i decreases as it causes more tasks to be preempted after priority assignment.

The overall procedure in TRAP is as follows: (1) task agent receives a task t_i, (2) tasks' priorities are updated through by Eq. (5.9), (3) tasks are resorted in the queue with respect to $pr_i(\tau)$, and then (4) SRAP is activated.

9.4.2 *Shared Resource Allocation Protocol (SRAP)*

Upon analyzing and assigning priorities of the tasks, SRAP is triggered for optimal allocation of resources to the tasks. Each resource in the system is managed by a corresponding resource agent. The task agent announces the task with the highest priority to the resource agents, taking into account their capabilities in processing certain tasks. Accordingly, each selected resource agent calculates its respective bid. The task agent then collects the bids and selects the best resource agent with the best bid. The steps in SRAP are as follows:

(1) The task agent announces t_i to the resource agents RA_j capable of processing t_i.
(2) RA_j resorts its queue considering t_i and updates the priorities of the tasks in its queue.
(3) RA_j calculates the bid for t_i as follows:

$$b_j = c^{ac} \sum_{i=1}^{q} \mu(pt_i) + c^{sc}, \tag{6.9}$$

where

b_j	The bid by RA_j	c^{ac}	Appreciation cost of RA_j per time unit
μ	Mean	pt_i	Processing time of t_i
q	Number of tasks in the queue of RA_j	c^{sc}	Setup cost of RA_j

(4) RA_j submits b_j for t_i to the task agent.
(5) The task agent selects RA_{j^*}, where

$$j^* = \arg\min_j \{b_j\}. \tag{7.9}$$

(6) The task agents awards t_i to RA_{j^*}, and t_i enters its queue with expected waiting time of b_{j^*}.

9.4.3 *Synchronization and Time-Out Protocol (STOP)*

In some situations, the current task (t_0) in process must be timed out and returned to the task queue of the task agent, in order to free the resources for more urgent tasks. Such decisions in TAP are handled by STOP, triggered in either of the following situations:

(1) *Excessive resource occupation*: Since faulty tasks need resources to be reworked, they may occupy the resources excessively and cause other tasks to be delayed. STOP checks excessive resource occupation (i.e., whether the current task uses the resource more than a certain time-out threshold) using the following formula:

$$TT_i = \mu(pt_i) + 2\sigma(pt_i), \tag{8.9}$$

where

TT_i	Time-out threshold for t_i	σ	Standard deviation

Nonetheless, there are two exceptions to the aforementioned time-out rule: (1) no task is waiting in the queue, and (2) the current task t_i may not satisfy its due date if timed-out.

(2) *Preemption by an urgent task:* Although the current task has been using the resource below the time-out threshold, an urgent task t_i in the queue may be late unless it is served before the process of the current task is completed. In this case, the current task is preempted by t_i, and its process will be resumed upon t_i is completed.

In summary, TAPs are composed of the three sub-protocols that enable efficient collaboration between service resources to process a number of tasks in a timely manner. The overall flowchart of TAPs is shown in Figure 9.19.

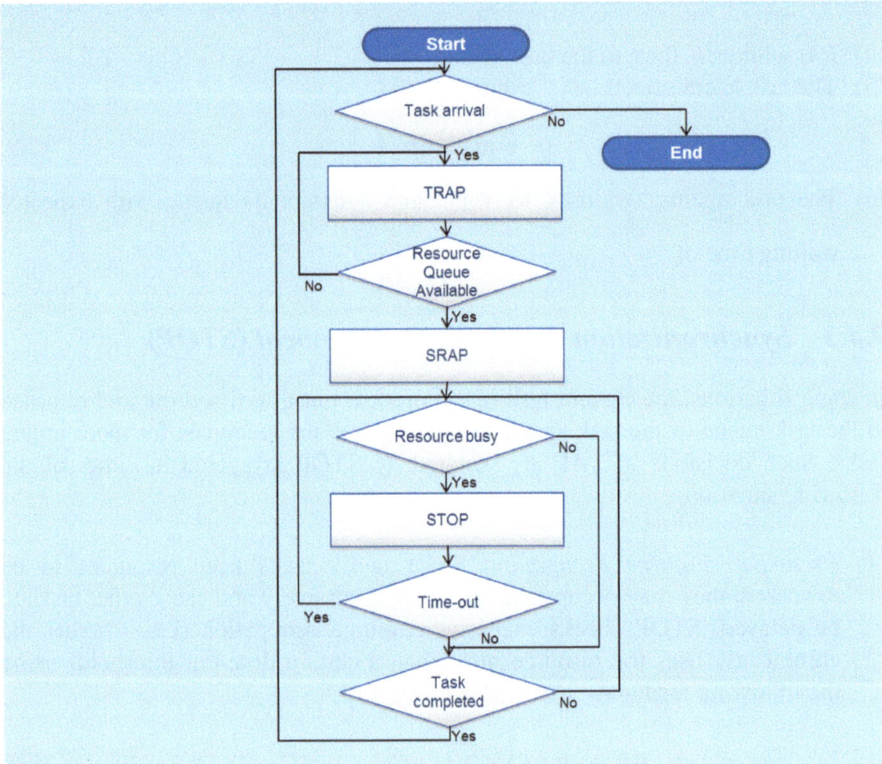

Fig. 9.19 The TAP general logic.

9.5 Collaborative ESP Design

ESP (e-Service Portal) refers to fully collaborative cyber-supported environments that enable remote access to large amounts of data in an e-Service system. ESP is formally defined as "*applications that enable companies to unlock internally and externally stored information, and provide users a single gateway to personalized information needed to make informed business decisions*" (Shilakes and Tylman, 1998). ESP enables users to collaborate over networks and provides customized services and various functions. Each participant in the system can access the portal, regardless of his/her/its current location, to update his/her/its assigned tasks and the status, and work with other peers through the tools and functions that the ESP provides. The roles of ESPs are to make these tools more accessible and unify a systematic access to these applications. The advantage is enhanced collaboration among users and systems and facilitated life cycle of knowledge in the e-Service system, from creation to exchange, retention, and reuse.

ESP is generally classified into four categories: (1) ESPs that deliver information and services to users, (2) ECP (e-Collaboration portal), which provides users with collaborative functions, (3) EXP (e-Expertise portal), which matches experienced people for decision-making, and (4) EKP (e-Knowledge portal), which supports various users with knowledge base. In all categories, ESP consolidates, manages, analyzes, and distributes information across and outside of an organization through the single gateway. It improves collaborative decision-making processes among users through providing customized services and various tools and information over electronic networks. In the process modeling stage of ESP design, it is critical to identify user and system requirements, e.g., workflow, business processes, involved users, input/output, and resources. Besides the design specifications, several fundamental functions need to be supported by ESP:

(1) *Content and document management*: ESP supports the full life cycle of content and document creation, and provides mechanisms for authoring, approval, version control, and scheduled publishing.
(2) *Collaboration*: ESP provides workspace where users communicate synchronously (through chat or messaging) or asynchronously (through threaded discussion, email, forums, and blogs).
(3) *Search and navigation*: This function is necessary to retrieve useful information, since retrieval becomes more difficult as the volume of information grows.
(4) *Personalization*: Users need to be able to customize and prioritize the layout, contents, and services in their work environment based on the users' attributes and metadata of the contents/services.
(5) *Entitlement*: This function limits user accessibility to specific types of contents and services, and protects intellectual property rights as well as privacy rights.
(6) *Integration*: ESP integrates various functions and data from multiple systems into a single interface.
(7) *Categorization*: Contents are categorized by metadata for efficient retrieval.
(8) *Single sign-on*: Users need to authenticate only once to access various contents and services.

The design specifications and required functions represent the overview of the ESP design, and define the interface requirements, type of services must be developed, and sources or services that must be connected to the ESP. A typical architecture for ESP design is depicted in Figure 9.20.

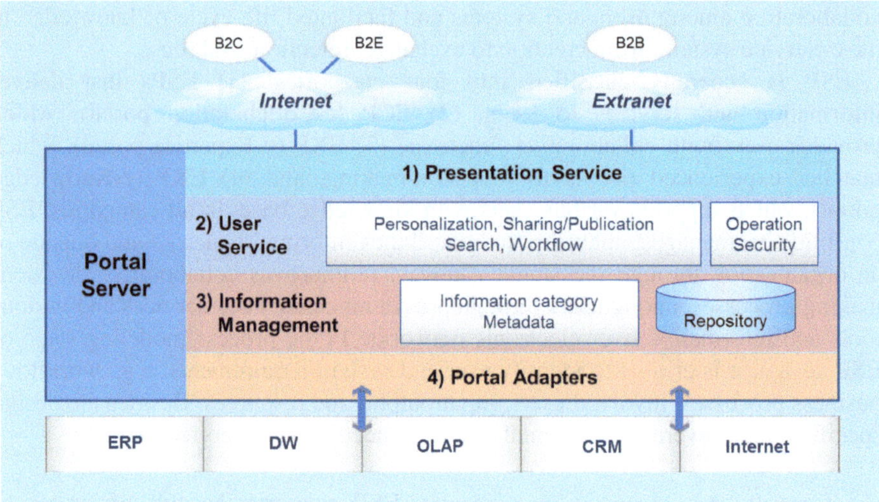

Fig. 9.20 General ESP architecture; (1) *Presentation service layer*: presents customized information to users; (2) *User service layer*: is invoked by the portal presentation layer and delivers services from the portal engine; (3) *Information management layer*: manages and categorizes the information from various sources with metadata and data repository; (4) *Portal adapter layer*: supports connection to legacy applications and other service systems (adapted from Lee, 2006).

Traditional ESPs, however, have some drawbacks due to lack of or low level of collaboration. Large amounts of information and task overloads diminish the quality of decisions made by the users without the help of smart collaboration functions. Hence, ESP needs to be augmented by the CSS functions that support high-level collaboration following the CCT (Collaborative Control Theory) design principles (see Chapter 2). The ESPs enhanced with such smart, sophisticated functions that enable high-level collaborative decision-making are called collaborative ESPs. In the following subsections, applications of some of the smart functions of CCT in design of collaborative ESPs for Service Decision Command Portal (SDCP) are presented. A SDCP is the front end of systems of systems, which delivers diverse information and services by harmonizing, integrating, aggregating/ disaggregating distributed systems, and thus satisfies the basic requirements for collaborative ESP.

9.5.1 CRP (Cooperation Requirement Planning) Function

(1) *Make information available*: SDCP acquires and avails enormous amount of data, scattered in multiple (internal and external) data sources. However, following CCT principles, it only forwards necessary information to the decision-maker.

(2) *Reduce information overload*: Users are overloaded by the massive amount of information. This issue is alleviated and controlled by information aggregation, categorization, proper search/navigation tools, and other relevant CCT principles.
(3) *Information aggregation*: Information makes more sense and is more easily accepted when grouped and represented at a proper level of aggregation.
(4) *Enterprise application integration*: SDCP enables users to enter data into different applications, tools, and systems that are integrated into the specific, customized workspace of each user.
(5) *Knowledge management*: SDCP manages the full life cycle of knowledge in the system with knowledge bases.
(6) *Collaboration*: SDCP provides traditional collaboration tools, e.g., e-mail, instant messaging, file sharing, group calendar, for better decision-making.
(7) *Remote access*: Users can access SDCP easily from remote locations.
(8) *Remote activation*: SDCP enables selection and authorization-based activation of decision support tools.

The SDCP is designed to satisfy all the basic requirements, from integration of multiple data sources and service systems (e.g., SAP, Oracle) to collection and visualization of necessary data at a proper level of aggregation in various formats (e.g., charts, maps, tables). It reduces information overload and enables rapid understanding of complex situations. For instance, a tactical manager may be interested in the information from a single regional service provider that he/she is in charge of, while a strategic manager may need the overall, aggregated information from the group of service providers in an entire district or state. However, if the strategic manager receives all the detailed information from every single regional service provider, he/she will be overloaded by a huge amount of information and be challenged to make effective, timely decisions out of it. The SDCP provides the information at the level of abstraction that the users require, but users can also retrieve the details and explore in the information, if required. At various levels of aggregation, users can share the necessary information with others and try to solve the problems together by applying collaboration tools supported by the SDCP (Figure 9.21-a). More importantly, the SDCP provides decision analysis tools customized to the users' needs at respective levels and authority (Figure 9.21-b). For example, users can run an online simulation to see the impacts of their decisions. They can change the values of input variables in the embedded simulation model and observe the results through the interface of the SDCP. These tools augment the users' ability to make better and timelier decisions by analyzing and predicting the envisioned outcomes of alternative users' decisions before implementing them. All the information, decisions, and analysis results can also be conveniently shared with other authorized SDCP users.

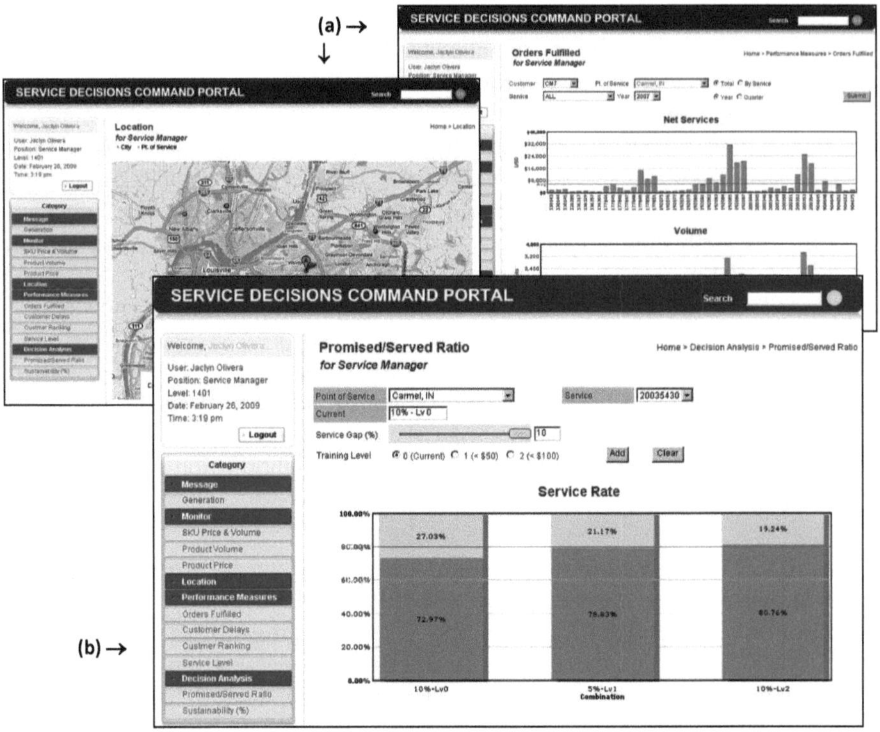

Fig. 9.21 SDCP; (a) chart and map examples to display aggregated information collected from multiple data sources in SDCP; (b) a decision analysis tool.

9.5.2 *ELOCC (Emergent Lines of Collaboration and Command) Function*

Upon receiving an alert about an exceptional situation and storing the information in the system, the corresponding users need to be notified. If all the information is distributed among all the users, they will be overloaded. For instance, if every tactical manager reports service delay and progress data to his/her supervisor, the supervisor cannot handle all the information on time. The information needs to be selectively delivered only to relevant persons and according to the level of urgency, i.e., the *need-to-know criterion*. As discussed in Chapter 2 in detail, BMP (Best Matching Protocol) is a promising coordination protocol that can be applied for matching and sharing the information with relevant decision makers considering its purpose and urgency. BMP uses logic of matching situations with decision makers, i.e., under each decision situation, specific matched organizational roles are identified. When urgent alerts are given and a decision has to be made, BMP sends the alert only to the relevant addressees based on the decision matching vector of attributes and the underlying workflow among them. All the information and decisions regarding a specific alert (i.e., time, corresponding user, subsequent decisions, information exchange procedure, and resolution mechanisms) can be specified under the BMP. BMP enables efficient collaboration to resolve urgent service issues. Figure 9.22 illustrates the BMP logic in collaborative ESP design (see Chapter 2 for further details).

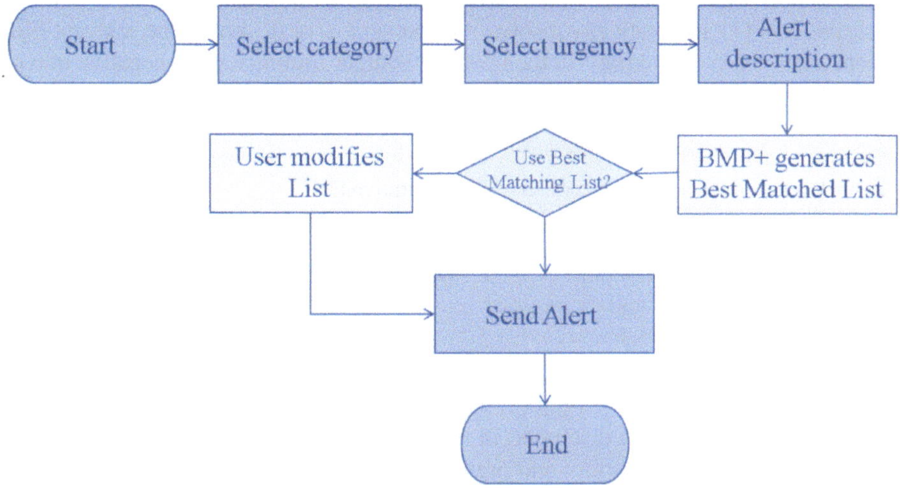

Fig. 9.22 BMP logic in collaborative ESP design.

In some situations, despite effective provision and allocation of relevant information to the users by the collaborative ESP, it is difficult for the users to detect and anticipate all the abnormal or undesirable facts. ESPs are also prone to errors and conflicts that may occur frequently due to various reasons, e.g., unexpected demand drop in a region, insufficient inventory of replacement parts, or lack of trained personnel. These errors/conflicts may result in an inability to provide maintenance and updating services during certain months. This issue is further elaborated in the next section.

9.6 Error Prevention and Conflict Resolution in e-Service

As discussed earlier in this chapter, performance of an e-Service system varies over time, during its life cycle. These performance variations may lead to errors by individual agents, e.g., faults, outages, breakdowns, etc. In general, an error is defined as *an act, assertion, or belief that unintentionally deviates from what is correct, right, or true. It is also the condition of having incorrect or false knowledge.* Collaboration, by nature, may also result in conflict situations, which arise due to complex interactions among collaborative agents. A conflict is defined as *a state of disharmony between incompatible or antithetical ideas, functions, or interests, that is, a clash* (See Chapter 2 for formal definitions of error and conflict in e-Work).

Errors and conflicts may escalate to higher levels of complexity and perpetuation, if not predicted, resolved, and prevented in a timely manner and through cost-effective and well-synchronized actions. Service performance problems caused by error and conflict situations may result in significant losses and service disruptions unless they are handled as early as possible. Collaborative ESP must provide early error and conflict detection capabilities to handle all the scenarios that may arise by human, system, and information. It should have the capability to detect an error as early as possible and (1) automatically resolve it or prevent it, or (2) notify the relevant ESP users about the errors to prevent them, solve them, or minimize their negative impact and damage. Some potential errors and conflicts in collaborative ESP are presented Table 9.3.

Table 9.3 Potential errors and conflicts in collaborative ESP.

Source	Human	System	Information
Error	- Data input - Redundant requests	Data corruption	Faulty/outdated/ redundant data
Conflict	- Rank/hierarchy - Scheduling	- System down - Time-out	Data integrity

In the SDCP example discussed in the previous section, the general procedure for early detection, resolution, and prevention of errors and conflicts regarding the capacity of service centers is illustrated in Figure 9.23. Data is collected, then the detection logic is applied, and finally alerts are generated (if any). Implementation of this system is based on agent-based technology, where intelligent agents collaborate and interact autonomously. The idea is to have the agents readily standing by in the system to identify any error/conflict once it happens, or even as it emerges. The software agent then finds the solution or sends an alert to human users whenever human intervention is needed to resolve the issue. Teams of SDCP agents can be designed to (1) monitor, check, and collect data from multiple sources, (2) detect an error by applying detection logic over the collected data, and (3) deliver the error and/or conflict information to relevant users of the SDCP. In this example, upon receiving the updated demand and/or capacity data, the agents check capacity data in multiple service centers considering open orders and appropriate service levels. If they cannot fulfill the demands or the capacity levels are too low, the agents trigger the alert in the SDCP automatically, so that responsible managers can notice and solve the problems promptly. Another important example is error and conflict detection and prevention in assembly and inspection, performed by integrated sensor-based control systems in flexible assembly cells. Information about previous errors and conflicts in assembly operations, and the corresponding recovery programs is stored in a knowledge base. The knowledge base provides support for both detection and recovery. Machine learning is an alternative approach, where diagnostics are emphasized as a necessary step after the detection and before the recovery through an integrated framework.

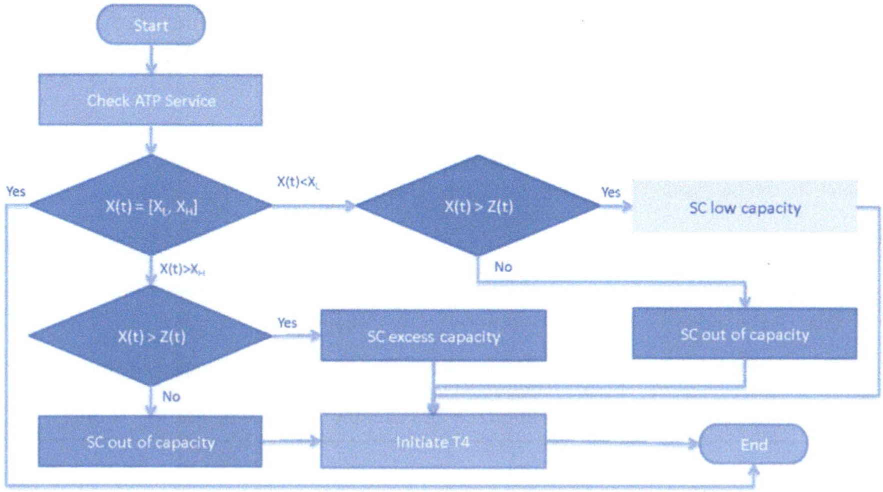

Fig. 9.23 Early error and conflict detection and smart alert mechanism regarding the capacity of a service center [X: Capacity; Z: Demand, t: Time].

As the use of computers and other automation devices, e.g., handwriting recognition and sorting machines in postal service, becomes increasingly popular, errors and conflicts can be effectively and automatically prevented and reduced to minimum in many service industries, including delivery, transportation, e-Business, and e-Commerce. In some specific service industries, e.g., healthcare systems, detection and prevention of errors and conflicts are very critical and thus, automated systems need to be developed to detect human errors and conflicts in the system. Much more research, however, needs to be conducted to advance the development of automated error and conflict detection and prevention in service industries.

An agent-based framework is discussed in this section, as an effective approach towards detection and prevention of errors and conflicts in the systems, in general. In the error detection procedure, an agent deployed at unit m executes the following expression to prevent errors:

$$\exists E(u_m), \quad \text{if } \left\{ U_m - L'_m < \sum_{i=1}^{m-1} Q_i \right\} \cup \left\{ L_m - U'_m > \sum_{i=1}^{m-1} Q_i \right\}, \tag{8.10}$$

where

$E(u_m)$	Error by unit m	Q_i	Performance of unit u_i
U_m	Upper limit for acceptable QoS after operation of unit u_m		
L_m	Lower limit for acceptable QoS after operation of unit u_m		
U'_m	Upper limit for acceptable performance of unit u_m		
L'_m	Lower limit for acceptable performance of unit u_m		

In case of multiple services, the following centralized algorithm can be used to predict errors in a sequential service line, which predicts at time 0 the potential errors that might occur at time τ:

$$\exists E[u_m(\tau)], \quad \text{if } \min_{i=1}^{m} \left\{ I_i(0) \times \prod_{i=1}^{m} \eta_i \right\} < \varphi_m(\tau), \tag{8.11}$$

where

$I_i(0)$	Quantity of available resources for unit u_i at time 0
η_i	Probability that a service being operated by unit u_i is within specifications
$\varphi_m(\tau)$	Required amount of qualified services after the operation of unit u_m at time τ

Eq. (8.11), however, often has low reliability, i.e., high false positive rates (errors are predicted but do not occur), or low preventability, i.e., high false negative rate (errors occur but are not predicted), because it is difficult to obtain accurate η_i in case of having several units in the system. To improve reliability and preventability, agent-based error prevention has been developed to prevent errors in the process of providing multiple services. An agent is deployed at specific unit while informed about the previous unit. Multiple agents deployed at different units can execute the following expression simultaneously to prevent errors. Each agent can have its own attitude, i.e., optimistic or pessimistic, toward the possible occurrence of errors. For further detailed information, the readers are referred to Chen and Nof (2010, 2012).

$$\begin{aligned} \exists E[u_m(\tau)], \text{ if } & \min\{I_m(\tau'), I_{m-1}(\tau') \times \eta_{m-1} + C_{m-1}(\tau') - N_m(\tau') - C_m(\tau')\} \\ & \times \eta_m + C_m(\tau') < \varphi_m(\tau), \quad \forall \tau' < \tau, \end{aligned} \tag{8.12}$$

where

$C_m(\tau')$	Number of cumulative conformities produced by unit u_m by time τ
$N_m(\tau')$	Number of cumulative non-conformities produced by unit u_m by time τ

The major conflict detection and resolution strategies are listed below, classified into subsequent stages. The general procedure is called MCR, which is an agent-based technique developed originally for conflict resolution in collaborative facility design (see Lara and Nof, 2003, 2009; Velasquez et al., 2008). The procedure can be generalized and applied in other similar instances, including conflict resolution in collaborative ESP and e-Service systems. The major methodologies and tools required for modeling and executing the agent-based conflict resolution procedure are heuristics, knowledge-based interactions, AI (Artificial Intelligence), and multi-agent systems.

Stage 1: Direct Negotiation

(1) Agent prepares resolution proposal and sends to counterparts.
(2) Counterpart agents evaluate the proposal. If they accept it, go to Step 5. Otherwise, go to Step 3.
(3) Counterpart agents prepare a counter-offer and send it back to the originating agent.
(4) The agent evaluates the counter offer. If accepted, go to Step 5. Otherwise, go to Stage 2.
(5) End of the conflict resolution process.

Stage 2: Third Party Mediation

(1) Third-party agent prepares resolution proposal and sends to counterparts.
(2) Counterpart agents evaluate the proposal. If accepted, go to Step 5. Otherwise, go to Step 3.
(3) Counterpart agents prepare a counter-offer and send it back to the third-party agent.
(4) The third-party agent evaluates the counter-offer. If accepted, go to Step 5. Otherwise, go to Stage 3.
(5) End of the conflict resolution process.

Stage 3: Incorporation of Additional Parties

(1) A specialized agent prepares resolution proposal and sends to counterparts.
(2) Counterpart agents evaluate the proposal. If accepted, go to Step5. Otherwise, go to Step 3.
(3) Counterpart agents prepare a counter-offer and send it back to the specialized agent.
(4) The specialized agent evaluates the counter-offer. If accepted, go to Step 5. Otherwise, go to Stage 4.
(5) End of the conflict resolution process.

Stage 4: Persuasion

(1) A third-party agent prepares persuasive arguments and sends to counterparts.
(2) Counterpart agents evaluate the arguments.
(3) If the arguments are effective, go to Step 4. Otherwise, go to Stage 5.
(4) End of the conflict resolution process.

Stage 5: Arbitration

(1) If the conflict management and analysis procedures result in *common proposals*, conflict resolution is achieved through management and analysis.
(2) If the conflict management and analysis procedures result in mutually *exclusive proposals*, conflict resolution is achieved, though *conflict confrontation*.
(3) If the conflict management and analysis procedures result in *no conflict resolution proposals*, conflict resolution must be used.

An integrated Conflict and Error Detection and Prediction Model (CEDPM) is illustrated in Figure 9.24, which comprises Conflict and Error Detection and Prediction Agents (CEDPA), Conflict and Error Detection and Prediction Protocols (CEDPP), and four components, developed to detect and predict conflicts and errors in distributed collaborative operational environments. The CEDPA is deployed at each collaborative unit to

(1) Detect and predict conflicts and errors by executing three functions, detection and prediction policy generation, error detection and prediction, and conflict prediction, which interact with and are supported by error knowledge base, and
(2) Communicate with other CEDPAs to send and receive the conflict and error announcements with the support of CEDPP.

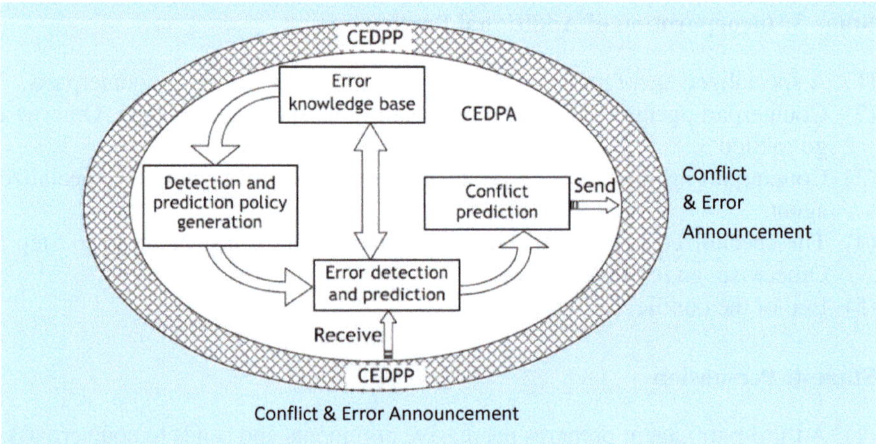

Fig. 9.24 The integrated CEDPM.

9.7 Emerging Trends and Challenges

The most challenging issue in the design, operation, and control of emerging e-Systems, in general, and e-Service, in particular, is the growing complexity caused by the overload and overabundance of data. e-Services are still overwhelmed by the influx of available data and incompleteness of information. The situation is further aggravated by cross-cultural and cross-disciplinary misunderstandings (see Chapter 11). The question that has remained open is how to "*engineer*" collaborative e-Service processes to ensure seamless, interactive, effective and efficient service through discovering and applying new integrated and multidisciplinary knowledge. The following research directions have to be taken into account in the design, evolution, and transformation of e-Service industries:

- *Incentive-based collaboration*: Design of incentive mechanisms to ensure effective processing of collaborative tasks when voluntary efforts are required (see Chapter 11 for details).
- *Smart search engines*: Revolutionizing search engines to incorporate more complicated functions such as mining, matching, and recommending target-filtered knowledge, beyond ''find anything and I will then sift through the ocean you found.''
- *Task assignment/distribution/sharing mechanisms*: Design of mechanisms that enable dynamic collaboration of participants in to accept or volunteer for task assignments, through reasonable sharing and pooling of resources.
- *Multicultural interactions*: Integration of multi-/cross-cultural issues in the expansion, refinement, innovation, globalization, and evolution of e-Services to ensure quality and sustainability of service to different client communities (see Chapter 11 for details).
- *Interactive, enhanced information/service visualization and experience*: Enabling enhanced visualization along with interactive technologies such as 3D avatars, tele-presence, tele-robotics, remote virtual labs, automated tag clustering, and semantic zooming for better comprehension, and better information and knowledge exploration.

References

Anussornnitisarn, P., Nof, S.Y., Etzion, O.: Decentralized Control of Cooperative and Autonomous Agents for Solving the Distributed Resource Allocation Problem. International Journal of Production Economics 98(2), 114–128 (2005)

Berger, R., Hohwieler, E.: Service platform for web-based services. In: Proc. 36th CIRP Int. Semin. Manuf. Syst. Produktionstech, vol. 29, pp. 209–213 (2003)

Bugge, J., Julian, D., Gundersen, L., Garnett, M.: Map of the future. ABB Rev. 2, 30–33 (2004)

Chen, X.W., Nof, S.Y.: Agent-based error prevention algorithms. Expert Systems with Applications 39(1), 280–287 (2012)

Chen, X.W., Nof, S.Y.: A decentralized conflict and error detection and prediction model. International Journal of Production Research 48(16), 4829–4843 (2010)

Claessens, S., Glaessner, T., Klingebiel, D.: Electronic Finance: Reshaping the Financial Landscape around the World. Journal of Financial Services Research 22(1-2), 29–61 (2002)

Fong, S., Meng, H.S.: A web-based performance monitoring system for e-government services. In: Proceedings of the ACM 3rd International Conference on Theory and Practice of Electronic Governance, pp. 74–82 (2009)

Jeong, W., Nof, S.Y.: Collaborative e-Work networks in industrial engineering. Computers and Industrial Engineering 57(1), 1–2 (2009a)

Jeong, W., Nof, S.Y.: A collaborative sensor network middleware for automated production systems. Computers and Industrial Engineering 57(1), 106–113 (2009b)

Jeong, W., Nof, S.Y.: Design of timeout-based wireless microsensor network protocols: energy and latency considerations. International Journal of Sensor Networks 5(3), 142–152 (2009c)

Ko, H.S., Nof, S.Y.: Design of protocols for task administration in collaborative production systems. International Journal of Computers, Communication and Control 1, 91–105 (2010)

Ko, H.S., Lim, H.J., Jeong, W., Nof, S.Y.: A statistical analysis of interference and effective deployment strategies for facility-specific wireless sensor networks. Computers in Industry 61(5), 472–479 (2010)

Ko, H.S., Nof, S.Y.: Modeling Task Administration Protocols for Human and Robot e-Workers. In: Duffy, V. (ed.) Handbook of Digital Human Modeling. Taylor & Francis (2008)

Lara, M.A., Nof, S.Y.: Computer-supported conflict resolution for collaborative facility designers. International Journal of Production Research 41(2), 207–233 (2003)

Lara, M.A., Nof, S.Y.: Conflict resolution in supply chain security. International Journal of Value Chain Management 3(2), 168–186 (2009)

Nof, S.Y., Morel, G., Monostori, L., Molina, A., Filip, F.: From Plant and Logistics Control to Multi-Enterprise Collaboration. Annual Reviews in Control 30, 55–68 (2006)

Ola, K.: Lifecycle management for improved product and system availability. ABB Rev. Spec. Rep. Ind. Serv. 29, 36–37 (2004)

Oliveira, R.R., Noguez, F.C., Costa, C.A., Barbosa, J.L., Prado, M.P.: SWTRACK: An intelligent model for cargo tracking based on off-the-shelf mobile devices. Expert Systems with Applications 40, 2023–2031 (2013)

Velasquez, J.D., Lara, M.A., Nof, S.Y.: Systematic resolution of conflict situations in collaborative facility design. International Journal of Production Economics 116(1), 139–153 (2008)

Wachs, J.P., Kölsch, M.: Vision-based hand-gesture applications. Communications of the ACM 54(2), 60–71 (2011)

Review and Challenge Questions

1. Compare and contrast, focusing on their purpose and their collaboration ability: (i) Service vs. e-Service; (ii) e-Service vs. Web service. (Illustrate your answers.)
2. Add another column to Table 9.1: For each entry, briefly explain an example of its use to augment collaboration in a given e-Service.

3. Consider two particular e-Services: (i) Retailing e-Books and music; (ii) e-Security. Specifically for each, identify the most significant challenges in this service industry, and the most promising emerging solutions.
4. For modeling service workflow, explain the relative advantages and limitations of (i) Petri Net diagrams; (ii) Reference Architecture diagrams; (iii) UML diagrams; (iv) OPM diagrams. Illustrate your answers with a particular e-Service.
5. Suggest several metrics to evaluate the success of alternative, competing e-Service companies in: (i) Cell phone service; (ii) cleaning service; (iii) real-estate service. (Apply e-Measures and e-Criteria discussed earlier in the book.)
6. What are some of the main benefits provided by e-Power services? How can they be rationalized?
7. Identify several common collaboration engineering features in e-Power and e-Healthcare. (Consider critical resource sharing; information sharing; demand and capacity sharing; privacy; sustainability; security; safety; performance metrics, etc.)
8. Historically, financial e-Services are among the oldest beneficiaries from e-Work, including commercial justification and rationalization of the Internet use for Business-to-Business (B2B) transactions. Identify several collaboration opportunities between and among multiple e-Services, (i) enabled by financial e-Services; (ii) Unrelated to financial e-Services. Illustrate your answer with specific examples.
9. Think and discuss how CCT design principles for augmenting collaboration can apply for your answers to Question 8.
10. Describe several collaborative multi-robot applications in e-Service, for instance: (i) e-Medicine; (ii) e-Delivery; (iii) e-Cleaning; (iv) e-Exploration; (v) e-Transportation; (vi) e-Security.
11. Illustrate the role, advantages and limitations in monitoring service performance with: (i) agents; (ii) GPS; (iii) flying robot swarms.
12. Explain the use of augmented collaboration in (i) e-Maintenance; (ii) e-Library; (iii) e-Laboratory experimentation.
13. Collaborative teamwork decision making is a common features of many e-Services, e.g., in design, planning, monitoring, evaluation, selection, recommendation, etc. (i) Discuss the role of collaboration engineering for collaboration optimization of decisions. (ii) Analyze your answer to (i) relative to (a) e-Design; (b) e-Healthcare; (c) e-Government.
14. Describe the Task Administration Protocols explained in this chapter, and how optimization algorithms are applied by each of them.
15. Explain how an ESP, e-Service Portal; ECP, e-Collaboration portal, and EXP, e-Expertise portal can each be useful for (i) e-Retail; (ii) e-Legal service; (iii) e-Library; (iv) e-Scheduling.
16. (i) Describe the main functions in the ESP architecture shown in Fig. 9.20. (ii) Explain how the architecture and its functions may or may not be different for ECP and for EXP.

17. Error Prevention and Conflict Resolution tools that are designed to operate automatically in parallel to an e-Work, an e-Business, or an e-Service system are beneficial in many ways. (i) Describe some of these ways, and explain how they are useful in the rationalization of those systems, e.g., by enabling them to scale up. (ii) Analyze and explain how these tools can also overlap functionally and collaborate with (a) CRP; (ii) ELOCC.
18. Explain the stages in the automated conflict resolution scheme discussed in this chapter, Section 9.6, and how they are designed to minimize the resolution costs.
19. Write a brief essay: "Is it a dream, or can we significantly improve the quality and timeliness of services, even government services, by revolutionizing collaboration?"

Lab Questions

20. Develop a simple Petri Net diagram for the e-Service case described in Figure 9.2.
21. Develop a UML use case diagram for modeling: (i) Scheduling co-design meetings; (ii) Archiving scientific data; (iii) Recommending video clips to an advertising campaign.
22. Following Question 10, explore on the Internet and describe several examples of implementing collaborative multi-robot systems for e-Services.
23. Develop the Task Administration Protocols TRAP, SRAP, and STOP, following the examples in the chapter, for the case of an alert system in a smart building.
24. Find (e.g., on the Internet) the following simple *"single machine scheduling"* models:
    a. Total Weighted Completion Time
    b. Maximum Lateness
    c. Number of Tardy Jobs
    d. Total (Weighted) Tardiness

    Then:
    (i) Analyze the similarities and differences of each model to/from the Task Requirement Analysis Protocol (TRAP).
    (ii) What are the unique functionalities of the TRAP, and (if any) their applications in real-life e-Service systems?
    (iii) Identify three real-life e-Service examples for SRAP (bidding) and STOP (preemption).
25. Consider an e-Service system with which you are familiar.
    (i) Identify and list several errors and conflicts in that e-Service system.
    (ii) Following your answer to (i), explain briefly the meaning of error and conflict propagation, and how each of the five stages described in Section 9.6 can be applied to resolve and prevent them.
    (iii) Develop a simple recovery tree for some of the errors and conflicts you described in (i) and (ii).
    (iv) Explain how the recovery tree developed in (iii) can be implemented through active middleware.

Chapter 10
e-Learning and e-Training

Abstract. Advances in ICT (Information and Communication Technologies), automation, and collaboration engineering have significantly revolutionized education and training over the past decades. The distributed and ubiquitous nature of Internet-based education and training, as remarkable instances of e-Service (Chapter 9), has transformed the emphasis of conventional education and training systems, from the transmission and acquisition to creation of knowledge, leading to further customized education and training. The emerging education and training systems are then more learner-oriented than instructor-oriented, i.e., concentrating more on the learning than the education aspect. The ever-increasing needs for knowledge update due to rapid advances in technology, the rising demand for postgraduate/continuing professional education of working adults, and limited capacities of educational institutions, along with high expenses of traditional education further justify the need for emerging e-Learning and e-Training systems. A major benefit of e-Learning and e-Training is the participative and collaborative infrastructure, as well as enabling engineered CCT aspects and functions of the learning process itself; just as in formation transfer in Internet 1.0 was mostly one-directional (from providers to users), Internet 2.0 enables bi-directional and multi-directional interaction and collaboration.

10.1 Automation and e-Work in Learning and Training

Nowadays, many universities and educational institutes have transformed their traditional education systems to dual mode education, offering both on-campus and off-campus Internet-based education. This evolution have caused the emergence of several corporate, for-profit online universities, which in turn lead to high competitions, a global lifelong learning economy (Taylor, 2002), and ultimately forces to the universities and educational institutions to change. This *revolution*, i.e., automation of learning and training, must be consistent with the educational psychology and technology, which is referred to as Instructional Design (ID). According to the Applied Research Laboratory at Penn State University (2009).

S.Y. Nof et al., *Revolutionizing Collaboration through e-Work, e-Business, and e-Service*,
Automation, Collaboration, & E-Services 2, DOI: 10.1007/978-3-662-45777-1_10

"ID, as a process is the systematic development of instructional specifications using learning and instructional theory to ensure the quality of instruction. It is the entire process of analysis of learning needs and goals and the development of a delivery system to meet those needs. It includes development of instructional materials and activities; and tryout and evaluation of all instruction and learner activities. ID as a discipline is that branch of knowledge concerned with research and theory about instructional strategies and the process for developing and implementing those strategies".

The ID models are generally classified into two categories of 1) Action, Design, Development, Implementation, and Evaluation (ADDIE), and 2) rapid prototyping, which are based on the ADDIE models (Figure 10.1). The main ID process design procedure for learning and training systems is composed of three steps of planning (analysis, design, and development), implementation, and evaluation.

In the first step (i.e., planning), the requirements of the system and the students, and the corresponding tasks are analyzed. The system is then designed and developed through determination of learning/training goals (e.g., target behaviors, evaluation conditions, and achievement levels), contents (e.g., knowledge, skills, and values), and methods (e.g., group/individual). In the second step, the generated plan is executed (e.g., tutoring, mentoring) and the outcomes are eventually evaluated at the third step through formative/comprehensive evaluations (e.g., reaction/learning/behavior/results).

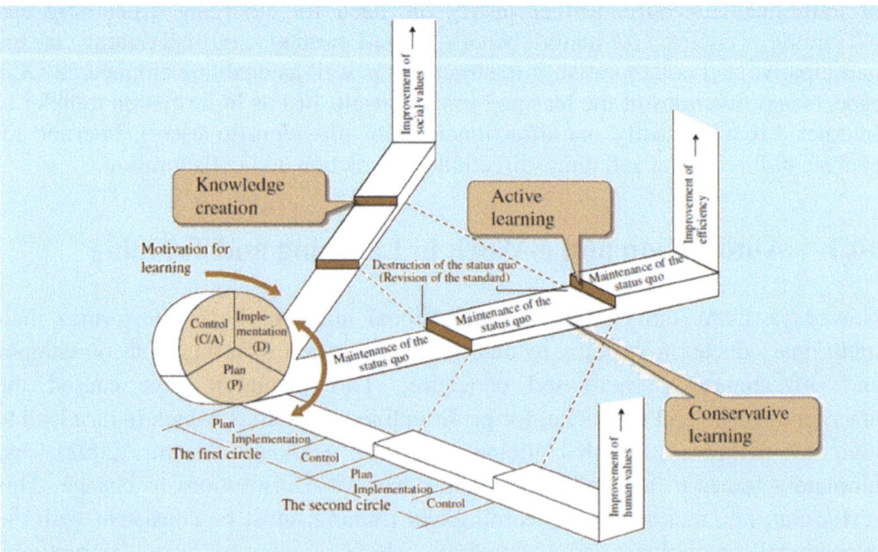

Fig. 10.1 The conceptual model for learning/training activities based on the plan/do/check management circle model, emphasizing *collaboration* for value creation by individuals (human values) and the entire system (social values) (after Ishii et al., 2009).

Learning and education programs are generally classified with respect to the learning goals, content, and formats, synchronous/asynchronous methods, and the limitations on time and educational resources. Well-known learning/education formats include (Ishii and Tamaki, 2009):

1) *Computer-Assisted Instruction (CAI)*: displays contents considering the level of understanding of each learner/trainee, and minimizes cost and time via computer assistance. Limitations: a) Instructor and learner must be at the same location; b) The learning/training progress is complied manually.
2) *Computer-Based Training (CBT)*: complies the contents into databases, allowing interactive setting of learning order. Limitation: Learning orders are fixed and determined before by manuals.
3) *Web-Based Training (WBT)*: transmits contents through Internet networks, enabling quick revisions and evaluations, and providing both synchronous and asynchronous learning/training services. Limitations: a) Unified management of learners/trainees is not possible; b) Is expensive; c) Modification and improvement of the contents is difficult.
4) *e-Learning/e-Training*: enables learners/trainees to study independently, optimizes utilization of ICTs, and customizes learning/education. Limitation: It is difficult to compile plans based on the evaluation results, history, and progress. Also known as *virtual Learning/Training*, e-Learning/e-Training systems model conventional *in-person* learning/training models through providing virtual environments for classes, contents, homework, tests, and assessments, using a *Content Management System (CMS)*.

The e-Learning and e-Training concepts encompass all the aforementioned learning and training formats. According to the American Society for Training and Development (2009), *"e-Learning* [and e-Training] *refers to all items provided, facilitated or transmitted by electronic technology that are clearly for the purpose of learning* [and training]". The purpose is to support individual learning and training, adjust the goals to the contents, improve interactions between the content provider and the learner/trainee, and enable all users to participate in the learning/training process via ICT (Figure 10.2). Emerging e-Learning and e-Training solutions are more cost-effective compared to the traditional learning and training method due to minimizing the indirect costs, and the costs associated with distribution of contents. They also enable timely efficient learning, feedback, and system performance management, as well as higher quality through enhancing update of contents, control of learning pace, access to historical data, and homogeneous distribution of educational contents.

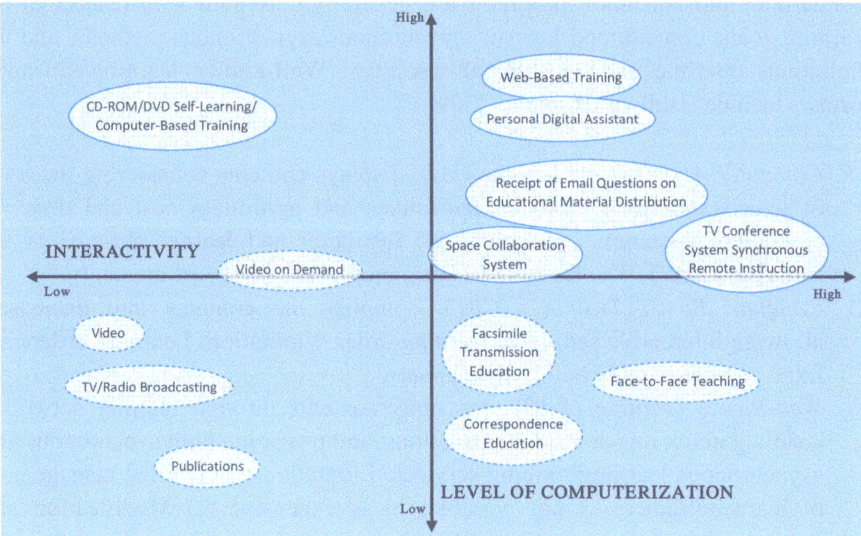

Fig. 10.2 e-Learning and e-Training formats according to the ranges of interactivity and computerization (automation); The solid lines characterize e-Learning and e-Training, while the dashed lines correspond to the formats out of e-Learning and e-Training (after Ishii et al., 2009).

There are two major platforms and standard tools for e-Learning and e-Training (see Table 10.1). The first platform is Learning Management System (LMS), which is a software application supporting administration, documentation, tracking, and feedback in e-Learning systems. The second platform is CMS, which provides a shared collaborative environment for developers, authors, learning/training system designers, and experts to create, store, update, control, and deliver contents from a central repository. The CMS platform is responsible for developing, controlling and publishing the contents delivered by the LMS platform. In the ADDIE model described above, for instance, CMS deals with the development tasks (e.g., authoring, accumulation and control of contents), while LMS is responsible for implementation (e.g., progress management, distribution of contents, mentoring and tutoring).

Table 10.1 Functionalities of LMS and CMS platforms (adapted from Kerschenbaum, 2013).

LMS	CMS
• Registration and administration	• Collaborative content development
• Event management (e.g., WBT delivery)	• Enhanced content management (e.g., indexing, reuse)
• Curriculum and certification management	
• Skill management and gap analysis	• Workflow integration
• Customized/individual plans	• Automated interface
• Record management	• Template-driven virtual spaces for users interactions
• Reporting and courseware authoring	

10.2 Learning via Computer Experiences and Games

Computer games are becoming promising solutions for next generation learning and education systems. Immersive learning simulations, known as *serious games*, are widely accepted methods, which emphasize education as the primary goal rather than entertainment. Serious games are growing rapidly and more organizations are adopting or expected to adopt them in their learning/training programs. Serious games comprise several characteristics that enhance learning and education in various areas such as government and NGO, defense, healthcare, education, and training (Figure 10.3). Various types of serious games have been developed so far. Examples include game-based education (e.g., learning and edutainment), production (e.g., machine behaviors), simulation (e.g., simulation games), messaging (e.g., marketing and advertisement), training and interfaces, health support (e.g., inform about diseases and risks), and advergames (e.g., social issues). Video games provide several unique characteristics addressing deficiencies of traditional learning and training systems (Mayo, 2007):

1) Unlimited access by massive groups of participants.
2) Experiential learning, i.e., navigation in the game scenarios, making decisions, and experiencing the consequences.
3) Inquiry-based learning, i.e., freeform exploration, discovery, and experimentation.
4) Self-efficacy, i.e., free trial-and-error process including bonuses and awards that encourage the player to try further/harder.
5) Goal-orientation.
6) Cooperation, especially in massive multi-player online games, where players teamwork to achieve a common goal.
7) Continuous feedback and tailored instructions.

Fig. 10.3 Serious game examples; (a) "America's Army": a powerful recruitment and training game for the U.S. Army (www.americasarmy.com); (b) Use of serious games in McDonald's restaurants for training the store personnel (www.3dsolve.com). (c) "Tragedy of the Tuna": a game that helps students, as tuna fishing fleets, to balance decisions and collaborate with their group with commonly shared resources (Courtesy of Learning Lab at The Wharton School of the University of Pennsylvania).

Serious games enhance learning and training through creating virtual worlds that enable development of situated experiences, social practices, and shared values, as well as higher levels of thinking in face of unanticipated situations, in computer-supported and collaborative environments. Such learning environments provide the players with a variety of authentic learning experiences that mimic those of real-world environments. The experiences and lessons learned from serious games improve future decisions, in case they are structured by specific goals. Moreover, immediate feedback mechanisms provide the players with a *trial-and error* feature and help them to recognize and evaluate their errors. Serious games also enhance collaborative decision-making, as the players need to learn from interpreted experiences and explanations of other people, both peers and experts (Gee, 2008). Two illustrative examples of serious games are presented in the following subsections.

10.2.1 Management Enterprise Resource Planning (MERP)

MERP is a powerful decision-making simulator that provides a *helicopter view* of a manufacturing enterprise and its environment, and a close to reality hands-on experience of complex managerial scenarios. This serious game simulates the day-to-day business operations of a fluctuating competitive environment and the enterprise information systems, along with the cross-enterprise effects of each decision. MERP is based on the concept of Enterprise Resource Planning (ERP) that allows users distributed users to work in a synchronized mode and create a competitive environment simulating real-life competitions using Web-based functions. The decisions range from multiple departments within an enterprise and the emphasis is on collaboration between the departments in decision-making. Examples include manufacturing guidelines such as EDD (Earliest Due Date) policy, marketing decisions such as advertisement, executive actions such as lay-offs or buying new equipment, and financial decisions such as setting the prices of the produced goods. MERP provides various business scenarios from simplest to complicated in which users have the opportunity to manage a manufacturing enterprise with five distinct departments (see Figure 10.4):

1) *Marketing*: links the manufacturing enterprise to the customers. The market of a MERP scenario is noted for the critical importance of maintaining an adequate due-date performance and clearly favors faster response time and reduced prices. Marketing policies are based on the commitment to ship every order within a Quoted Lead-Time (QLT). Demand is increased when the QLT and/or the selling price are reduced. Uncertainty is incorporated in the market demand, i.e., the number of orders received is random. The global logistics policy is a strict *make-to-order* policy, i.e., only units ordered by customers are produced. There is no safety net of products' stock. Safety stocks are kept to consider the amount of unassigned quantity to be produced and maintained. A redline time is a control mechanism to alert *stockouts* to the user (Figure 10.4(a)).
2) *Operation*: is a detailed shop floor that transforms raw materials into finished goods. The entire production planning and control is provided by an internal information system, which produces work orders based on the actual orders received and the planning parameters as set by the users. Operations policy is based on make-to-order policy without safety stock. Uncertainty is incorporated in the timing of the production machine operations. Work orders are generated weekly based on incoming orders. All new customers' orders for a product are combined into one work order. No extra units are produced.

Machine operators load their work orders sequentially, according to the time that the orders reach their workstation, i.e., FIFO (First In, First Out) discipline (Figure 10.4(b)).

3) *Purchasing*: controls raw material inventory, replenishment policies, and supplier selection. Usually there is more than a single vendor, and the faster vendors are typically more expensive. Purchasing policy is purchase-to-stock and managed according to maximum-order levels. Uncertainty is incorporated in the response times by the suppliers. The main inventory policy is (s, S), i.e., re-order point, order level (Figure 10.4(c)).
4) *Finance*: tracks financial status of the manufacturing enterprise without incorporating any policies or actions. Customers pay upon receipt of goods, and suppliers are paid on delivery. Interest is earned on the cash held at the bank. A monthly profit and loss statement informs the user how he/she is doing financially (Figure 10.4(d)).
5) *Management*: provides a real-time graphical summary of the simulation, in terms of several KPIs such as profit, cash, and reputation (as the three main KPIs), along with other KPIs such as sales and backlogs, on a daily/monthly basis. The significance of reputation (percentage of on-time orders) is that customers are unhappy if they receive their orders later than the promised date. When orders are late or missed, the company's reputation declines, and thus, future market demand declines (Figure 10.4(e)).

In MERP, managerial success is measured both by numerical results (i.e., main KPIs) and by the actions that the users take. The profit earned (the more, the better) is the primary measure of success, while the others serve as qualifiers (at the end of the simulation, users must have positive cash on hand and minimum 75% reputation). The quality of the decision-making is also evaluated to avoid successes based on random acts not supported by an analysis. MERP experience improves performance of the users in decision-making through exercising business skills, gaining business intelligence and insights, dealing with and handling uncertainty, practicing how to think in the world of systems and systems of systems, combining theory and practice, and cause-and-effect thinking.

As a serious game, MERP is also useful for e-Learning and e-Training of e-collaboration among students and trainees of e-Business and e-Service (Fig. 10.4(f)). As such, it has also been applied for research on and experimentation with different collaboration augmentation levels and techniques; cultural diversity in e-Collaboration; and the impact of best practice shared knowledge on collaboration (e.g., Matsui et al., 2003; Fujii et al., 2005, Yoon et al., 2011).

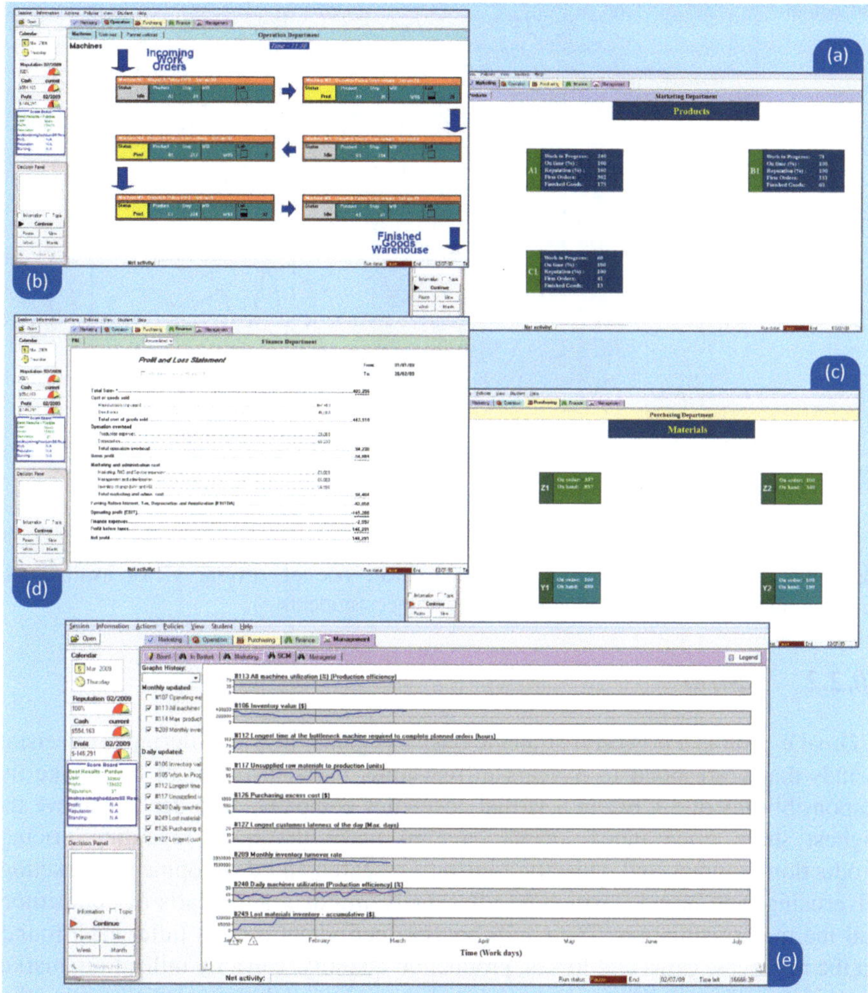

Fig. 10.4 MERP departments: (a) Marketing, (b) Operation, (c) Purchasing, (d) Finance, and (e) Management (Courtesy of MBE-Simulations Ltd.).

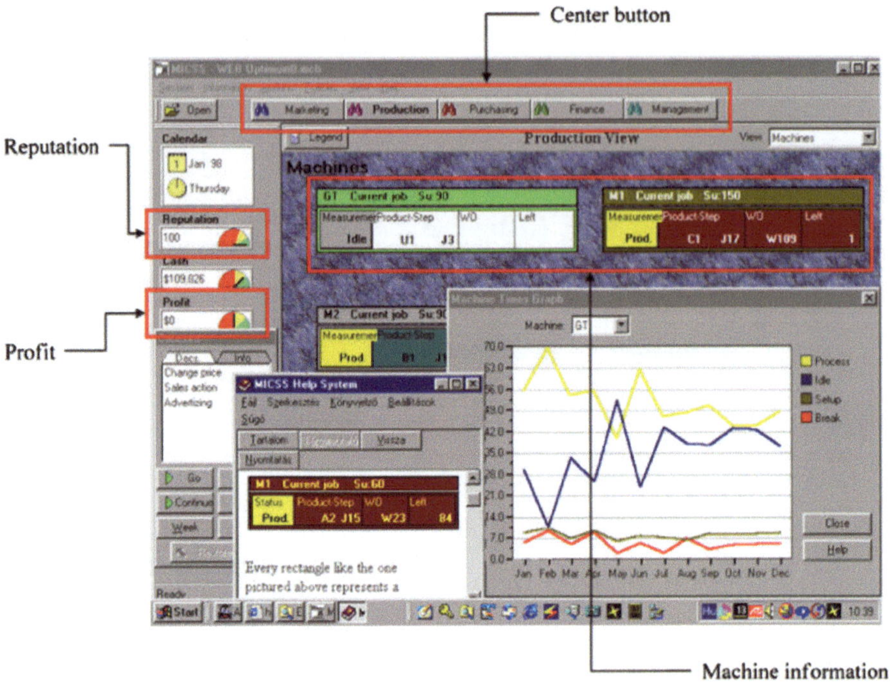

Fig. 10.4 (f) Production Center view in MERP, one of several views available to collaborating or non-collaborating managers, trainees/students.

10.2.2 e-Strat

L'Oreal e-Strat is a powerful strategic business game, in which during six decision rounds that correspond to 3 years in real-time, users in teams of three take the responsibilities of a CEO in a virtual cosmetics company. In order to achieve the highest share price index, teams have to make decisions regarding pricing, production volumes and capacity, logistics, research and development, marketing, advertising, and brand positioning after analyzing data and charts of competitors and the market situation. The CEOs analyze periodical reports (after each round) on the revenues, gains and losses, production quantity, capacity utilization, market research studies, competitors, brand perceptions, and benchmarking with other competitors. Accordingly, decisions are made in the following areas of the company's value chain: research and development, production planning, brand management, distribution, and corporate social responsibility. The decisions are limited by a budget, determined by the performance of the last round.

In e-Strat, reputation, demand, and profit are sensitive to the research and development decisions. In the production-planning department, there are policies for pricing, advertising, distribution, research and development that can be made for changing the market behavior (i.e., demand). For example, high prices for

SKUs (Stock Keeping Units) and/or low prices for competing SKUs may lead to a lower demand rates. In order to satisfy the demand of all customers and in order not to produce products to stock, the management has to forecast market demand and incorporate the forecasts in the production planning. Low levels of production capacity may lead to stockouts, which in turn has negative impacts on the revenues of the enterprise and Shared Price Index (SPI). In the e-Strat environment, there are four distribution channels. CEOs have to allocate distribution and trade marketing budget to each retail channel depending on the shopping habits of the target group. Among other considerations, the e-Strat structure includes corporate social responsibility. Management can decide if the company is willing to invest in activities such as water reduction, career tracking, and kindergartens for employees. Such investments improve the enterprise's reputation and the SPI. e-Strat is a complex business simulation that covers all areas of a company's value chain.

Serious games provide close to reality replications of a reduced size business that needs to be managed by the users. A unique characteristic of such environments is the *learning business by doing* concept that enable to experience real-world scenarios in a virtual environment. MERP and e-Strat have been introduced in this section, as two well-known solutions that represent different approaches to serious games in business management. Table 10.2 presents a comparison between these two types of serious games.

Table 10.2 Comparison between MERP and e-Strat (adapted from Vastag and Yerushalmy, 2009).

Criteria	MERP	e-Strat
Decision type	Realistic, daily-based events (operational)	Strategic (investments, a new market, new product release, etc)
Decision making	Dynamic; on time and type of Decisions considering KPI status	Mandatory periodical (i.e., quarterly) set of decisions
Competitive nature	Based on actual benchmark results in Turbulent/uncertain daily business life	Direct competition based on profit or SPI, dynamic competition cycles
Learning effectiveness	Reinforced learning, i.e., repeated runs of the same enterprise	The game runs once, cannot be repeated
Learning scopes	Strategy execution abilities, business and managerial insights, cause-and-effect thinking under uncertainty	Strategic learning

10.3 Collaborative e-Learning Design and Applications

This section presents three examples of design and application of collaborative e-Learning systems. The first case introduces *grid-computing techniques* to support collaborative learning in a high school using an e-Lab (Bardeen et al., 2006). The second case provides an e-Learning environment based on *virtual reality* to support stimulations, interactions, and collaborations in learning (Monahan et al., 2008). The last case introduces an example of the development of e-Learning systems in Japanese manufacturing industry (Ishii and Tamaki, 2009).

10.3.1 Grid Collaborative Learning e-Lab

The QuarkNet/Grid e-Learning system supports interactive and collaborative learning in high schools, where students explore, gather, and upload data through an e-Lab. The QuarkNet system deploys a network of classroom cosmic ray detectors (Figure 10.5) that capture the time and location of the arrival of cosmic rays and save the data to local computers. The detectors are set up to function as an array that covers an area including several schools. Upon uploading data to a central server, students can collaborate through investigating data from one or a combination of schools. The techniques used for data exploration in the e-Lab are virtual data transformations, workflows, metadata cataloging/indexing, data product provenance and persistence, and job planners for local and grid execution. In order to acquire knowledge and improve problem-solving skills, students must participate in a network of users, which requires access to computer cycles, storage capacity, shared data, community-defined analysis tools, and communication tools for collaboration.

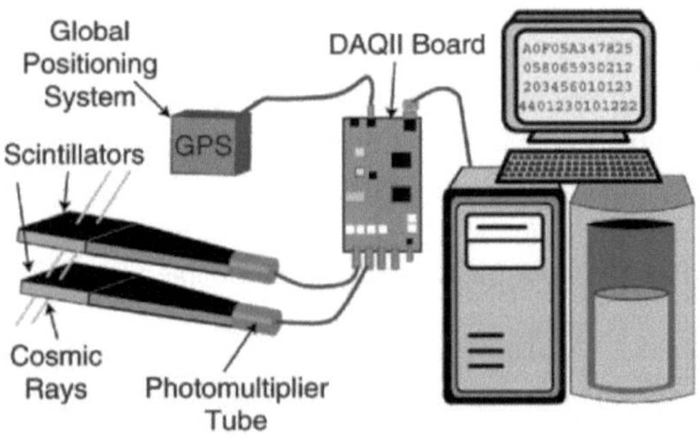

Fig. 10.5 The school-based cosmic ray detector architecture.

The Grid e-Lab provides a collaborative e-Learning environment where students are able to investigate in real-world problems through online projects, share knowledge with synchronous and asynchronous users, and publish their work. In the case of classroom cosmic ray detectors, for instance, students from different schools upload their experimentation data to the server, and share with the students from other schools (who may/may not have their own cosmic ray detectors) through the e-Lab. The e-Lab enables students to discuss about their own questions in forums, manipulate the parameters on built-in analyses, and get support and insights from expert physicists (Figure 10.6). This system is supported by websites for both students and teachers, with the capabilities of accessing to the teams' e-Logbooks, tutorials, animations, relevant URLs, experts, and contact information for all participating teams, as well as searchable database, data validation, accessible data files, reproducible computer workflow, comment sections on data files, and graphs and posters. The e-Lab also provides teachers with online student performance tracking and assessment options using e-Logbooks.

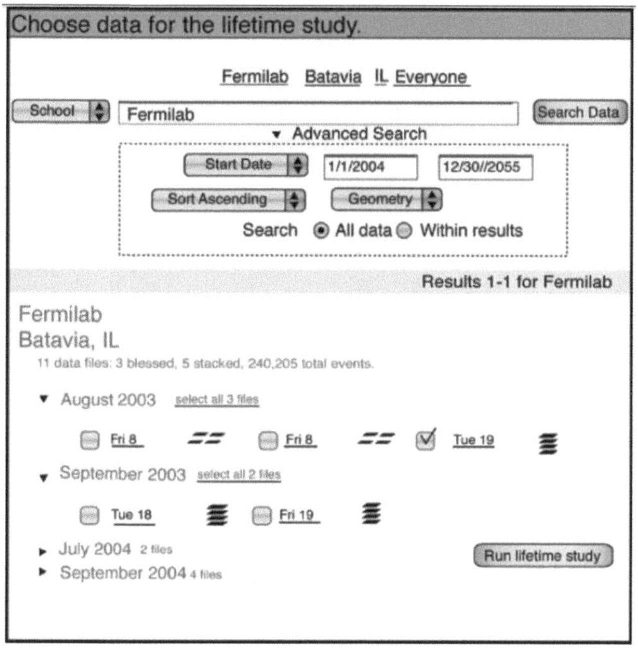

Fig. 10.6 Student interface for searching and selecting uploaded data files.

The e-Lab architecture is shown in Figure 10.7. The system is based on the GriPhyN virtual data system and the Chiron portal. The features and implementation characteristics are summarized below:

1) *System architecture*: consists of Apache Tomcat web server with JavaServer pages, GriPhyN virtual data system API and libraries, backend databases for the Virtual Data System, customized application software, GraphViz for graph visualization, and workflow engines for local execution. The jobs can be submitted to either the local or the grid planner. The site enables security (supporting SSL), interactivity (through DHTML and Javascript), and lessons tracking.
2) *Virtual Data System*: supports data-intensive applications to share data, data products, and the required resources, as well as deriving data products with Virtual Data Languages (VDL), storage of metadata in Virtual Data Catalogs (VDC), and query functionalities that facilitate the discovery and reuse of resources.

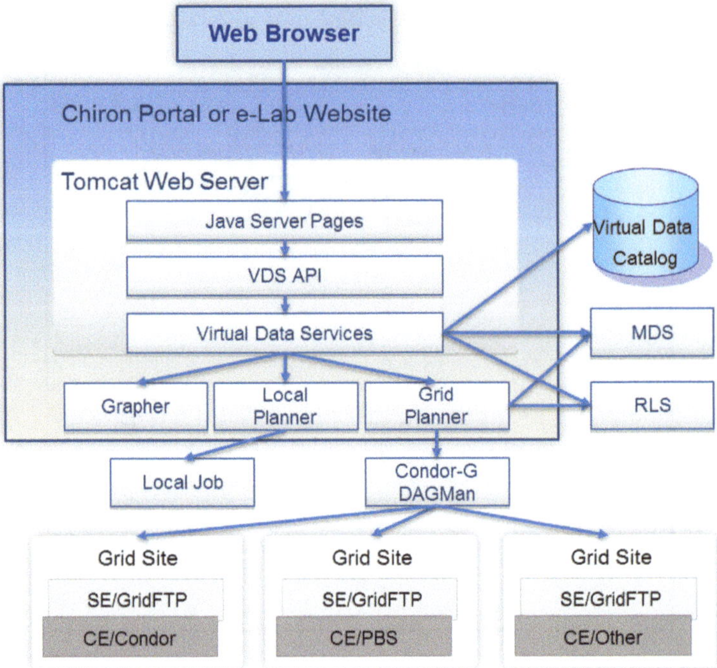

Fig. 10.7 The e-Lab architecture.

3) *Chiron portal*: provides interfaces to the services provided by the GriPhyN virtual data toolkit to enhance defining and executing workflows, publishing and sharing data products, entering metadata (i.e., information about the data and definitions) into the VDC, and provenance (i.e., provision of audit trail for the derivation of a data product). The workflows are executed either on the local server or in the grid.

10.3.2 CLEV-R: Collaborative Learning Environment with Virtual Reality

The CLEV-R system provides a web-based environment based on virtual reality, 3D graphics, and communication tools to support collaborative e-Learning. CLEV-R enables students to interact, learn from each other, take part in lectures, and have online group meetings. The CLEV-R system is composed of the following e-Learning concepts:

1) *Course Management Systems*: provide web-based environments that enable course management tasks such as uploading course material (e.g., texts, images, URLs) by the teachers, interactions between groups of users and teachers (e.g., email, online forums), and customized synchronous/ asynchronous learning. Widespread examples of CMS are Blackboard (www.blackboard.com), Web Course Tools (www.webct.com), and Moodle (www.moodle.org). CMSs support *constructionist learning*, i.e., student-oriented learning systems that enable learning from experiences (e.g., serious games).
2) *Collaborative Learning Environments (CLE)*: enable web-based collaborative learning. Original studies in this area are
   - Virtual European Schools project (Bouras et al., 1999): enhances CLE in secondary schools using computers, 3D graphics, and animations in an interactive and collaborative environment (e.g., information sharing and navigation, text-chats).
   - Intelligent Distributed Virtual Training Environment (INVITE) project (Bouras et al., 2001): provides a collaborative 3D environment for industrial e-Training, enhanced with voice-chat communications and whiteboard applications for interactions and information/knowledge sharing.
   - C-VISions project (Chee and Hooi, 2002): is designed for science education (e.g., Physics), and helps students carry out experiences on mass, velocity, acceleration, conservation of momentum, friction and coefficient of restitution in a virtual, multi-user, and collaborative environment.
   - Educational Virtual Environments (EVE) project (Bouras et al., 2003): besides INVITE functionalities, creates a virtual 3D character or *persona* for each student, who recognizes the other personas during the virtual experience. EVE enables students to access course and user information, upload and download files, view and reply to personal messages, and manage their profile in virtual classrooms.
3) *Mobile Learning*: varies from simple mobile functions such as SMS (e.g., alerting students about deadlines, providing administration information) to MMS and m-Learning (e.g., interactive games, streaming audio and video).

Following the e-Learning concepts and initiatives discussed above, the CLEV-R system offers more interaction for students, tutors and their peers, through text- and voice-chats and video streaming and broadcasting in a virtual 3D environment. The system comprises a virtual lecture room (Figure 10.8(a)) along with several meeting rooms (Figure 10.8(b)) that model a real university, and enhances social interactions, rapports, and friendships among students. The mobile CLEV-R system also offers an interactive 3D learning environment based on ICTs, which enables collaboration with other system users, logged in on desktop PCs or mobile devices (Figure 10.8(c)). The mobile CLEV-R environment supports asynchronous and *anytime-anywhere* access to the e-Learning system.

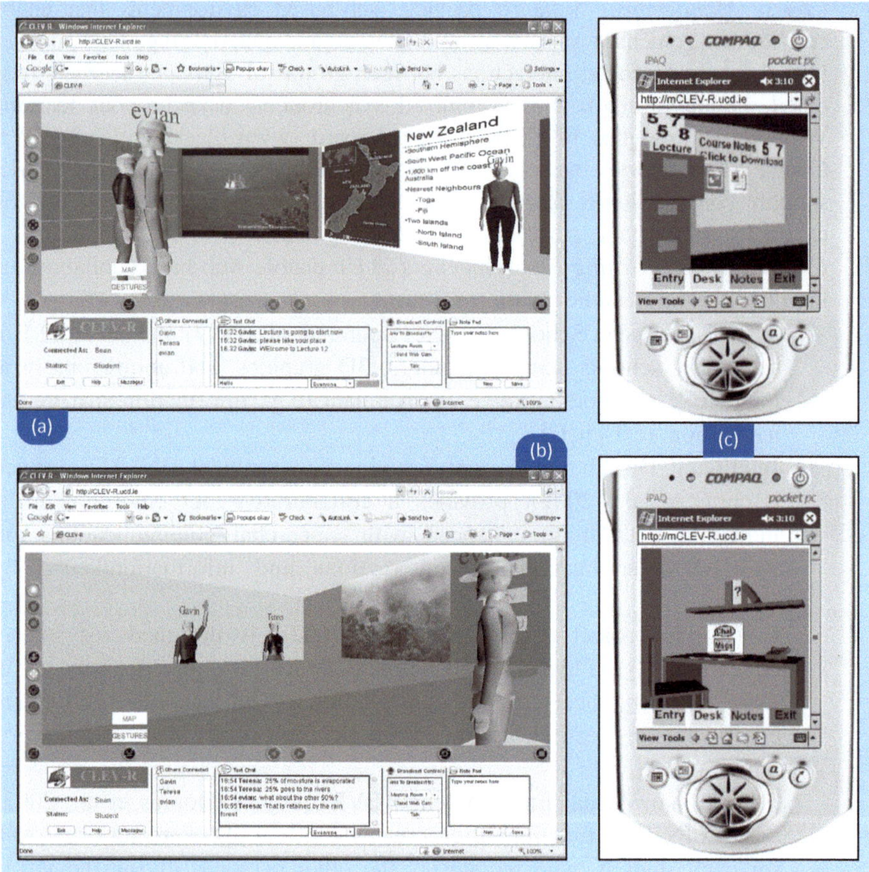

Fig. 10.8 CLEV-R environment; (a) A live online lecture, (b) Students' group meeting, (c) Mobile CLEV-R 3D environment (after Monahan et al., 2008; with permission from Elsevier B.V.).

10.3.3 Blended Learning in Cyber Manufacturing

The program is initiated at The Research Center for e-Learning Professional Competency in the Research Institute of Aoyama Gakuin University in Japan, as a joint program between higher education and industry. The cyber manufacturing courses include four course units on engineering processes and four course units on operational management processes (Figure 10.9). The e-Learning system for each course is characterized by a *blended-learning style* between lectures and practices supported by a Learning Management System (LMS) and simulation software. The instructional strategies used are

- *Case Method Learning (CML)*: is a learning method based on a case study in a virtual experience model of the real world, assisted with 3D computer graphic simulation (Figure 10.10).
- *Problem-Based Learning (PBL)*: helps students to acquire practical skills through practice using numerical simulation software in a university-industry cooperative environment.
- *Computer Supported Collaborative Learning (CSCL)*: provides a collaborative e-Learning environment as discussed in the previous section.

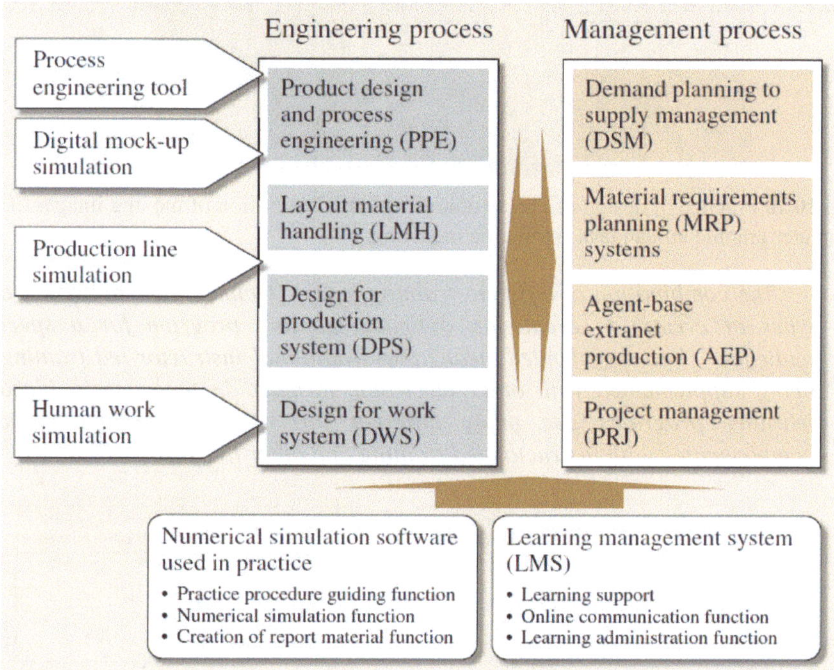

Fig. 10.9 Learning map for each course unit composed of cyber manufacturing core curricula.

An instructional strategy of *comprehensive learning style* is applied in each course unit through combining CML, PBL, and CSCL, in order to conduct the practice corresponding to each lecture (Figure 10.11). The aim of each lecture is to provide the theoretical knowledge about the engineering and management processes. The purpose of each lecture practice is to improve practical skills through 3D computer graphics and simulation, numerical simulation software, and an asynchronous Bulletin-Board System (BBS), and communication functions. This system results in a *practical e-Learning environment* through blended learning. The goal of blended learning is to synthesize learning media into an integrated mix, which can be tailored to create a high-impact, efficient, and exciting learning program. Blended learning is defined as (Bersin, 2004)

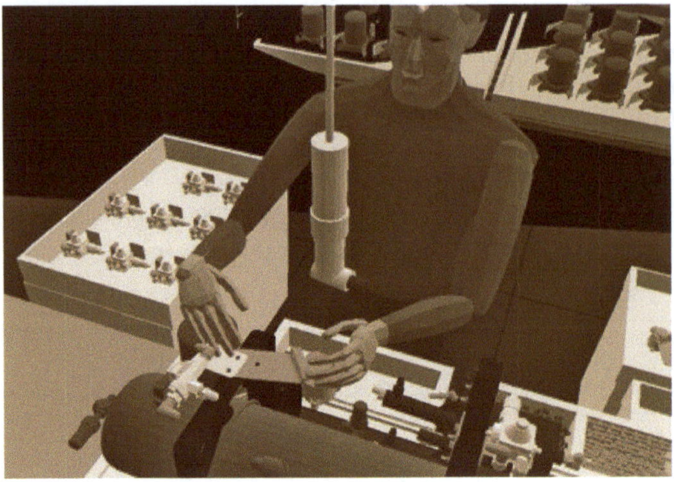

Fig. 10.10 Practice of design of course unit using task observation of moving images of 3D computer graphic human task simulation modeling.

> *"the combination of different training media (technologies, activities, and types of events) to create an optimum training program for a specific audience. The term blended means that traditional instructor-led training is being supplemented with other electronic formats. In the context, blended learning programs use many different forms of e-Learning, perhaps complemented with instructor-led training and other live formats".*

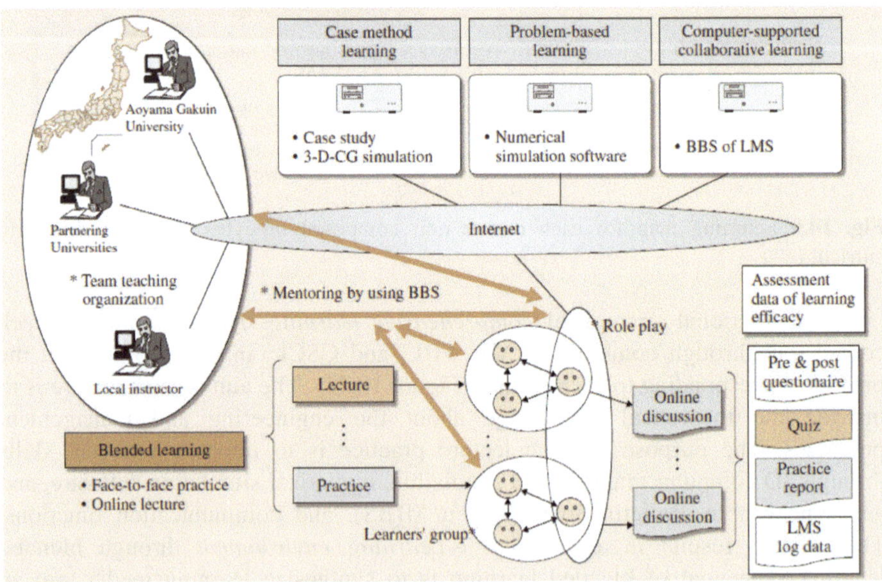

Fig. 10.11 The comprehensive blended learning architecture combined with PBL, CML, and CSCL.

10.4 Collaborative e-Training Design and Applications

This section presents three examples of design and application of collaborative e-Training systems. The first case discusses computer-based collaborative e-Training for enhancing security awareness, alertness, and emergency responsiveness in the State Departments of Transportation (DOT) (Velasquez et al., 2010). The second case introduces an e-Training program for manufacturing managers based on *kaizen* (continuous improvement) knowhow in Japanese manufacturing industry (Ishii and Tamaki, 2009). The third case presents a knowledge-based architecture for collaborative problem solving in military e-Training (Liao and Ho, 2008).

10.4.1 Collaborative e-Training in State DOTs

The terrorist attacks over the past decade and threats in the United States and around the world have forced government agencies to reassess their preparedness in protecting citizens, securing critical infrastructure, and effectively responding to and recovering from emergencies. The State DOTs are therefore in charge of securing their transportation network to ensure the safety of the traveling public and critical infrastructure, and improving their awareness, alertness and responsiveness accordingly. The employees of the State DOTs are then challenged to recognize, assess, and respond to potential or ongoing security threats. Training is one of the most fundamental and effective solutions through which transportation agencies and their employees can be better prepared to identify threats and deal with emergencies. Such training programs, however, face several challenges in practice, due to limitations in time and financial resources required for training. For example, State DOTs often lack the resource and budgetary flexibility to commit their staff's time and travel to training activities. In addition, the percentage of transportation agency workers eligible for either retirement or early retirement is higher than ever. This *workforce-aging phenomenon* is not only a State DOTs problem, but also a rising dilemma of several industries, especially western industries, which is anticipated to intensify over the next decade, to one third of workforce being over 50 years of age by 2040 (see Figure 10.12). It is therefore of the highest importance to address these gaps and formulate a stronger and more cohesive training policy and direction in the State DOTs (and other industries as well).

A Computer-Based Collaborative Training (CBCT) prototype is designed for the Indiana Department of Transportation (INDOT) for better preparation of their staff on emergency response in a safe, low-cost environment (Velasquez et al., 2010). The CBCT prototype functions are identified using field observations and interviews with members from both INDOT Operation and Research Department and the Indiana Department of Homeland Security. The CBCT prototype is based on client-server architecture (Figure 10.13). The CBCT server functions as a training controller connecting several distributed clients, which are in turn controlled by a specific set of staff during the security exercises. The CBCT prototype provides three main functions: (1) as an e-Training database and information warehouse for decision analysis, (2) as a collaborative decision support tool during the security exercise, and (3) as a collaborative environment by which to simulate the necessary intra and interagency decision making process during emergencies. The objectives are to effectively and efficiently.

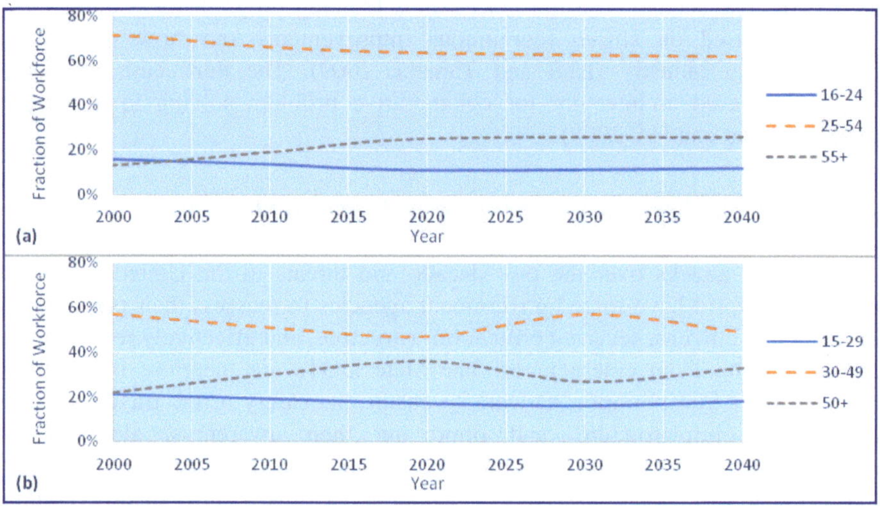

Fig. 10.12 Future structure of workforce in the US (Source of data: Toossi, 2012) and Germany (Source of data: Thun et al., 2007). After a peak of old age (>50) workers in 2020 there is a lack of balance between: Practical knowledge, experience, work ethic, discipline, quality awareness, reliability, loyalty, punctuality, and social competence associated with older workforce, and: Flexibility, willingness to learn, aptitude, reactivity, psychological and physical problems associated with younger workforce.

1) Communicate and collaborate with representatives from other agencies.
2) Identify, secure and procure all needed equipment from all districts and agencies to respond to the emergency.
3) Reference Standard Operating Procedures to guarantee appropriate response efforts and chain of command.
4) Inform the public of any on-going emergencies via dynamic messages signs to aid the routing of traffic and support of first responders.

The CBCT server provides participants with a global view of the emergency scenario and the responses being carried out by other training participants. It receives all the decisions and actions made by each team and combines the emergency events, decisions, and actions together, and transmits them to the CBCT clients after streamlining the information. This global view provides the system with some experiential learning since each team can observe the performance of other teams and learn from their actions and decisions (Velasquez et al., 2010). On the CBCT client side, the participants have multiple-function windows for decision-making and access to decision support system tools (Figure 10.14). The decision-making window enables participants to access historical emergency information, previous actions/decisions by other teams, including their own, for their references, and decision/action execution (Figure 10.14(a)). During the decision-making process, each team can refer to the various types of decision support systems (DSS) embedded in the CBCT client. The DSS is divided into several critical areas for emergency response, such as resources, communication, maps, intelligent transportation, and an archive of emergency plans and procedures:

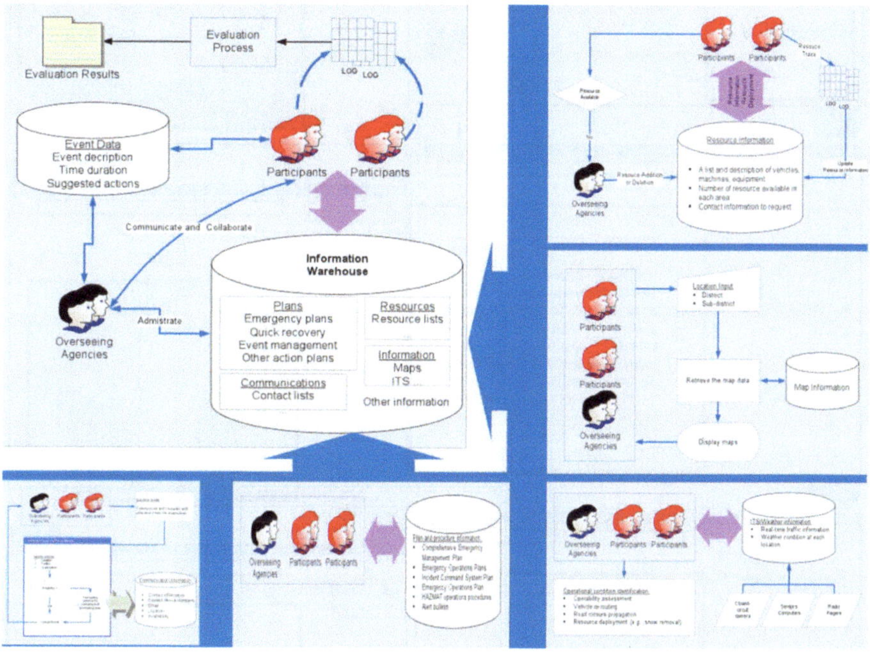

Fig. 10.13 The transportation security collaborative e-Training architecture.

1) *Resource information*: allows participating teams to access the agency resource information from the CBCT database, categorized by location, equipment functionality, and quantity at a sub-district level. It expedites processes in securing and procuring the resources in emergencies and enables the teams to simulate emergency resource deployment, modify the number and type of resources requested, assigned, released, or being transferred as well as deploy the number of workers needed to operate the resource deployed (Figure 10.14(b)).
2) *Communication information*: enables participating teams to access lines of communication, including roles, contact information, and availability, and search multiple levels of the communication channels to identify additional lines of communication in case of unavailability of critical emergency responders (Figure 10.14(c)).

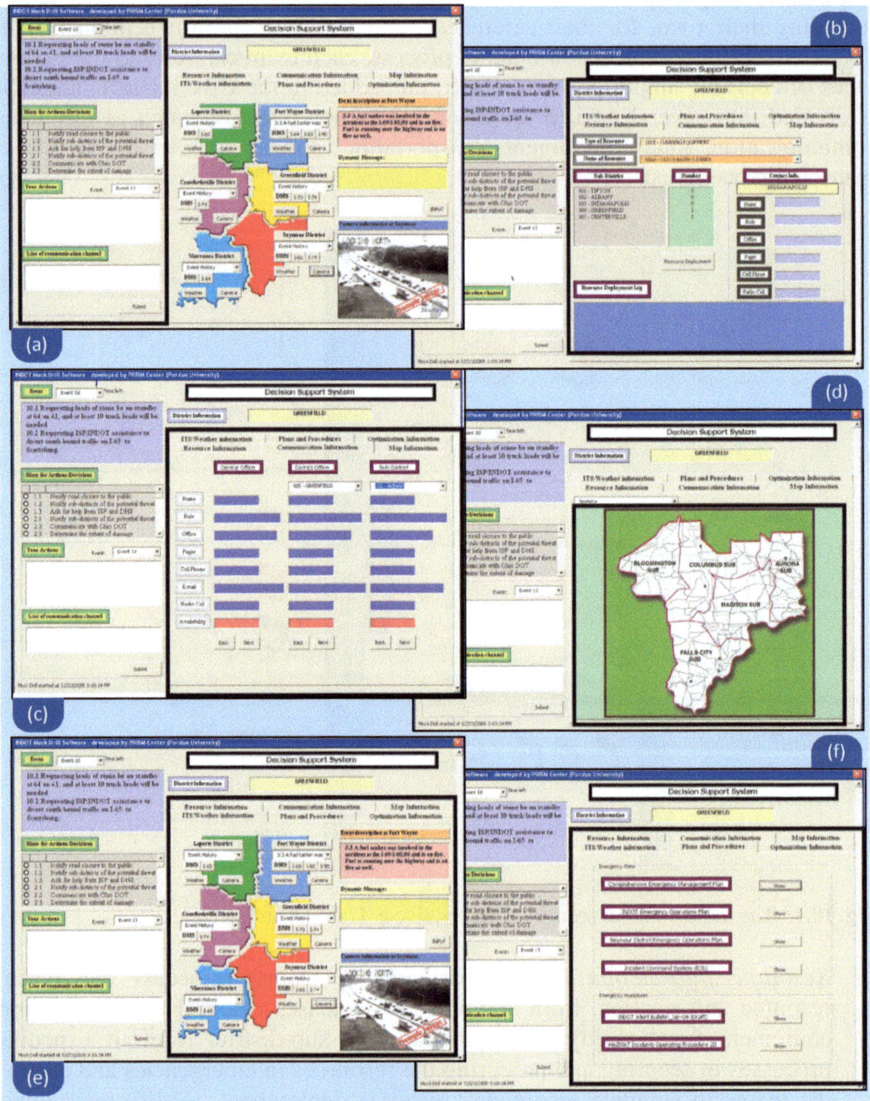

Fig. 10.14 CBCT interfaces (a) Decision making; (b) Resource allocation DSS; (c) Collaborative communication DSS; (d) Map DSS; (e) ITS and collaborative DSS; (d) Collaborative plans and procedures DSS.

3) *Map information*: allows participating teams to access different types of road maps in different levels of detail. It is useful in rerouting the traffic and reallocating the resources. The maps also help each team identify emergency locations to determine accessibility and alternative detours, identify areas that might be affected by the events, and notify local governmental agencies that need to be informed (Figure 10.14(d)).
4) *Intelligent Transportation System (ITS) information*: allows participating teams to identify the operability of emergency responses, such as accessibility, traffic control, and public communication (Figure 10.14(e)). Examples of ITS information are weather information, camera feed of critical infrastructure, and dynamic message signs.
5) *Archive of emergency plans and procedures*: enables a team access to regularly available and used emergency plans and procedures, e.g., comprehensive emergency management plans, emergency operations plans, incident command system plans, district specific emergency operations plans, HAZMAT operations procedures and alert bulletins (Figure 10.14(f)).

10.4.2 e-Training for Manufacturing Managers

This e-Training system aims at providing managers with the knowledge and skills required for managing manufacturing processes in Small-to-Medium-sized Enterprises (SME). The course contents include: 1) Make-to-order manufacturing and flexible manufacturing systems controlling cost, quality, and lead-times, 2) Process optimization (knowledge chain management), and 3) Continuous process improvement (manufacturing knowhow and creativity). The e-Training instructional design follows a workflow called *circuit model* (Figure 10.15). Practical exercises are designed to enhance the ability of managers to implement *kaizen* knowhow and problem solving that are often complicated processes and involve unconventional approaches. The first two processes are therefore *learning* and *understanding* the unconventional approaches. The third process is *creating*. This learning cycle is repeated several times during the course and when the participants complete the program, they repeat the learning cycle once again to evaluate their understanding.

IT educational tools, also known as *IT jigs*, are used in design of the e-Training system to help managers during production planning process, using different types of bill of materials, master files, layout editors, loading analyzers, and flexible manufacturing system simulators (Figure 10.16). Through powerful computer interfaces, managers exercise creating process masters (type, program, requirements, times, etc.), equipment masters (identification and allocation of equipment, layout design, equipment information analyses, etc.), part masters (process and delivery schedules, process selection, etc.), and master production schedules, in a collaborative environment.

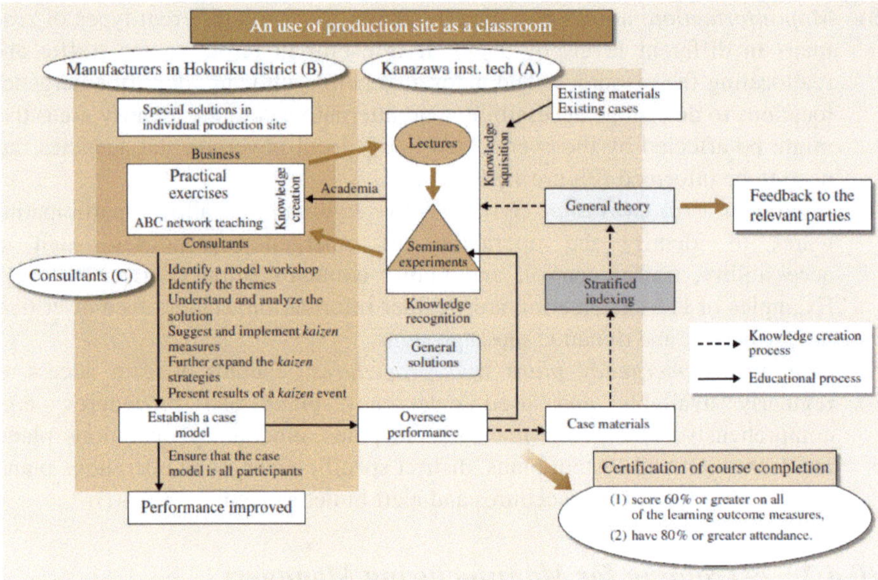

Fig. 10.15 Circuit model for e-Training instructional design.

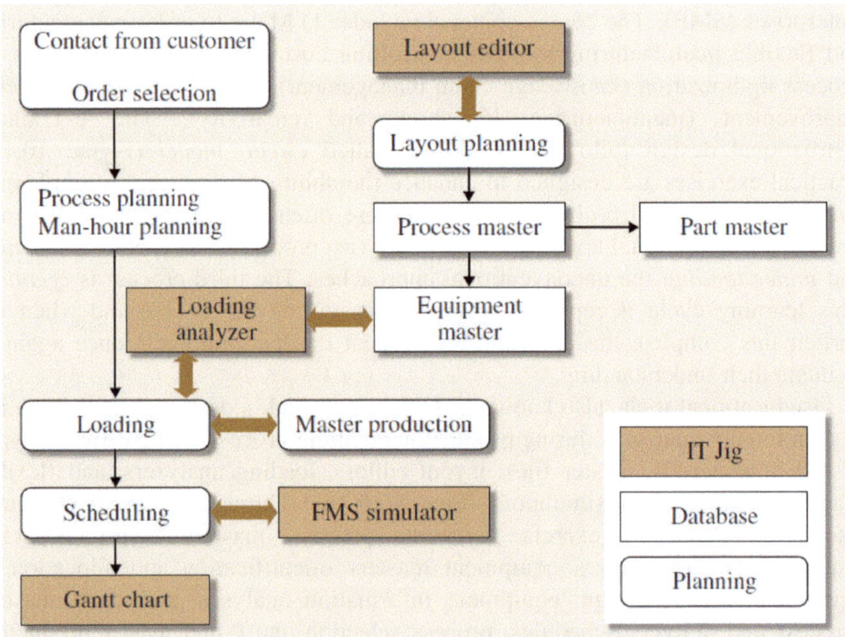

Fig. 10.16 IT educational tool (IT jig) for production planning and control including a series of activities from customer contact to master planning and scheduling (after Liao and Ho, 2008; with permission from Elsevier B.V.).

10.4.3 Collaborative Problem Solving in Military e-Training

Collaborative military e-Training systems provide environments that support collaborative decision-making and problem solving in military training programs, e.g., simulation of battlefields, situation analysis, plan generation, decision-making under uncertainty, and information/knowledge sharing. The case presented here belongs to the Taiwanese Military programs, based on knowledge-based system architecture that enables functions such as military training materials, scenario case base, database, and Geographical Information Systems (GIS). The e-Training program focuses on integrated and collaborative problem solving considering the Command and Control (C^2) processes. C^2 enables officers to analyze situations, make and execute decisions and plans, and accomplish different missions.

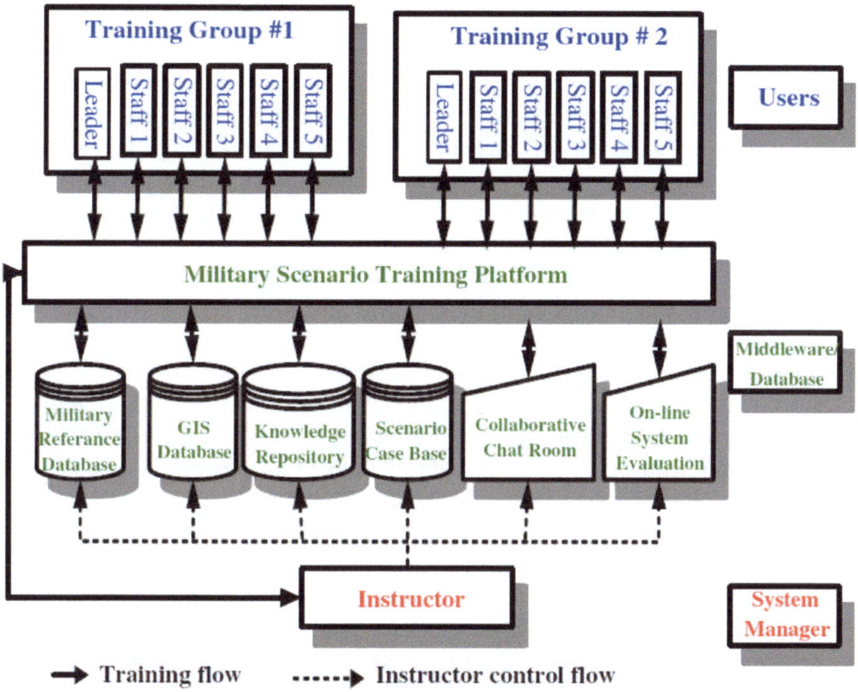

Fig. 10.17 The military e-Training knowledge-based system architecture (after Liao and Ho, 2008; with permission from Elsevier B.V.).

Following the knowledge-based system architecture, the trainees access a network training homepage through the client terminal, middleware, and database to employ the training scenarios. The instructors, on the other hand, use the server terminal to join the network, supervise, and induce the overall training procedure (Figure 10.16). The system architecture is composed of three layers: 1) Users, 2) Middleware/database, and 3) System manager. The Internet and Intranet protocols used are TCP/IP (Transmission Control Protocol/Internet Protocol), FTP (File Transfer Protocol), HTTP (Hypertext Transfer Protocol), SMTP (Simple Mail Transfer Protocol), IRC (Internet Relay Chat), NNTP, (News Network Transfer Protocol). The middleware uses ASP (Active Server Pages) and IMS (Internet Map Server) to connect to the database and respond to the queries. The interface layer uses GIS software, as an object-oriented tool that supports graphic interface operation, query, display, analysis, data or picture design, message production and interaction.

The collaborative military e-Training scenario is composed of the following nine steps (see Figure 10.17):

1) *Team formation*: Student officers are divided into different training groups with respect to the military branches, specialties, experiences and/or backgrounds.
2) *Training preparation*: Student officers study the reference materials and prepare for the scenario exercises before starting the training.
3) *Geographical data study*: Student officers exchange their knowledge and ideas regarding the scenario geographical characteristics to improve the knowledge of the group members. The required materials are usually principles, handbooks, guidelines, and textbooks.
4) *Scenario outline analysis*: Each group carries out training operational issues based on the training scenario outline (e.g., location of the enemy, resources available on each side, etc.).
5) *Special situation and assignment identification*: The scenario contents, strategic/tactical concepts, and operational plans are defined in this step.
6) *Assignment accomplishment and presentation*: Student officers interact and collaborate with each other on special situations through a brainstorming process. Each student officer acts as a commander or takes one of the supporting roles of *analyzing situations*, *deliberating possible plans*, *opening discussions*, or *making decisions*.
7) *Reference solution distribution*: The instructor evaluates the progress of each student officer and his/her suggested reference solution(s).
8) *Instruction and discussion*: The instructor may stop the situation, discuss and brainstorm the reference solutions and possible conflicts in solutions.
9) *Conclusion*: After discussions of the exercises, the instructor updates (if required) the scenario through integration of opinions from different reference solutions. Knowledge creation and development of new scenarios take place in this step.

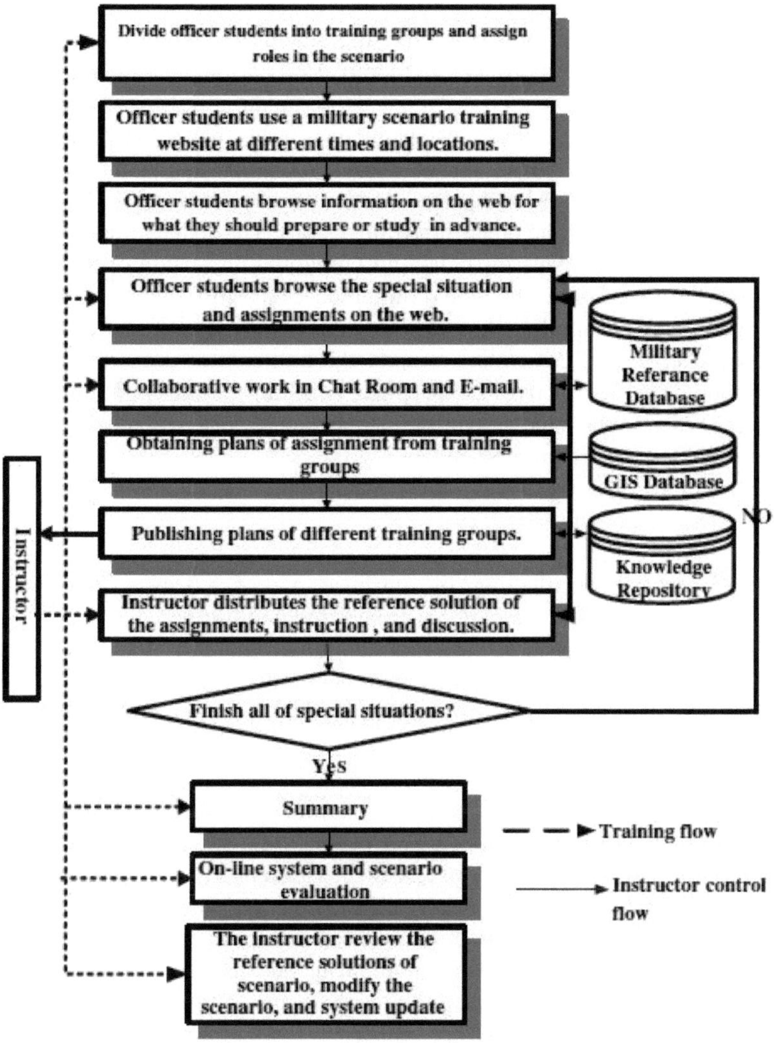

Fig. 10.18 The collaborative military e-Training architecture (with permission from Elsevier B.V. after Liao and Ho, 2008).

10.5 Performance Impacts, Monitoring, and Evaluation

This section deals with issues on realizing, monitoring, and evaluating the performance of e-Learning and e-Training systems. Performance monitoring and evaluation is a critical part of computerized and online education systems, especially for detecting malfunctioning individuals/groups (e.g., not participating), resolving potential conflicts, and improving the processes, communications, interactions, and learning management. In order to monitor and evaluate the performance effectively, however, the first step is to define the corresponding criteria. The computer-supported collaborative e-Learning and e-Training systems discussed in the previous sections impact traditional educational systems from a variety of perspectives. For instance, the performance impacts of online discussions in CSCL systems can be analyzed in four major dimensions (Henri, 1992): 1) *Participative*, 2) *Social*, 3) *Cognitive*, and 2) *Interactive*. The participative dimension encompasses the other three dimensions. The social dimension is defined as the ability of individuals in social and emotional projection of themselves in a learning/training community, and communication with other individuals. The cognitive dimension implies that the individuals must also be able to obtain and create knowledge in a critical community through continual communications and interactions. The interactive (also called teaching) dimension emphasizes on enhancing the social and cognitive dimensions for realizing effective and customized learning/training processes.

The methods for monitoring and evaluation of e-Learning and e-Training performances can be classified as

1) Self-assessment (reflecting on intrinsic and/or extrinsic feedback) often provided by the electronic/online environment in use.
2) Peer-to-peer assessment through collaborating on the solutions, and articulating and negotiating with other peer individuals/ teams.
3) Assessment by teacher, performed on various solutions provided by different individuals/ teams.

Besides solutions, monitoring and evaluation must be performed on *participation* of the individuals (e.g., frequency of messages within a lapse of time), *productivity* (e.g., number of documents generated), *reactivity* (e.g., average delay/latency in replying to a query/fulfilling an assignment), and *impact* (e.g., degree of interactions with consecutive concatenate answers).

The e-Learning and e-Training monitoring and evaluation processes require efficient *web tracking* on the individuals/teams performance. This can be performed through *Social Network Analysis* (SNA) on collective actions (e.g., extent/aggregation of individuals' neighborhood or direct contacts, central or peripheral role in information exchange/ transmission, and participation in subgroups). A social network is typically characterized as follows:

$C_D(i)$	Degree centrality of node i	$C_C(i)$	Closeness centrality of node i
$D(i)$	Density of node i	$C(i)$	Centralization of node i
$C_{DI}(i)$	In-degree centrality of node i	$C_{DO}(i)$	Out-degree centrality of node i
$C_B(i)$	Betweenness centrality of node i	$C_{CO}(i)$	Out-closeness centrality of node i

The SNA technique enhances automated monitoring of e-Learning/e-Training systems, and detecting and alerting malfunctioning individuals in collaborative environments (Marcos-García et al., 2009):

1) *Isolated individuals*: refer to those who do not participate in collaborative activities, characterized by an absolute lack of interactions. In the SNA context, this isolation incident is characterized as

   $$C_D(i) = C_{DI}(i) = C_{DO}(i) = 0.$$

   In the corresponding *sociogram*, the node corresponding to the isolated individual is disconnected from the entire social network. In Figure 10.18, the individual $x18$ is isolated and thus $C_D(18) = C_{DI}(18) = C_{DO}(18) = 0$.

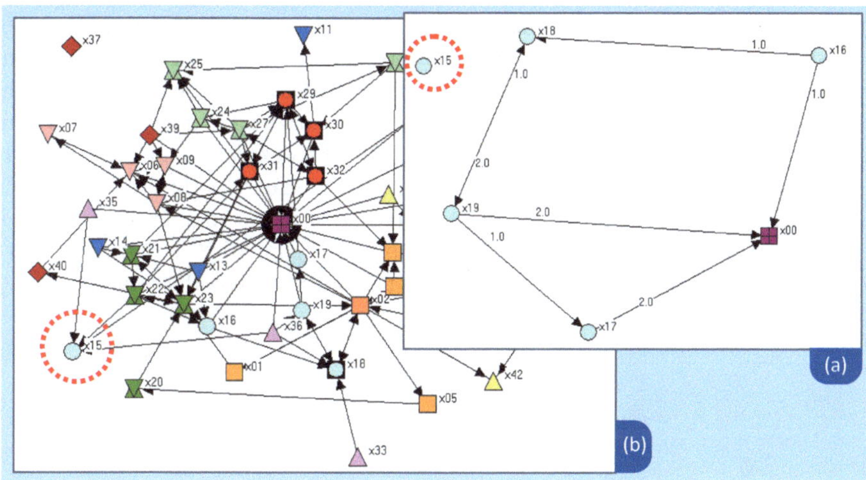

Fig. 10.19 Sociograms representing (a) $x15$ as an isolated individual within her group, and (b) the entire social network where $x15$ is not an isolated individual.

2) *Intra-group conflicts*: refers to intra-group isolated individuals, i.e., those who do not interact with their group members. An individual isolated in her group, however, may not be an isolated individual in the entire social network. For instance, individual $x15$ in Figure 10.19 is isolated from the rest of her group (see Figure 10.19-a), but is not an isolated individual (see Figure 10.19-b), since $C_{DI}(15) = 5$ and $C_{DO}(15) = 3$.

3) *Group malfunctioning*: refers to negative situations caused by the *group coordinator*, who is also a learner/trainee organizing group activity, initiating interactions and discussions, and managing and maintaining the relationships with other groups. A negative impact, for instance, could be lack of interaction and collaboration between the group members (excluding the group coordinator). A group coordinator, in general, must meet some specific requirements, as shown in Table 10.3.

Table 10.3 Indicators and relative requirements for the group coordinator role (adapted from Marcos-García et al., 2009).

Indicator	Threshold	Significance	Relevance
$C_{DO}(i)$	> 75%	High: Represents high control power over the group member activities	First level
$C_{CO}(i)$	> 75%	High: Represents high proximity of the group members to each other	First level
$C_B(i)$	> 75%	High: Represents good connection to the rest of the group members	First level
Position in sociogram	Near the center	Centered: Represents prominence over the rest of individuals	Second level

4) *Instructor dependent individuals*: refers to undesirable collaborative situations where autonomy of an individual is diminished due to high dependency on the instructors' interventions and suggestions. The ideal situation occurs when the instructor's role changes from *guide* to *collaborator* over time. If this situation does not happen for some group, this is regarded as a malfunction. In Figure 10.20(a), for instance, the instructor (the square node) has maintained her *guide* role, as she is the most centered node in the sociogram. Clearly, if the instructor is excluded from this group (see Figure 10.20(b)), it turns into a disconnected network without effective interactions and collaborations.

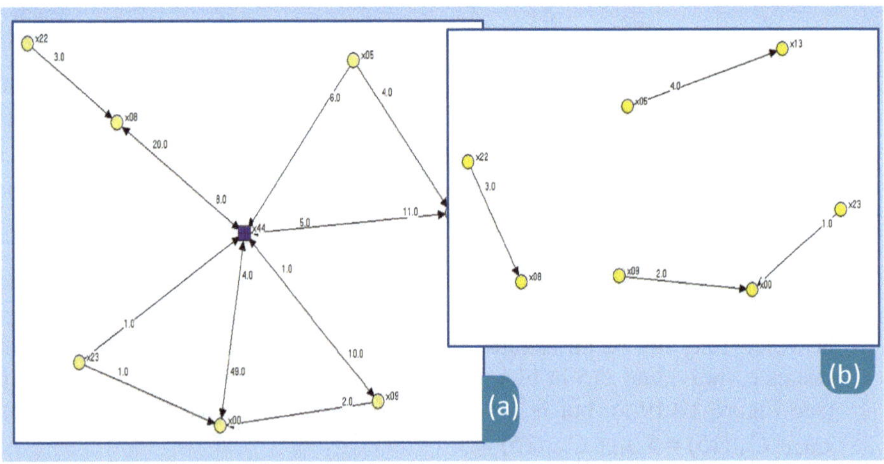

Fig. 10.20 (a) Sociogram of an *instructor-dependent* group; (b) Disjoint sociogram of the same group excluding the instructor.

10.6 Emerging Trends and Challenges

This chapter introduced emerging e-Learning and e-Training systems and presented an overview of the design guidelines and techniques based on ID (Instructional Design), automation, computers and communication technologies. Various examples of serious games for learning and computerized collaborative e-Learning/e-Training techniques were reviewed and their impacts along with execution and evaluation procedures were discussed. Current e-Learning and e-Training systems are still maturing, and there is a variety of open questions and challenges from different viewpoints that must be addressed effectively.

- *Systematic perspective*: Future Scientists, Technologists, Engineers, and Mathematicians (STEM) must be well prepared for the complex technical, social, and ethical challenges raised by emerging disciplines and technologies. Successful e-Learning and e-Training systems are required to prepare STEMs with systematic perspectives and interdisciplinary leadership capabilities to handle complex modeling, analysis, design and control tasks.
- *Student-oriented education*: Learners and trainees must be able to take more responsibilities in designing the educational systems and collaborate with instructors in designing their customized e-Learning and e-Training processes. The system must be developed under actual contracts between student and instructor/educational system (e.g., syllabus). The system must enable each individual to build his/her career (i.e., having options), and interact and collaborate with other peer individuals during the learning/training process.
- *Social value creation*: e-Learning and e-Training systems have the potential to link *personal value creation* to *social value creation* through bridging the gap between educational systems and business/society. For instance, they can be integrated with knowledge management, sales force management, human resources management, enterprise information portals, electronic performance support systems, and community services.
- *Game-orientation*: Advances in virtual reality, groupware, and simulation technologies will make the future managers and employees more game-oriented. Serious games and simulation experiences enhance abilities of students in analyzing/taking risks, making decisions in uncertain environments, and handling challenging situations. Current serious games require effective assessment and monitoring tools for evaluating the performance of individuals, their progress/deficiencies, and achievement of different learning/training objectives.
- *Collaboration*: The systematic perspectives and interdisciplinary nature of next generation educational systems will require further developments in collaborative e-Learning and e-Training to address the holistic realm of a learning/training subject. Collaborative e-Learning and e-Training systems can be further improved following the design principles of CCT

(Collaborative Control Theory) and collaborative e-Work (see Chapter 2). Examples include protocols for 1) designing and planning the distributed collaborative e-Learning/e-Training networks (CRP-I principle), 2) executing and revising the collaborative networks (CRP-II, AD principles), and 3) detecting, resolving, and preventing potential conflicts and errors during e-Learning/e-Training communications, interactions, and collaborations (EPCR principle). The concept of *"teaching is learning"*, i.e., instructor-student partnership and collaboration, should also be further investigated through public disclosure of best practices, continuous instructor training, and development of new methods.

- *Human-machine systems design:* LMSs and CSCL methods must be improved further from a variety of aspects such as securing and maintaining motivation of learners/trainees, detecting deficiencies of processes and interfaces, building *open systems*, and constructing dynamic and automated evaluation systems.
- *Cost-effectiveness*: Issues of *intellectual property* for e-Learning and e-Training systems must be effectively resolved. In addition, social capital development and cost rationalization issues must be clearly addressed through effective collaboration between educational service providers, institutions, universities, governments, and private sectors.

The universal and essential role of education and training can benefit significantly from the collaboration revolution. Affordability of bi-directional and multi-directional information and knowledge flows and exchange offers engagement a sense of active participation that can enrich the experience and quality of learning and training.

References

American Society for Training and Development: Official homepage (2009), http://www.astd.org/astd

Bardeen, M., Gilbert, E., Jordan, T., Nepywoda, P., Quigg, E., Wilde, M., Zhao, Y.: The QuarkNet/Grid Collaborative Learning e-Lab. Future Generation Computer Systems 22, 700–708 (2006)

Bersin, J.: The Blended Learning Book: Best Practice, Proven Methodologies, and Lessons Learned. Pfeiffer, Somerset (2004)

Gee, J.P.: Learning and games. In: Salen, K. (ed.) The Ecology of Games: Connecting Youth, Games, and Learning, Foundation Series on Digital Media and Learning, pp. 21–40. MIT Press, Cambridge (2008)

Fujii, Y., Chen, X., Yang, C.L., Yamada, T., Matsui, M., Nof, S.Y.: Enterprise Decision Support with Evolutionary Organizational Knowledge. In: Proc. of ICPR-18, Salerno, Italy (August 2005)

Henri, F.: Computer conferencing and content analysis. In: Kaye, A.R. (ed.) Collaborative Learning Through Computer Conferencing: The Najaden Papers, pp. 115–136 (1992)

Ishii, K., Tamaki, K.: Automation in Education/Learning Systems. In: Nof, S.Y. (ed.) Springer Handbook of Automation, ch. 85, pp. 1503–1527 (2009)

Ishii, K., Ichimura, T., Kondoh, S., Hiraki, S.: An innovative management system to create new values. International Journal of Technology Management 45(3/4), 291–305 (2009)

Kerschenbaum, S.: LMS Selection Best Practices (White paper). Adayana Chief Technology Officer, 1–15 (2013)
Liao, S.H., Ho, Y.P.: A knowledge-based architecture for implementing collaborative problem-solving methods in military e-training. Expert Systems with Applications 35, 976–990 (2008)
Marcos-García, J.A., Martínez-Monés, A., Dimitriadis, Y., Anguita-Martínez, R., Ruiz-Requies, I., Rubia-Avi, B.: Detecting and Solving Negative Situations in Real CSCL Experiences with a Role-Based Interaction Analysis Approach. In: Daradoumis, T., Caballé, S., Marquès, J.M., Xhafa, F. (eds.) Intelligent Collaborative e-Learning Systems and Applications. SCI, vol. 246, pp. 129–146. Springer, Heidelberg (2009)
Mayo, M.J.: Games for science and engineering education. Communications of the ACM 50(7), 30–35 (2007)
Matsui, M., Aita, S., Nof, S.Y., Chen, J., Nishibori, Y.: Analysis of Cooperation Effects in Two-Center Production Models. Int. Journal of Production Economics 84(1), 101–112 (2003)
Monahan, T., McArdle, G., Bertolotto, M.: Virtual reality for collaborative e-learning. Computers & Education 50, 1339–1353 (2008)
Taylor, J.C., Jahrestagung, G.I.: Automating e-Learning: The Higher Education Revolution. LNI, pp. 64–82 (2002)
Thun, J.H., Größler, A., Miczka, S.: The impact of the demographic transition on manufacturing: Effects of an ageing workforce in German industrial firms. Journal of Manufacturing Technology Management 18, 985–999 (2007)
Toossi, M.: Projections of the labor force to 2050: a visual essay. Monthly Labor Review, 3–16 (2012)
Vastag, G., Yerushalmy, M.: Automating Serious Games. In: Nof, S.Y. (ed.) Springer Handbook of Automation, ch. 73, pp. 1299–1311 (2009)
Velasquez, J.D., Yoon, S.W., Nof, S.Y.: Computer-based collaborative training for transportation security and emergency response. Computers in Industry 61, 380–389 (2010)
Yoon, S.W., Matsui, M., Yamada, T., Nof, S.Y.: Analysis of effectiveness and benefits of collaboration modes with information- and knowledge-sharing. J. Intelligent Mfg. 22(1), 101–112 (2011)

Review and Challenge Questions

1. Explain the evolution of the role of computers and communication in education; from computer-assisted instruction to e-Learning and e-Training.
2. Following your answer to Question 1, identify and describe several advantages and limitations of each stage of this evolution compared with traditional educational methods.
3. Besides interactivity and level of computerization (automation), what other e-Criteria may be involved in the design of e-Learning and e-Training? Explain.
4. Recall the "MIT Beer Game" discussed in Chapter 4. (i) How can this game be played, by either humans or software agents, through MERP? (ii) What additional understandings and insights beyond those with the MIT Beer Game are enabled by MERP?
5. Answer Question 4 for e-Strat.

6. (i) Explain briefly the real-life collaborative e-Learning systems discussed in Section 10.3. (ii) What CCT design principles can be applied to improve each system? Justify your answers.
7. Answer Question 6 for the real-life collaborative e-Training systems discussed in Section 10.4.
8. Compare your answers to Questions 6 and 7 in terms of CCT design principles. Do you notice any significant difference between the collaborative processes involved in e-Learning and e-Training?
9. List and explain the most important criteria for monitoring, evaluating, and adjusting the performance of e-Learning and e-Training systems. Comment on the need to customize and personalize the learning and training experience, and on the need to leverage team experiences.
10. Explain some emerging challenges in e-Learning and e-Training. How does e-Education, in general, influence the respective collaborative activities?
11. Write a brief essay: "Is it a dream, or can we revolutionize learning and training by revolutionizing collaboration?"

Lab Questions

12. Computer experiences and serious games:
    (i) Search the Internet and find several serious games with different e-Learning and e-Training purposes.
    (ii) Explain the unique functionalities and relative advantages of each game compared with the equivalent traditional method of learning/training.
    (iii) Suggest a serious game for an e-Learning/e-Training area for which no serious game has been developed yet. Justify your answer.
13. Consider the application of SNA (Social Network Analysis) for performance monitoring and evaluation:
    (i) Explain the benefits provided by sociograms in monitoring and evaluation of an e-Learning/e-Training system.
    (ii) How can the SNA method be applied for performance monitoring and evaluation of a multi-agent system? Explain your answer with examples.
    (iii) What other methods do you know that can apply for performance monitoring and evaluation of multi-agent systems, in general?
    (iv) How can the methods described in (iii) be applied for monitoring and evaluation of the performance of an e-Learning/e-Training system?
    (v) Explain how the SNA method can be applied for modeling the emergence and evolution of collaborative networks (Hint: Consider AD and ELOCC design principles of CCT).

Chapter 11
Emerging Trends and Research Challenges

Abstract. The methods by which distributed agents, from micro- and nano-sensors to computers, machines, humans, and enterprises, interact and collaborate are being rapidly transformed. The concepts, theoretical foundations, and applications of collaborative e-Work discussed in this book are both enablers and direct results of automation of collaborative processes, activities, and interactions among the distributed entities. The future of production and service systems depends highly on the effective design and engineering of collaborative e-Work, and their integration into their design, i.e., automated and collaborative systems. The principles, models, and techniques of CCT, the Collaborative Control Theory (Nof, 2007), as explained in the previous chapters, provide insights for design, and supports execution and control through optimized e-Activities. The future challenges are also concerned with the evaluation and assessment of new e-Criteria, such as integrity, viability, collaboration-ability, survivability, and so on. This chapter summarizes and briefly discusses the emerging trends and challenges of next generation e-Work, e-Business, and e-Service systems (simply e-Systems) from a variety of perspectives. The objective is to bring up new open questions and research challenges, and provide organized insights for future research and development.

11.1 Computer and Communication Security

Emerging e-Systems call for higher levels of dependability and security, enabled by structured backup and recovery mechanisms for information, computer, and communication systems. The focus should be on preventing and eliminating any conceivable errors, failures, and conflicts, and sustaining critical continuity of operations. The critical dimensions of risk are mainly concerned with privacy invasion, electronic surveillance, accuracy and integrity, intellectual and physical property protection, accessibility issues, and confidentiality. Following the trends towards highly distributed and networked computer and information systems, security is being enabled by autonomous agents. Examples include 1) self- and

S.Y. Nof et al., *Revolutionizing Collaboration through e-Work, e-Business, and e-Service*,
Automation, Collaboration, & E-Services 2, DOI: 10.1007/978-3-662-45777-1_11

situation- awareness, 2) self-configuration, 3) self-explaining and self-rationalizing, 4) self-healing and self-repair, 5) self-optimization, 6) self-organization, and 7) self-protection for security.

Such transformation, however, has changed the traditional scope and definition of e-Systems security, which in turn necessitate new cyber-supported mechanisms for system protection against threats. In traditional, less integrated systems, where computer and information systems were decoupled, or used interconnected binary I/O signals, system security was limited to physical security, i.e., to prevent unauthorized people from accessing the system by physical means. With transformation to distributed and autonomous e-Systems supporting the aforementioned e-Activities, cyber-security of information, computer, and communication systems is now a crucial challenge for emerging e-Work, e-Business, and e-Service. For instance, denial-of-service attacks influencing the financial systems of a company or even a country's infrastructure through spyware or phishing attacks (Terwiesch and Ganz, 2009) impose significant damages and losses to the system. Another vital example is network security in wireless sensor networks, operating in hostile environments (e.g., battlefields). To protect such networks against intrusion and spoofing, the foremost concerns that must be addressed are:

- Level and type of security.
- Data authentication mechanisms.
- Detection and isolation of misbehaving nodes.
- Prevention of false information exchange.
- Trade-off between security and required resources (e.g., energy).

In any e-System, in general, security involves computer and controller software and hardware, information and knowledge bases collected via the Internet, imaging satellites, sensor networks, etc., and automation devices, appliances, systems, networks, and other platforms. Major challenges and open questions regarding the security of information, computer, and communication systems are summarized below (Ramaswamy and Joshi, 2009):

- Identification of major vulnerabilities security that affect privacy, property, quality, and accessibility of the system.
- Development of protection and recovery procedures, and administration of relevant tasks.
- Development of procedures for tracking, tracing, and blocking the instigators and initiators of security problems.
- Decisions on how to maintain, share, and guarantee system's security.
- Assignment of responsibilities for different security breaches, such as trespassing, espionage, sabotage, information extortion, data acquisition attacks, cyber-terrorism, cyber-crime, compromised intellectual property, private information theft, etc.

- Identification of breaches caused by unintentional human error versus malicious, unethical acts.
- Identification of sources and causes of software attacks, such as software viruses, worms, Trojan horses, denial of service, phishing, spamware, and spyware attacks.

An important contribution of collaborative mechanisms as described and explained in this book, e.g., for error and conflict detection and prevention, and for emergent lines of collaboration and control, is their ability to detect irregularities faster and alert defensive protocols in time to engage recovery and prevent propagation. Another emerging response to this challenge is the collaborative fault tolerance by teaming, which has also been discussed in previous chapters.

11.1.1 Information Assurance

In line with the concerns for security of information, computer, and communication, next generation collaborative e-Systems must be attentive to the information assurance, i.e., assuring information and managing risks associated with the use, processing, storage, and transmission of information among distributed entities. Information assurance is critical for ensuring the confidentiality, integrity, availability of information and knowledge-based system. To protect industrial systems against hackers, spies, and disgruntled insiders, Multari (2004) mentioned four major challenges based on the best practices at the Boeing Information Assurance R&D group:

- Design of complete and verifiable user access control mechanisms the, following the corporate access policies, to detect malicious authorized users.
- Reduction of false alarms generated by the intrusion detection systems, associated with the previous attacks.
- Development of predictive behavior analysis mechanisms to improve system availability, without any need for redundancy.
- Mechanisms for controlling information sharing among users, considering its sensitivity and criticality.

The information assurance challenges in mobile networking, considering variable connectivity, waveforms, and topology are to design networks that are self-defending, self-monitoring, and self-evolving, enhanced with robust defense, situation awareness, trust-based dynamic credentialing, and trace-back and attribution tools (Ghosh, 2005).

The realization of benefits such as flexibility and agility associated with service-oriented architectures relies on securing service-oriented environments through effective information assurance strategies. The critical elements of service-oriented architectures to be secured against threats are architectures, web services (including composites), and applications, due to the following challenging issues associated with the service-oriented environments (Simanta et al., 2009):

- *Ubiquitous interface style*: The standard definition of interfaces, which are beneficial to the service consumers, makes it possible for intruders to penetrate the system via few widely publicized techniques.
- *Commercial products*: The commercial back-end systems behind the interfaces are vulnerable to attacks hidden inside messages handled by a web service.
- *Black boxes*: Services are black boxes to the consumers/integrators, which prevent them from performing static analyses on the system's vulnerabilities, and thus reduce their trust, especially in applications with sensitive data.
- *Deployment configurations*: The variability and complexity associated with the configuration of hardware/software infrastructure make the system's identity, location, configuration, etc. vulnerable against hackers.
- *Dynamism*: Dynamic employment, orchestration, and implementation of services along with lack of real-time knowledge provision to the clients in intensify system's vulnerability.
- *Untrustworthy networks*: A challenging issue is to establish trust across a distributed and large network of web and application servers, service registries, service repositories, individual services, etc.
- *Distributed security zones*: Mechanisms are needed for performing end-to-end local security testing in distributed security zones, following their specific security and privacy policies.

11.1.2 Ethics

Besides security and information assurance, emerging e-Systems encounter new challenges to individuals and organizations from a broader perspective, the ethics. By definition, ethics refers to a set of principles that an entity, from an individual to an organization, applies while making decisions influencing its behavior and environment (Ramaswamy and Joshi, 2009). The emerging collaborative e-Systems have transformed ethical challenges and undesirably, have enabled and simplified unethical behavior compared to traditional systems (Figure 11.1). Common examples are privacy violation and anonymous access to individual property, recording conversations and proprietary knowledge, cyber-crime, cyber-terrorism, information hiding or obscuring, etc. Major challenges concerning ethics in collaborative e-Work, e-Business, and e-Service are:

- Challenges caused and intensified by technology and automation, e.g., cyber- and robo-ethics.
- The impacts of unethical behaviors on the entities and the system, e.g., privacy, property, quality, and accessibility.
- Security threats and vulnerabilities caused by automation, e.g., information security, technical failure, cyber-crime, cyber-terrorism, cyber- or robo-warfare, etc.

- Development of collaborative mechanisms for rational, acceptable, and sustainable ethical sensitivities and behaviors supported, maintained, and monitored by autonomous entities.
- Gathering, mining, and observing user information-seeking behavior, while protecting individuals' privacy and rights.
- Marketing and information dissemination in digital world.
- Issues of privacy, copyrights and intellectual properties, and cyber crime.
- Investigating the global ethical impact of computers and communication, online communities and social networks, and virtual reality.
- Development of collaborative protocols for rational process control, equality of access under authorization procedures, and trust and authentication agreements.
- Design of self-repair and self-recovery systems mitigating the risks associated with unexpected failures and disasters that cause vulnerabilities to unethical behaviors.
- Automation solutions for automatic resolution of unethical scenarios, e.g., copying and file sharing, infringement of intellectual property, security risks and threats, automation-assisted forgery, identity theft, unethical employee surveillance, etc.
- Questions on how to ethically control autonomous systems and restrict their level of autonomy in different situations.

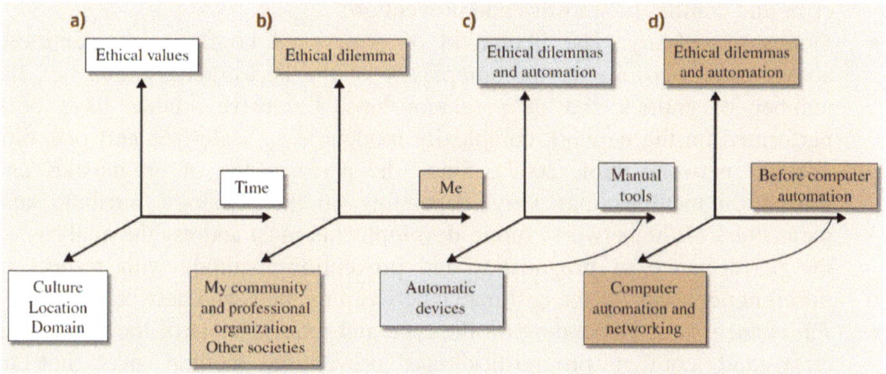

Fig. 11.1 Ethics challenges: (a) Transformation over the time, cultural changes and technological evolutions; (b) Ethical dilemmas as conflicts between individuals/groups and community, organization, and society; (c) Impacts of automation on the emerging ethical dilemmas, e.g., remote imaging, banking automation, and Internetworking; (d) Role of computer, network, and communication technologies in emergence of new ethical dilemmas (after Ramaswamy and Joshi, 2009).

11.1.3 Error Detection, Reduction, Prevention, and Recovery

The increasing trends towards highly distributed e-Systems with decentralized intelligence and autonomy have brought up new challenges regarding potential errors and conflicts in collaborative processes. The autonomous entities (e.g., intelligent agents) are not necessarily error-proof and the increasing rates of interactions among entities are proportional to the likelihood of conflicts among them. The distributed and autonomous nature of emerging e-Systems, on the other hand, enhances efficient and timely detection, reduction, prevention, and recovery of errors and conflicts in the systems. As explained in the fourth design principle of CCT (see Chapter 2), under proper coordination protocols, distributed and autonomous agents are able to detect, identify and isolate errors, and communicate effectively to diagnose, resolve, and prevent conflicts. The primary challenge, however, is the design of effective and robust protocols for ensuring early detection, recovery, and prevention of errors and conflicts in the system, considering:

- *Autonomy and intelligence*: Future error and conflict prognostics and prevention systems need to be enhanced with tacit knowledge and high-level decision-making capabilities, inspired by human intelligence, for dealing with complex and dynamic systems. The emerging trends in this area is the application of fuzzy logic, pattern recognition, and artificial neural networks in design of self-learning, self-improving artificial intelligence systems for error and conflict prognostics and prevention.
- *System complexity*: The likelihood of errors and conflicts, as mentioned above, is proportional to the complexity of the networked system, i.e., the number of entities and their relationships. Extensive studies have been performed on the network complexity models, e.g., scale-free network, bio-inspired network (Nof, 2007). Since the performance of prognostics and prevention methods may vary, depending on the topology, attribute, and parameters of the network, future developments must address the analysis on the performance of prognostics and prevention methods with respect to different networks for the best match between methods and networks.
- *Financial issues*: Depending on the type and requirements of the system, an error and conflict prognostics and prevention method may not be economically justifiable. In other words, the damages caused by errors and conflicts may be much lower than the cost of designing and implementing the conflict prognostics and prevention system. Robust economic rationalization models then must be developed for analysis of short-term, long-term, and intangible effects and damages caused by errors and conflicts compared to the costs of detection, reduction, prevention, and recovery.

11.1.4 Sustainability

Sustainability, including the three main pillars of economy, environment, and society, is an emerging concept concerning design of collaborative e-Systems (Figure 11.2). Major challenges are about strategic reduction of waste, reduction of industrial pollution, energy consumption and hazards, and environmental disasters caused by the system (environmental), employee satisfaction, quality of life, expenditure on peripheral development (social), and sustaining profitability and viability, transportation and civil activities, turnover ratio, gross margin, and capital investment (economic). Design of sustainable e-Work, e-Business, and e-Service then necessitates the design of technical and organizational solutions regarding business processes, supply networks, product design and lifecycle management, process innovation, and technology advancements (see Chapter 3). Beyond the inherent challenges, the complexities associated with sustainable design are intensified due to the following sustainability planning and control challenges (Seok et al., 2012), which call for advanced control techniques for effective solutions:

- *Cognitive reasoning*: Refers to the complexities and dynamic characteristics associated with real-time monitoring and updating of changes such as new constraints, policies, regulations and parameters. In addition, the dynamic changes in the interactions within and between organizations are other critical factors for successful and reasonable sustainability decision processes.
- *Decision support systems*: Design and selection of appropriate methodologies for the decision process, from recognition of the characteristics and objectives of the sustainability issue to suggestion of appropriate methods and procedures, is another challenging issue concerning sustainability planning, and control.
- *Conflict resolution*: Due to potential conflicts among participants, an effective control mechanism must be designed and integrated into the decision support system to keep the balance between the organizations.
- *Sustainability criteria*: In addition to the conflicts among organizations and participants, there are conflicts among sustainability problems and objectives due to high variations of criteria, factors and perspectives. Hence, effective multicriteria decision-making mechanisms must be designed and incorporated in the sustainability decision support system to solve the trade-off among multiple sustainability criteria and objectives.
- *Customized and dynamic mechanisms*: More analytical considerations are required to deal with specific types and levels of disruptions in different systems, considering volatile customer demand, government policy, and market condition. The major question is how to handle the modeling of different kinds of industries with their different constraints and criteria.

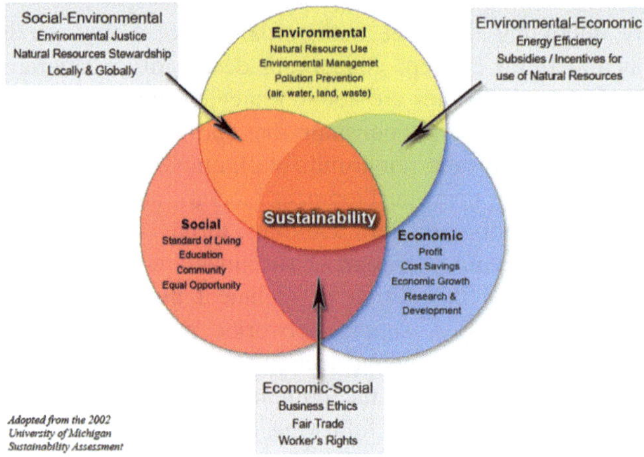

Fig. 11.2 Three sustainability areas (Sustainability, 2002).

11.1.5 Ability to Collaborate More Effectively

The increasing abilities for ubiquitous and real-time access to information and knowledge are enabling e-Systems to collaborate along the entire spectrum of the distributed networks of entities and provide all entities with the information they need to work towards common goals. Major applications of such features are in the emerging e-Commerce systems. Examples include electronic brokerage and dealerships, online B2B and B2C collaboration, communication services such as EDI, financial e-Services, online education, and digital authentication/copyright services, and technology infrastructures such as public and private communication utilities, and Multimedia management capabilities (Zwass, 2003; Yoon and Nof, 2008; Velasquez and Nof, 2009; Seok and Nof, 2011; Seok et al., 2012). Other trends in applications of ubiquitous and real-time information technologies are:

- *Mobile commerce*: As an emerging trend in e-Commerce, mobile commerce, also known as m-Commerce, is increasingly facilitating the traditional fixed-wire Internet-based commercial systems with wireless and ubiquitous access using mobile devices (e.g., smart phones, laptops). Major challenges and open questions regarding mobile commerce are 1) integration of existing systems into mobile environments, 2) development of easy-to-use devices with minimum requirements for modification in the existing system, 3) value addition over and above that provided by traditional e-Commerce, and 4) provision of relevant, inexpensive, and customized services (Pagiavlas et al., 2005).

- *Telemedicine (e-Health)*: Telemedicine refers to the application of telecommunications for exchanging medical information (e.g., x-rays, tomography scans, magnetic resonance images, digital images of pathology specimens) between remote locations (ATA, 2007). Design of control mechanisms for effective collaboration and conflict resolution in telemedicine processes, especially human-robot interactions, is a major challenge in this area.
- *Fee-based information delivery*: These services are designed to provide customers with specific information for a payment. The motivations and challenges regarding the development of such systems are concerned with money, time, learning, decision-making, and quality of the information provided (Wang et al., 2005; Lopes and Galletta, 2006).
- *Ubiquitous computing*: The ability of accessing to information from anywhere, and at anytime, has enabled ubiquitous computing capabilities, i.e., using any device, in any location, and in any format for performing specific computing tasks. An emerging application is in the area of robotics. Future robots will be characterized by *ambient intelligence*, enabled through embedding them into emerging information technology environments, along with the growing spread of ubiquitous computing and communications, and of ad hoc networks of sensors (Figure 11.3).

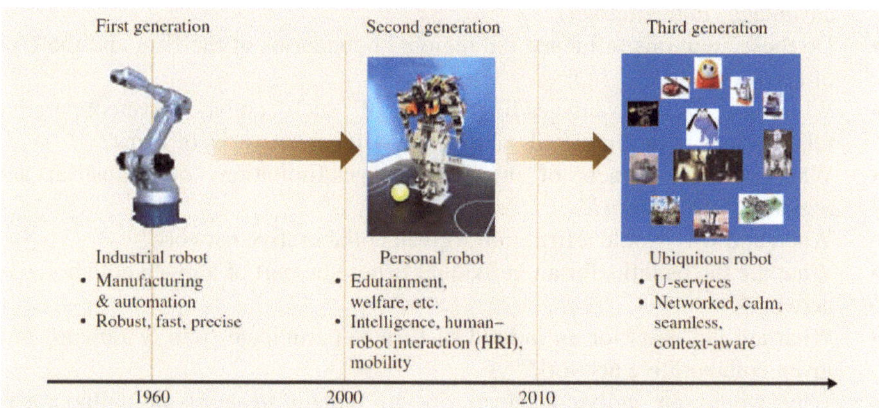

Fig. 11.3 Development trends in robotics and emergence of ubiquitous robots (after Kim, 2006).

11.1.6 Incentive-Based Collaboration

Since 1980's, global competition has emerged as a challenging concern for the modern economy, which, from a protectionist point of view, may seem contradictory with the concept of collaboration. The key questions are then why to share, and not to compete, what are the role and optimal level of collaboration to ensure the competitiveness of a firm? Traditionally, competition was known as the preferred type of behavior, maximizing the firm's profit. The structural evolution of markets empowered by the emerging e-Work, e-Business, and e-Service solutions along with the industry dynamics, however, increasingly impelled firms towards collaboration (Golich, 1992). Nevertheless, collaboration, by itself, may not guarantee the success. According to a study by Fraunhofer Society on technology transfer, it was concluded that the dynamic competition and collaboration interactions among firms ensure the sustained technological competitiveness of the German economy (Harding, 2002). Hence, firms may need to establish effective competitive, collaborative, or cooperative activities (known as the 3C's) to remain competitive in the market.

Profit is definitely the main factor for taking the optimal strategy out of the 3C's, but not the only one. In automotive industry, for example, firms make collaborative agreements, even with their competitors, in order to protect their competitive edge and control the market (Schoenberger, 1994). In general, a firm must evaluate all economic, social, and political aspects of the 3C's, in order to define whether to collaborate/cooperate and to what extent (Figure 11.4). These issues are the target of the AD (Association and Dissociation) and ELOCC (Emergent Lines of Collaboration and Command) principles of CCT, dealing with the following challenges of CNO (Collaborative Network Optimization) (Polenske, 2004; Yoon and Nof, 2008; Prins, 2010; Seok and Nof, 2014a, b):

- Do collaborative/cooperative solutions provides the firm with competitive advantages in the market?
- Do these solutions influence the regional boundaries of the firm and the type of networks?
- What are the advantages/limitations of such collaborative/cooperative relationships regarding local, national and global networks of firms?
- What are the impacts of these advantages/limitations on industrial and regional restructuring?
- When and why would a firm join a given collaborative network?
- What are the benefits for an individual firm to be part of a given collaborative network?
- What are the costs for an individual firm to participate (join or remain) in a given collaborative network?
- Why would an individual firm opt to remain in a given collaborative network?
- What are the criteria and performance evaluation measures to characterize and assess the effectiveness and performance of the collaborative network?
- What are the asymmetries (e.g., power, resources, social network) and paranoid concerns (e.g., fear to lose one's identity, mutual stereotyping)?
- How can collaborative relationships (e.g., trust, mutual learning) be developed?
- What are the strategies for dealing with uncertainties, openness, emergence, and complexity of the collaborative network?

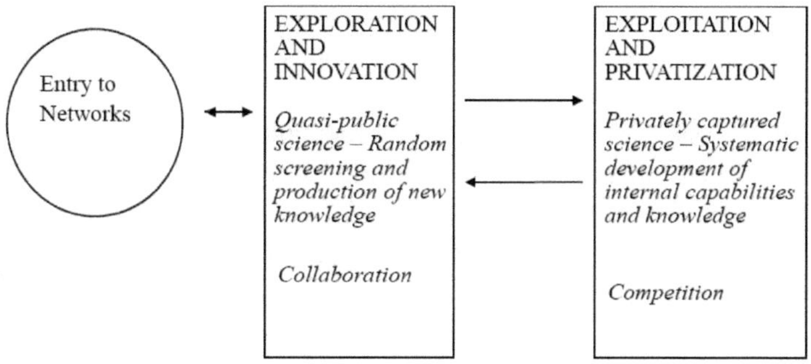

Fig. 11.4 The collaboration-competition dynamic symbiosis (after Oliver, 2004; with permission from Elsevier B.V.).

Table 11.1 Alternative restructuring models considering the 3C's (adapted from Polenske, 2004).

Attribute	Success		
	Small firm	Competitive firm	Large multinational firm
Competition	Limited forms exist	- Even on shop floor - Few long-term contracts - Control over time and space - Very competitive in a global marketplace	- Driving force of the global firm - Control over the entire supply chain of large and small suppliers
Collaboration	Limited joint work on projects	Long-term contracts between large and small firms	Employer-employees collaboration
Cooperation	- Prevalent behavior - Constant product-process innovation - Regional organizations: credit unions, R&Dtraining, and IT	Limited forms exist	Limited forms exist
Networks	Among small firms in an industrial district	Small firms cluster near large firms to achieve JITand information sharing	Important for firms of all sizes
Regional	- Industrial district - Strong regional density	- Small and large firms cluster - Governmental influences regarding location decisions	- Innovators and collaborators in the core region - Low-cost firms in the peripheral region - Networks cross spatial boundaries

It is proven that all type of firms, from small innovative and cooperative firms to competitive and large multinational firms can benefit from CNO for guaranteeing their success in competitive market (Table 11.1). A good example is the firms in the Silicon Valley that form regional networks for collaborative research and development, social networking, labor sharing, etc. The critical questions are then how to optimally restructure the networks of firms and define the boundaries among the 3C's with respect to a proper set of criteria. Moreover, in many cases, even competing firms can benefit from collaboration. For instance, Okumura (2012) analyzed the formation of collaborative networks among firms that are located in a circular city and engage in price competition, and its impact on the stability and social efficiency of collaborative networks. He proved that a price equilibrium exists, at which there is no conflict between the stability and social efficiency of collaborative networks.

11.1.7 Multi-cultural Interactions

With the increasing trends towards globalization, the design of collaborative e-Systems is becoming more concerned with cross-cultural issues, i.e., communication, interaction, and collaboration among individuals from a variety of cultures who must function together for achieving common goals. The challenging question is how to integrate and harmonize the knowledge and expertise dispersed among several disciplines and research areas to avoid cultural conflicts. Thus, it is necessary to assess and incorporate cultural contexts in all stages of design and development of products and services. The increased dimensions of reachability and connectivity within the collaborative e-Systems further intensify the exposition, meeting, and potential collisions/enrichment of cross-cultural interactions, conflicts, and bridges. Hence, a critical issue is the level of the knowledge of the multi-cultural decision-makers about the cross-cultural aspects of the product/service under development. In addition, distribution of the cross-cultural knowledge among the entities, computer-supported collaborative tools and interfaces is another dilemma concerning the design of products and services. Furthermore, design of customized products and services addressing consumers cross-culturally and incorporating consumer expectations in innovation are other issues that are worthy of attention (Table 11.2).

Accordingly, a *precondition* for successful development of future e-Systems is to recognize the cultural differences, such that the creativity of collaborating individuals from unfamiliar intellectual traditions is guaranteed (Leung et al., 2008). In fact, the focus must be on the elimination of creativity obstacles, i.e., mental sets, caused by different cultural experiences (Figure 11.5), and improvement of multicultural mindset and interdisciplinary collaboration. The major challenges and trends for cross-cultural design of e-Work, e-Business, and e-Service systems are (Proctor et al., 2011):

Table 11.2 Collaborative vs. intellectual-centric responses (adapted from Proctor et al., 2011).

Property	Collaborative response	Intellectual-centric response
Information search	Distribute attention to technological and user benefit information among decision - makers	Focus attention on technological information among designers and user benefit information among marketers
Idea sampling	Sample information across disciplines for marketable innovation	Sample information primarily from one's own disciplines for marketable innovation
Idea generation	Generate ideas from multidisciplinary perspectives	Generate of ideas primarily from one's own disciplinary perspective
Idea editing	Open to suggested revisions of one's own ideas from people with different disciplinary backgrounds	Resist suggested revisions of one's own ideas from people with different disciplinary backgrounds
Dissemination stress	Experience low stress when presenting ideas to an audience with different disciplinary backgrounds	Experience heightened stress when presenting ideas to an audience with different disciplinary backgrounds
Managing conflicts in work team	Attribute conflicts to disputants' personal qualities or experiences; prefer problem - solving strategies	Attribute conflicts to clash of disciplinary perspectives; prefer competitive strategies

- *Provision and utilization of cultural data*: 1) Meta-analysis of existing studies to extract cultural factor loadings, and 2) Enrichment of operational data through interaction and collaboration with service providers.
- *Big data and complex network analysis*: Development of data mining methods analyze network data and capture real-time interactions of entities, including consumers.
- *Assessment of psychological and cultural features*: 1) Analysis of relations between culture, technology, and decision-making, 2) Measurement of culture and multiculturality (including identity and competence), and 3) Modeling changes in culture and multiculturality (i.e., growth mixture modeling).
- *Analysis of psychological processes of cross-cultural collaboration*: Controlled experiments to reveal the impact of culture on the psychological processes in decision-making, and to understand the causal roles of awareness of cultural differences, multicultural learning mindset, epistemic security, and accountability.

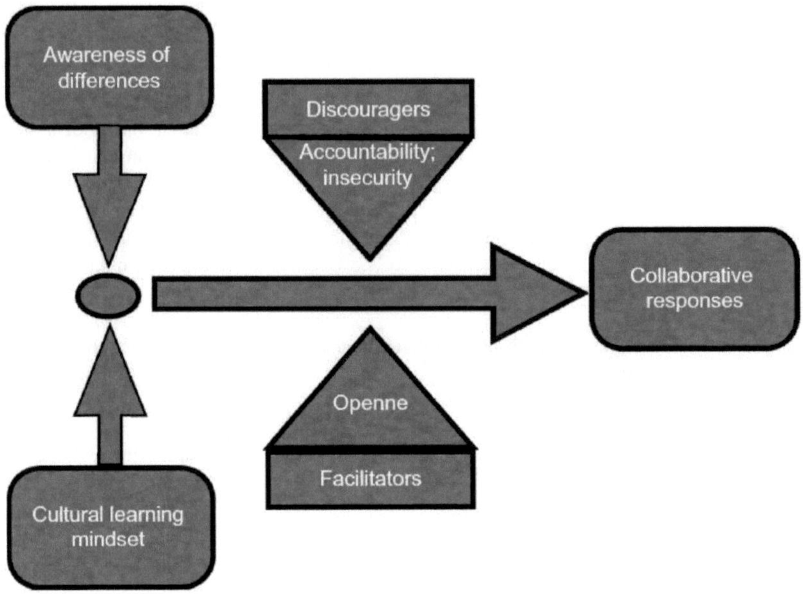

Fig. 11.5 A process model for collaborative responses in cross-functional teams (after Proctor et al., 2011).

11.1.8 Cost Reduction in Service Processes

Financial efficiency is a classic design consideration for any system, which must be addressed in the design of service processes and delivery systems as well. To elaborate, however, a low-cost system does not necessarily imply poor technology performance, but enhanced affordability, and lower lifecycle and ownership costs. In other words, cost reduction strategies aim at developing innovative and intelligent solutions that enable achievement of the same performance as sophisticated systems but with lower costs. Hence, cost-oriented strategies must be developed taking into account the cost of ownership associated with the lifecycle of the system, i.e., 1) design, 2) implementation, 3) operation, 4) reconfiguration, 5) maintenance, and 6) recycling. The cost of ownership can be used to support acquisition and planning decisions for a wide range of assets that have significant maintenance or operating costs throughout their usable life (see, e.g., Figure 11.6). Examples include computing systems, vehicles, laboratory and test equipment, and manufacturing equipment. The critical issues and challenges regarding cost-effective realization of service processes and delivery systems are (Erbe, 2009):

- Application of field robots in various domains such as manufacturing plants, building, offices, agriculture, mining, etc. for reduction of operational costs.
- Enterprise integration, support for networked enterprises, and trending towards Small and Medium-Sized (SME)-oriented agile manufacturing for cost reduction.
- Condition-monitoring and e-Maintenance to reduce maintenance cost and avoid downtimes machines and equipment (see examples in Chapter 9).
- Distributed and collaborative engineering and control of ongoing work from remote locations, for lower-cost processes, e.g., tele- or web-based maintenance.

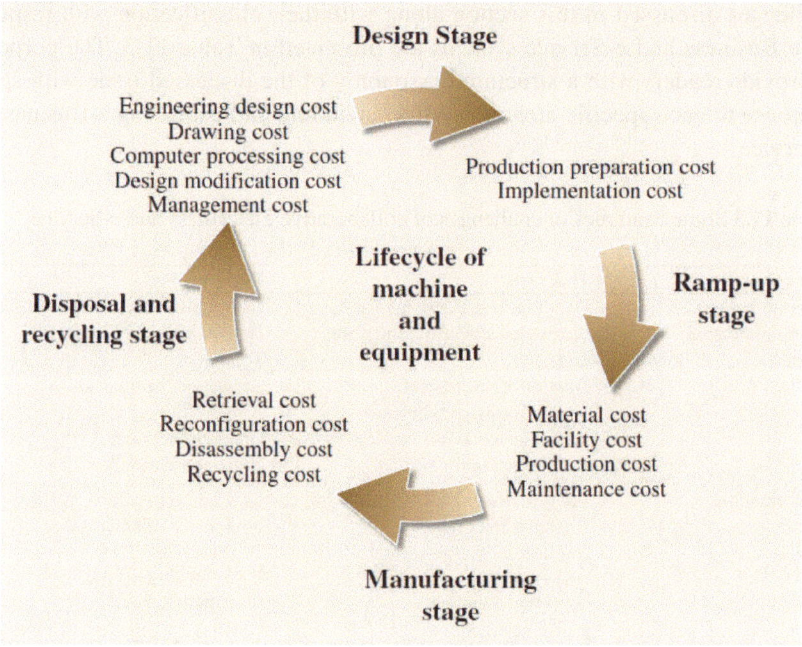

Fig. 11.6 Cost contributions of the life cycle of machine and equipment (after Erbe, 2009).

- Integrated product-process development, enhanced with effective information / knowledge sharing, collaborative decision-making, and interoperability.
- Mixed-reality concepts and virtual manufacturing supporting integrated product and process development.
- Smart devices (e.g., micro- and nano-sensors) with specific level of autonomy and intelligence for local information processing for cost reduction in various fields such as automotive, robotics, mechatronics, and manufacturing.

- Shop-floor-oriented technologies (e.g., job-shop controls) focusing on agile manufacturing, which create synergetic effects to master the manufacturing processes, and support execution of tasks and decision-making with lower costs.

11.1.9 Summary and Classification

Recalling from Chapter 1, collaborative e-Work is defined as the collection of e-Operations, e-Functions and e-Supports that enable e-Systems, including e-Business and e-Service. In other words, it can be argued that e-Business and e-Service are subsets of e-Work. Hence, a summary of the emerging trends and challenges discussed in this section along with their classification with respect to the e-Business and e-Service systems are presented in Table 11.3. The purpose is to provide readers with a structured taxonomy of the discussed issue with special reference to each specific area of e-Work, including and excluding e-Business and e-Service.

Table 11.3 Some examples of challenges of collaborative e-Business and e-Service.

Trend	e-Business	e-Service
Computer and communication security	- Data loss caused by malware, e.g., viruses, worms, software bugs - Data/information theft - Vulnerability of wireless networks to eavesdropping - Authentication	- User anonymity, and traceability - Data confidentiality - Trust - Security dependencies - Auditability - Denial of service - Non-repudiation
Information assurance	- Encryption and digital signature - Business integrity	- Data and transaction integrity - Financial settlements
Ethics	- Intellectual property theft - copyright trolls - Baits and switches	- Poor service - Web tracking - Information theft - Fraud
Sustainability	- High performance and quality - Cost-effective production - Reconfigurable, adaptive, and interoperable processes - Mass customization	- Intelligent transportation system - Smart energy distribution - Standardization and government regulations/support - Cultural issues
Ability to collaborate more effectively	- Mobile commerce	- Tele-medicine - Fee-based information delivery
Cost reduction in service processes	- Robotics application in plants, offices, mining, etc - Enterprise integration - Agile manufacturing - e-Maintenance - Intelligent shop floor devices	- Robotics application in healthcare, delivery, transportation, etc - Online education - e-Banking

11.2 Theoretical Issues in Collaborative e-Work, e-Business, and e-Service

11.2.1 Network Physics and Bio-Inspired Techniques

Network models have been extensively studies and applied to describe the nature and design of smart networked entities in the context of e-Work, e-Business, and e-Service. These models are applicable to different instances of networks, from extended enterprises and networks of SMEs to micro- and nano-sensors and robots. Recent studies on neuroscience have uncovered significant opportunities for enabling bio-inspired and adaptive control mechanisms for collaborative e-Work, e-Business, and e-Service. The enablers of collaborative control, e.g., adaptivity, dependability, interoperability, decision supports, active middleware, etc. are increasingly influenced by the natural mechanisms such as competitive animal behaviors and brain models (Figure 11.7). Further research is expected to inspire from the nature's design and control mechanisms for optimization of collaborative e-Work, e-Business, and e-Service systems, and understand the limits of such network models in highly complex environments, taking into account the following challenges and open questions (Nof, 2007):

- *Join/Leave/Remain (JLR) decisions*: Best matching and timing of JLR in distributed networks of sensors, robots, SMEs, etc. (e.g., which organization to join, when, and how?)
- *Cluster formation*: When an entity (e.g., organization) wishes to expand by collaborating with other entities, and create multi- or super networks, taking into account best matching and scalability of service.
- *Cluster optimization*: Optimal shrinkage or expansion of clusters, in accordance with dynamic changes in the network characteristics.
- *Profitability vs. sustainability*: Balance between the measures of reward and sustainability in terms of economic survival, especially in non-profit enterprises.
- *Legal limits and penalties*: Agreements, contracts, and legal binding protocols, along with their corresponding limitations and penalties, which influence the design of network models.
- *Topology-based performance*: The impact of network topology in terms of looseness, redundancy, and contract-binding terms on the productive performance of individuals and the entire network.

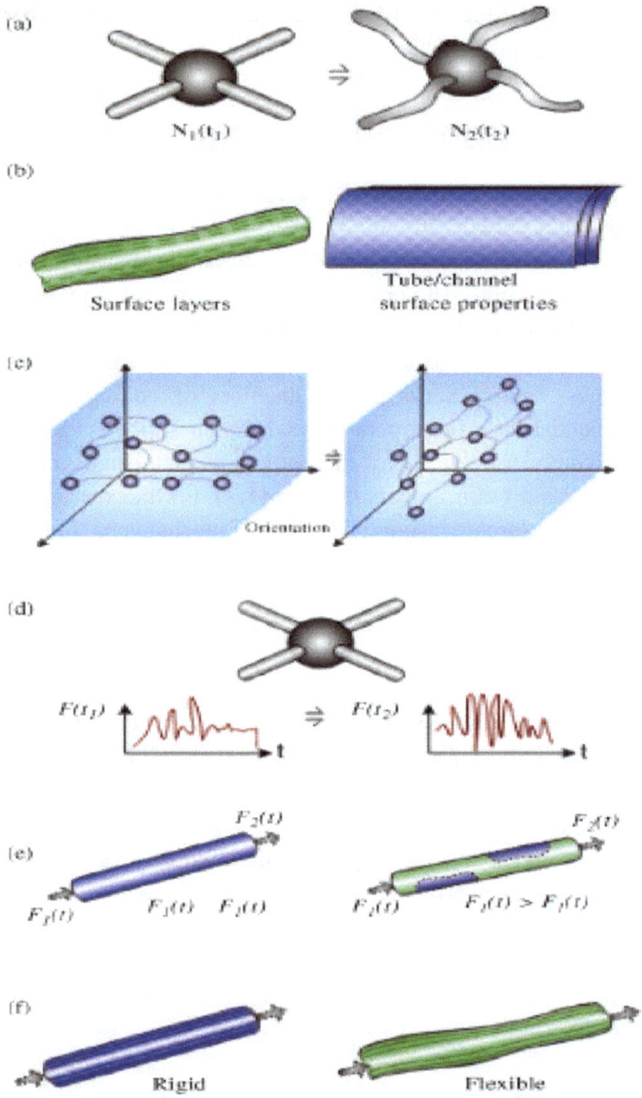

Fig. 11.7 Bio-inspired network models of e-Work, e-Business, and e-Service systems: a) Channels modify their shape and plasticity over the time. b) The channels may have multiple surfaces, and their surfaces may change dynamically. c) Network's spatial orientation may change over the time. d) Time-variable electromagnetic channel waves (as observed in nano-wires in mice). e) Network learning, structural plasticity, channel rigidity, and blockage, f) Rigid vs. flexible channels in network learning.

11.2.2 Collaborative Computing

Advances in collaboration and interaction models have enabled geographically distributed and well-organized computing services for dealing with the increasing complexity and dynamic behavior of emerging e-Systems. Such collaboration grids facilitate rapid sharing, selection, and aggregation of distributed computing resources for solving large scale and data intensive problems. A computational grid provides dependable, pervasive, consistent, and inexpensive access to high-end computational resources and enables its users to perceive heterogeneous resources as homogeneous (Foster and Kesselman, 2003). Grid computing infrastructures provide *virtual* computing capabilities, empowered by diverse collaborative computing resources, which support virtual service-oriented environments to schedule the data resources considering applications and computing engines (He and Castro-Leon, 2009). The critical challenges of transition to virtual service-oriented grid environments are:

- Trustability of distributed computing resources, and their reliability in fulfillment of their allocated tasks with the expected quality.
- Advanced networking, interfaces, interaction paradigms, and interoperable and application-specific tools.

11.2.3 Complexity Handling

In the context of collaborative e-Work, complexity is characterized by the uncertainties, nonlinearities, and time-delays associated with systems with large numbers of variables and networks of interconnected subsystems. A complex system is formally defined as "*A system which is composed of a number of smaller constituents, which serve particular functions, share common resources, are governed by interrelated goals and constraints and, consequently, require more than one controller*" (Mahmoud, 1977). The emergence of distributed network paradigms such as extended and virtual enterprises over the last decades has intensified the complexity of complex systems, due to (Mårtenson, 1999)

- The increasing number of interacting subsystems performing various functions and utilizing resources and technologies from different domains.
- Communication and interaction problems among systems agents (e.g., experts).
- Emergence of new tasks and routines, and differences in the level of skills, cultures, etc.

Accordingly, next generation systems are expected to comprise higher levels of flexibility in responding to the dynamic needs of subsystems and network elements in a timely manner (Nof et al., 2006). There is a growing demand for the design and control of complex *super-systems* or *systems of systems*, which have emerging applications in space exploration, healthcare, computer networks, software integration, air and auto transportation systems, energy management, etc. In general, handling systems complexity requires effective decentralized control and coordination mechanisms analogous to the *divide and conquer* algorithm, i.e., decomposition of the complex system into two or more manageable subsystems (Filip and Leiviskä, 2009). The emerging challenges and directions for further investigations in this area are[1]:

- *Stigmergy*: as an indirect mechanism for coordination between agents and actions, this principle refers to the stimulation of the actions performance by the traces of the previous actions left in the environment, and is a modern form of coordinating the actions of intelligent agents (Hadeli et al., 2004).
- Development of advanced decentralized control mechanisms such as *Internet congestion control* (Srikant, 2004), *aerial vehicles* (Stipanovic et al., 2004), and *flexible structures* (Bakule et al., 2006) for handling the complexities of large-scale complex systems.
- Development of tools for solving various large-scale control problems, such as *integrated/embedded control*, *distributed control of communication networks*, *hybrid/discrete-event systems*, and *autonomous systems* (Filip and Leiviskä, 2009).
- Unification of humans, numerical models, and artificial intelligence-based tools using numeric optimization and simulation models, agent-based systems, etc.
- Enabling *virtual* collaborative decision-making capabilities through mobile communications and web technology.
- Issues in social sciences, such as (San Miguel et al., 2012).

  - Data gathering by large-scale experiments.
  - Participatory sensing and social computation.
  - Managing large-scale, distributed, dynamic, and heterogeneous databases.
  - Dynamical models, cause-effect relationships, ensemble modeling and data assimilation.
  - Multi-level (micro and macro) systems of systems.
  - Prediction, forecasting, and risk, especially in intelligent systems.

11.2.4 Automation Lifecycle Planning

Design of human-machine systems in an automation project involves a formalized plan composed of five individual phases (Figure 11.8):

---

[1] See Chapter 4 for examples of bio-inspired agent collaboration techniques. See the book by DiMario (2010) on *System of Systems Collaborative Formation.*

1. *Pre-run*: includes processing the system's concept. The objectives are (a) identification of the current problem, (b) evaluation of the need for an automation system, and (b) applications of the system.
2. *Definition*: involves deconstruction of the problem into sub-functions and subsystems. The objectives are (a) development of the specifications, including characteristics, performance features, and human/technology requirements, and (b) development of the layout and evaluation of alternative solutions for each subsystem.
3. *Development*: refers to the realization of the detailed design of the subsystem's hardware and software. The objective is to integrate the developed solutions into a functional system.
4. *Acquisition*: includes development of the system's hardware and software, technical evaluation of the subsystem's functionality, integration of mechanical components, and development of instructions.
5. *Use*: refers to realization of the usage of the system. Improvement, modernization, and redesign feedback are generated based on operational experiences for the design of similar systems in the future.

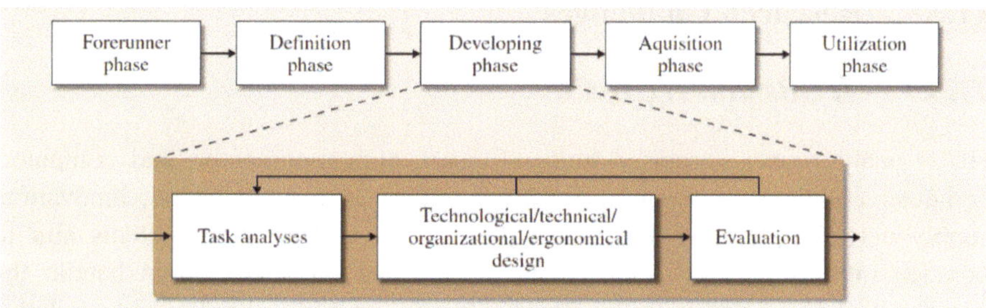

Fig. 11.8 Lifecycle of an automation project (after Spath et al. 2009).

At the end of each phase, a review must be performed to evaluate its maturity and decide upon the start of the next phase. There are several concerns regarding the accomplishment of an automation project and the lifecycle of the system:

- The *cost of ownership* (as explained in Section 11.1.8) is a critical issue in the life cycle management of an automation system. According to Blasi and Puig (2005), "*automation is not useful by itself: there are a lot of additional requirements to accomplish its function as needed, not only putting automation into the factories, putting automation just where it is needed and economically justified*".

- *Ecological constraints* must be addressed by the legal, ethical, strategic and operational decisions through modification of the cost functions in different decision levels of an automation project.
- Optimal balance between the complexity of systems, information uncertainty, and individual decision subjectivity must be made to address *dependability* concerns, caused by the economic constraints and the demands for safer, lower risk technological controls.
- The coherence of the individual and local decisions along the entire life cycle chain of events, and in the global context must be considered, in order to harmonize the regional decisions with overall system's goals.
- Certain traditional automatic control techniques must be incorporated in the development of control and automation support mechanisms in large scale, complex industrial and other systems to prevent catastrophic situations.
- Team collaboration supported by *computer supported collaborative work* must be performed to address multidisciplinary/multicultural aspects in the concurrent design and engineering of an automation project.

11.3 Research Challenges

11.3.1 HUBs and HUB-CI

HUBs enable cyber-supported human-human, human-computer, and computer-computer collaborative systems for applications such as education, innovation, supply network planning and control, etc. Such collaborative systems aim to broaden interdisciplinary science and engineering knowledge, and handle the complexities associated with the integration and control of distributed knowledge for collaborative understanding. HUB-CI, as a collaborative initiative due to Purdue University's school of Industrial Engineering, is a high performance computing platform that enables distributed entities (e.g., students, researchers, clients) to remotely collaborate and dynamically monitor the processes in a virtual environment. HUB-CI is an extension of Purdue's HUBzero, which is a platform developed by Purdue researchers to create websites for scientific research and educational activities, supporting extensive analysis, allowing access to a wide array of application tools, and enabling researchers to publish and share research software and materials on the Web. HUB-CI integrates CI (Collaborative Intelligence) and CCT principles to optimize human, system and research collaboration (Figure 11.9). The primary advantage of HUB-CI over the existing HUBs is its focus on improving human collaboration through CCT collaboration support (Nof, 2011; Seok and Nof, 2011; Seok et al., 2012).

The most important applications of HUBs, and the associated challenges and research questions targeted by HUB-CI are as follows:

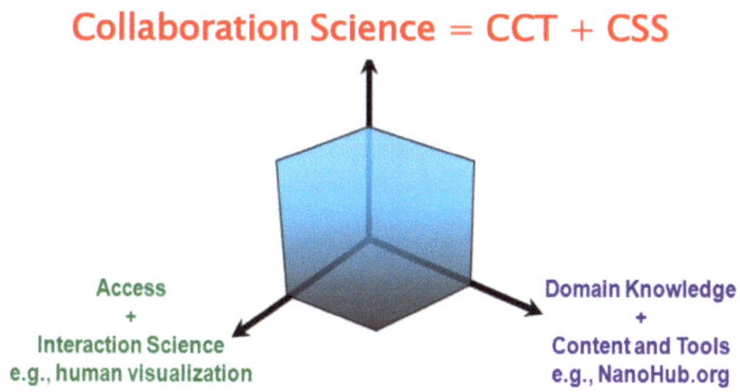

Fig. 11.9 The concept of CI Collaboratorium (CSS: collaboration support systems).

- *Innovation network HUBs*: refer to the emerging global networks (hubs/clouds) to trade, adapt, engage, and learn diverse ideas through collaboration. The research challenges, however, are cross-cultural capabilities, multi-cultural interaction and infrastructures, challenged web-based applications, and asynchronous multimedia (Figure 11.10).

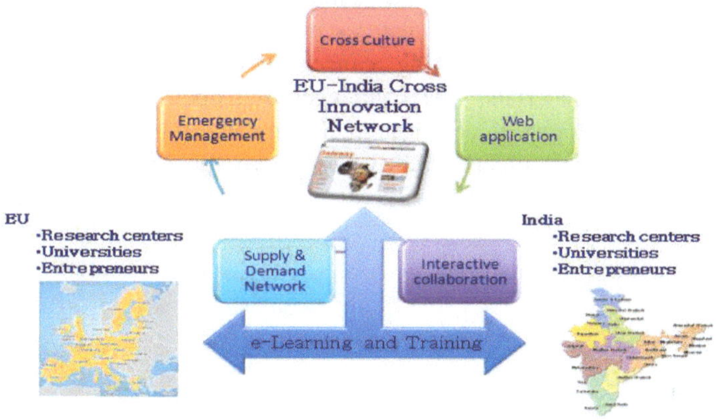

Fig. 11.10 Challenges of EU-India Cross Innovation Network targeted by HUB-CI.

- *Educational network HUB (nanoHUB)*: is a HUB for research, education, and training in nano-technology. NanoHUB enables collaboration and resource sharing without any need for installation, and access to plenty of simulation tools. Nevertheless, there are several challenges concerning nanoHUB, among them are interactive resource visualization, tag-clustering analysis, knowledge/resource/group best matching protocols, intuitive interaction techniques (Figure 11.11).

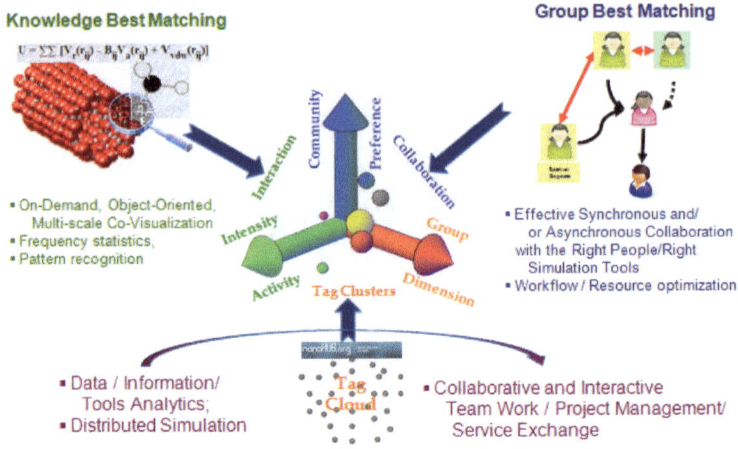

Fig. 11.11 Challenges of nanoHUB targeted by HUB-CI.

- *Healthcare network HUB (cceHUB)*: is a HUB for Cancer Care Engineering (cce-HUB), which provides a combination of clinical, scientific and engineering disciplines. The features of cce-HUB are exposition of current paradigms of cancer, sharing cancer related research, and effectively exchange any materials. The challenges and open questions, on the other hand, are development of research document repository and smart search engine, task assignment ability, selective posting and meeting, and coordinating tools (Figure 11.12).

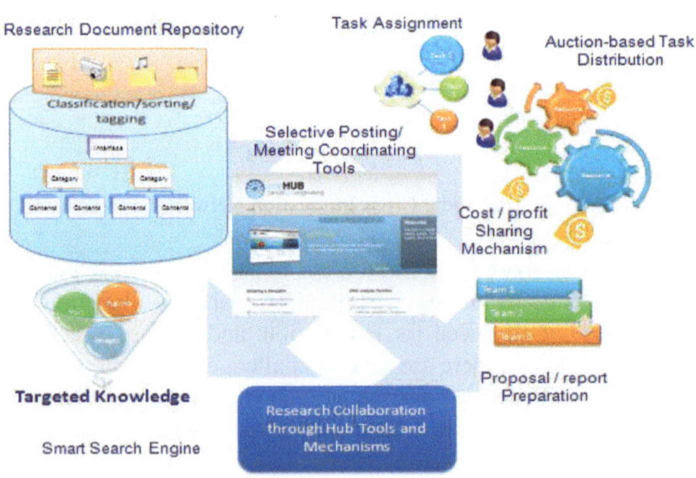

Fig. 11.12 Challenges of cceHUB targeted by HUB-CI.

- *Supply network HUBs*: provide a combination of marketing, manufacturing, and logistics engineering disciplines, which enable selection of best models of supply, helping consumers' regarding their needs, and effective information exchange. The challenges of supply network HUBs, however, are development of knowledge repository, need for smarter search engines, task scheduling ability, and automatic knowledge matching through HUB tools (Figure 11.13).

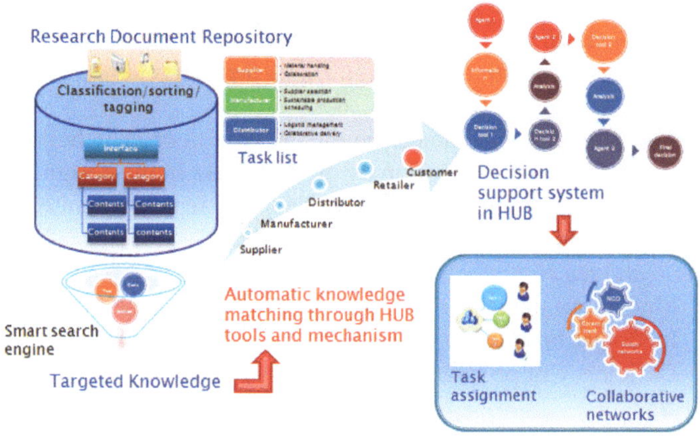

Fig. 11.13 Challenges of Supply HUBs targeted by HUB-CI.

In general, the purpose of HUB-CI is to enable faster and better achievement of the collaboration goal by automated support, in order to (a) enhance synthesis and integration of knowledge and discoveries, (b) understand the dynamics of interactive collaborative research work, and (c) support timely delivery of critically needed discoveries and shared knowledge. These objectives can be realized by the following emerging CCT-based HUB-CI tools (see Chapter 2):

- *CNO (Collaborative Network Optimization)*: optimizes the networking by using CRP (Collaboration Requirement Planning), CFT (Collaborative Fault Tolerance), AD (Association and Dissociation), ELOCC (Emergent Lines of Collaboration and Command) principles.
- *KBMP (Knowledge Best Matching Protocols)*: suggests appropriate knowledge and tools to the users by CRP and EPCR (Error Prevention and Conflict Resolution) principles.
- *Co-Vis (Collaborative Visualization)*: uses dynamic matching of analytics, applying KISS (Keep It Simple System) principle.

11.3.2 New Ways of Interacting in the World with e-Work

In addition to increasing the systems' complexity, the emerging e-Work technologies are gradually undermining the skills of human operators in their day-to-day tasks. Hence, human-centered systems, as an emerging area initiated by Mike Cooley (1982), aim at preserving or even improving human skills for better interactions with the advancing technologies. The emerging areas of research and major open questions concerning these systems are (Bullinger and Nof, 2010):

- *Communication*:
  - Dynamic formations of design teams for different project.
  - Custom-made products.
  - Complexities associated with the integration of IT in buildings.
  - Weak/lack of participation by the end-users in the design process.
  - Multicultural/multidisciplinary planning teams (architects, consultants, and end-users), and the resulting conflicts.
- *Data management*:
  - Multi-level data.
  - Incompatibility of software tools.
  - Deficiency of product-process integration.
- *Privacy*:
  - Securing systems against fraud.
  - Incompatibility of software tools.
- *Ambient intelligence*:
  - Development of interfaces and integration into mobile environments without the need for programming.
  - Accessibility of the developed systems through multi-modal combination of interaction devices and techniques without reducing the use of rich graphics.
  - Efficient usability assessment lifecycle through the design-and-play approaches.

References

American Telemedicine Association: Defining Telemedicine. ATA, Washington (2007), http://www.americantelemed.org/i4a/pages/index.cfm?pageid=3333 (accessed February 9, 2009).

Bakule, L., Rodellar, J., Rossell, J.M.: Robust overlapping guaranteed cost control of uncertain steady-state discrete-time systems. IEEE Transactions on Automatic Control 12, 1943–1950 (2006)

Blasi, A., Puig, V.: Conditions for successful automation in industrial applications. A point of view. In: Proceedings of 15th IFAC World Congress. Elsevier, Oxford (2005)

Bullinger, H.J., Nof, S.Y.: Special Issue in Honor of Professor Gavriel Salvendy. Computers in Industry 61(4), 309–389 (2010)

DiMario, M.J.: System of Systems Collaborative Formation. World Scientific, Systems Research Series, vol. 1 (2010)

Erbe, H.H.: Quality of Service (QoS) of Automation. In: Nof, S.Y. (ed.) Springer Handbook of Automation, ch. 41. Springer (2009)

Filip, F.G., Leiviskä, K.: Large-Scale Complex Systems. In: Nof, S.Y. (ed.) Springer Handbook of Automation, ch. 36. Springer (2009)

Foster, I., Kesselman, C. (eds.): The Grid: Blueprint of a New Computing Infrastructure. Morgan Kaufmann, San Francisco (2003)

Ghosh, A.: Challenges in mobile networking and information assurance. In: Proceedings of Fourth IEEE International Symposium on Network Computing and Applications, NCA (2005)

Golich, V.L.: From competition to collaboration: the challenge of commercial-class aircraft manufacturing. International Organization 46, 899–934 (1992)

Hadeli, H., Valckenaers, P., Zamfirescu, C.B., Van Brussel, H., Germain, B.S.: Self-organizing in multi-agent coordination and control using stigmergy. In: Di Marzo Serugendo, G., Karageorgos, A., Rana, O.F., Zambonelli, F. (eds.) ESOA 2003. LNCS (LNAI), vol. 2977, pp. 325–340. Springer, Heidelberg (2004)

Haarding, R.: Competition and collaboration in German technology transfer. European Management Journal 20(5), 470–485 (2002)

He, J., Castro-Leon, E.: Automation under Service-Oriented Grids. In: Nof, S.Y. (ed.) Springer Handbook of Automation, ch. 24. Springer (2009)

Kim, J.W.: Unpublished Transparencies. Summerschool, Kaist, Daejon (2006)

Leung, A.K.-Y., Maddux, W.W., Galinsky, A.D., Chiu, C.-Y.: Multicultural experience enhances creativity: The when and how? American Psychologist 63, 169–181 (2008)

Lopes, A., Galletta, D.: Consumer perceptions and willingness to pay for intrinsically motivated online content. Journal of Management Infrastructure Systems 23, 203–231 (2006)

Mahmoud, M.S.: Multilevel systems control and applications. IEEE Transactions on Systems, Man, and Cybernetics SMC–7, 125–143 (1977)

Mårtenson, L.: Are operators in control of complex systems? In: Proceedings of 13th IFAC World Congress, vol. B, pp. 259–270 (1990)

Multari, N.J.: Information Assurance technical challenges. In: Proceedings of the ACM SIGMOD International Conference on Management of Data, p. 899 (2004)

Nof, S.Y.: Collaborative control theory for e-Work, e-Production, and e-Service. Annual Reviews in Control 31(2), 281–292 (2007)

Nof, S.Y.: Cultural Factors: Their Impact in Collaboration Support Systems and on Decision Quality. In: Proctor, Nof, Yih (eds.) Cultural Factors in Systems Design: Decision Making and Action. Francis & Taylor (2011)

Nof, S.Y., Morel, G., Monostori, L., Molina, A., Filip, F.: From Plant and Logistics Control to Multi-Enterprise Collaboration. Annual Reviews in Control 30, 55–68 (2006)

Okumura, Y.: Spatial competition and collaboration networks. International Journal of Game Theory 41, 455–472 (2012)

Oliver, A.L.: On the duality of competition and collaboration: network-based knowledge relations in the biotechnology industry. Scandinavian Journal of Management 20, 151–171 (2004)

Pagiavlas, N., Stratmann, M., Marburger, P., Young, S.: Mobile business – comprehensive marketing strategies or merely IT expenses? A case study of the US airline industry. Journal of Electronic Commerce Research 6, 251–261 (2005)

Polenske, K.R.: Competition, Collaboration and Cooperation: An Uneasy Triangle in Networks of Firms and Regions. Regional Studies 38(9), 1029–1043 (2004)

Prins, S.: From Competition to Collaboration: Critical Challenges and Dynamics in Multiparty Collaboration. The Journal of Applied Behavioral Science 46(3), 281–312 (2010)

Proctor, R.W., Nof, S.Y., Yih, Y., Balasubramanian, P., Busemeyer, J.R., Carayon, P., Chi-Yue, C., Farahmand, F., Gonzalez, C., Gore, J., Landry, S.J., Lehto, M., Pei-Luen, R., Rouse, W., Tay, L., Vu, K.-P.L., Sang, E.W., Salvendy, G.: Understanding and Improving Cross-cultural Decision Making in Design and Use of Digital Media: A Research Agenda. International Journal of Human-Computer Interaction 27(1-3), 151–190 (2011)

Ramaswamy, S., Joshi, H.: Automation and Ethics. In: Nof, S.Y. (ed.) Springer Handbook of Automation, ch. 47. Springer (2009)

San Miguel, M., Johnson, J.H., Kertesz, J., Kaski, K., Díaz-Guilera, A., MacKay, R.S., Loreto, V., Érdi, P., Helbing, D.: Challenges in complex systems science. The European Physical Journal Special Topics 214(1), 245–271 (2012)

Schoenberger, E.: Competition, time and space in industrial change. In: Gereffi, G., Korzeniewicz, M. (eds.) Commodity Chains and Global Capitalism, Praeger, Westport, pp. 51–66 (1994)

Seok, H.S., Nof, S.Y., Filip, F.G.: Sustainability Decision Support System based on Collaborative Control Theory. Annual Reviews in Control 36(1), 85–100 (2012)

Seok, H.S., Nof, S.Y., The HUB-CI.: initiative for cultural, education and training, and healthcare networks. In: Proceedings of the 21st International Conference in Production Research, Stuttgart (August 2011)

Seok, H.S., Nof, S.Y.: Dynamic coalition reformation for adaptive demand and capacity sharing. Int. J. Production Economics 147(A), 136–146 (2014a)

Seok, H.S., Nof, S.Y.: Collaborative capacity sharing among manufacturers on the same supply network horizontal layer for sustainable and balanced returns. Int. J. Prod. Res. 52(6), 1622–1643 (2014b)

Simanta, S., Morris, E., Balasubramaniam, S., Davenport, J., Smith, D.B.: Information assurance challenges and strategies for securing SOA environments and Web services. In: IEEE International Systems Conference, p. 6 (2009)

Spath, D., Braun, M., Bauer, W.: Integrated Human and Automation Systems. In: Nof, S.Y. (ed.) Springer Handbook of Automation, vol. 34. Springer (2009)

Srikant, R.: The Mathematics of Internet Congestion Control. Birkhäuser, Boston (2004)

Stipanovic, D.M., Inalham, G., Teo, R., Tomlin, C.J.: Decentralized overlapping control of a formation of unmanned aerial vehicles. Automatica 40(1), 1285–1296 (2004)

Sustainability, The Triple Bottom Line (2002), http://www.sustainability.com/philosophy/triple-bottom/tbl-intro.asp

Terwiesch, P., Ganz, C.: Trends in Automation. In: Nof, S.Y. (ed.) Springer Handbook of Automation, ch. 8. Springer (2009)
Wang, C., Zhang, Y., Ye, L., Nguyen, D.: Subscription of fee-based online services: what makes consumer pay for online content. Journal of Electronic Commerce Research 6, 304–311 (2005)
Yoon, S.W., Nof, S.Y.: Demand and capacity sharing decisions and protocols in a collaborative enterprise network, PRISM Center Research Memo, Purdue University (2008)
Velasquez, J.D., Nof, S.Y.: Collaborative e-Work, e-Business, and e-Service. In: Nof, S.Y. (ed.) Springer Handbook of Automation, ch. 88. Springer (2009)
Zwass, V.: Electronic commerce and organizational innovation: aspects and opportunities. International Journal of Electronic Commerce 7, 7–37 (2003)

Review and Challenge Questions

1. Identify a number of computer and communication security issues which are critical concerns for (i) e-Work, (ii) e-Business, and (iii) e-Service. Illustrate your answers.
2. Explain how information assurance concerns impact: (i) Factory sensor networks; (ii) Collaborative e-Work that relies on these sensors.
3. Understand the ethics challenges described in Fig. 11.1, and describe how they influence the rationalization of a particular e-Business.
4. Can the emerging techniques for error and conflict detection, reduction, prevention, and recovery/resolution have direct or indirect impacts on the ethics concern in e-Businesses and e-Services? Explain, illustrate and justify your answer.
5. Reflect on your answer to Question 4 specifically in the case of: (i) Insurance claims; (ii) Safety; (iii) Security; (iv) Maintenance; (v) Emergency response.
6. How is the ability to collaborate more effectively through engineered, augmented collaboration enabling better (i) Design for resiliency; (ii) Design for sustainability; (iii) Innovativeness; (iv) Agility; (v) Competitiveness?
7. Describe briefly some of the emerging technologies for more effective B2B, B2C, and e-Commerce collaboration.
8. Research has already proven that optimized collaboration is not "full cooperation" or "no cooperation"; "full sharing" or "no sharing", but somewhere in between, depending on a combination of conditions. Is this finding in conflict with incentive-based collaboration techniques?
9. (i) Explain some of the multi-cultural factors influencing the performance of e-Businesses and e-Services. (ii) Illustrate your answer from your own experience.

10. Describe some of the theoretical issues that may enable further innovations in augmenting collaborative e-Work.
11. (i) List the three dimensions (axes) of the Collaborative Intelligence (CI) Collaboratorium, and illustrate them for a particular e-Business or e-Service. (ii) Explain some of the features for augmented collaboration that are enabled by HUB-CI, and illustrate their emerging benefits for your answer in (i).
12. Write a brief essay: "Is it a dream, or can we revolutionize our life, or at least our work, business, and services by revolutionizing collaboration?"

Index

D

E

N

O

P

Q

R

T

U

V

W

MIX
Papier aus verantwortungsvollen Quellen
Paper from responsible sources
FSC® C105338

If you have any concerns about our products,
you can contact us on
ProductSafety@springernature.com

In case Publisher is established outside the EU,
the EU authorized representative is:
Springer Nature Customer Service Center GmbH
Europaplatz 3, 69115 Heidelberg, Germany

Printed by Libri Plureos GmbH
in Hamburg, Germany